Bureau of Flora and Fauna, Canberra

ZOOLOGICAL CATALOGUE OF AUSTRALIA

Volume 3

ARACHNIDA:

MYGALOMORPHAE, ARANEOMORPHAE IN PART, PSEUDOSCORPIONIDA, AMBLYPYGI AND PALPIGRADI

Australian Government Publishing Service
Canberra 1985

ISBN for series: 0 644 02840 8
ISBN for Volume 3: 0 644 03988 4

Zoological Catalogue of Australia

The compilation of the *Zoological Catalogue of Australia* is conducted under the auspices of the Bureau of Flora and Fauna, B. J. Richardson, Assistant Director (Fauna).

D. W. Walton, Executive Editor

Vol. 3

ARACHNIDA : Mygalomorphae

by

Barbara York Main
Department of Zoology
University of Western Australia
Nedlands, W.A.

ARACHNIDA : Araneomorphae (in part)

by

Valerie Todd Davies
(R.J. McKay for **The Lycosidae)**
Queensland Museum
Fortitude Valley, Qld.

ARACHNIDA : Pseudoscorpionida, Amblypygi, Palpigradi

by

Mark S. Harvey
Australian National Insect Collection
C.S.I.R.O. Division of Entomology
Canberra, A.C.T.

Printed at Brown Prior Anderson Pty Limited, Victoria

PROF. V.V. HICKMAN (1894–1984)

This volume of the *Zoological Catalogue of Australia* is dedicated to Professor V.V. Hickman. It is with deep gratitute that this treatment of some arachnid orders is offered as a tribute. Professor Hickman's great and lasting contribution to the study of Australian arachnology has served as a stimulus and incentive to all other workers in this field.

CONTENTS

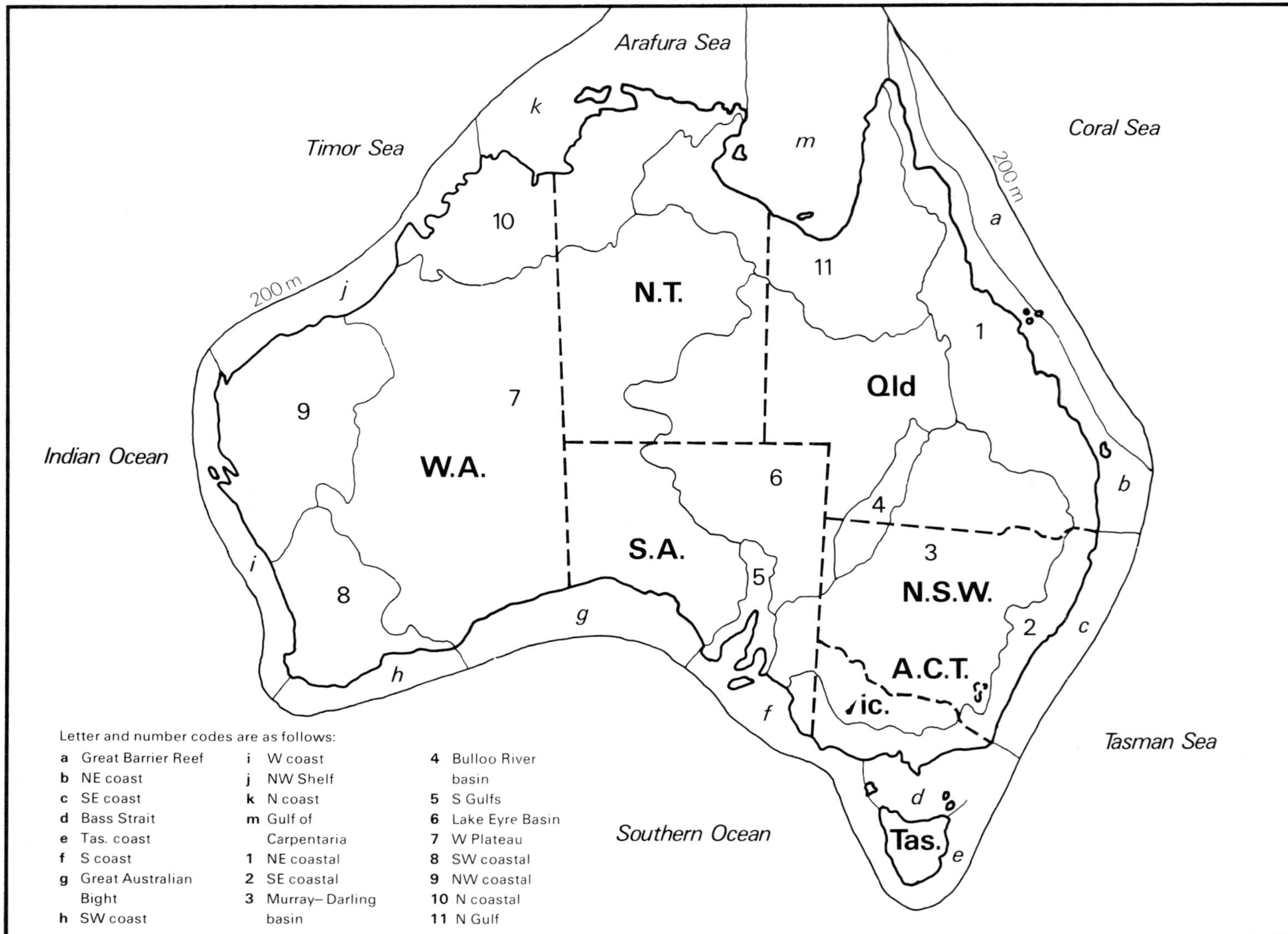
Arafura Sea
Timor Sea
Coral Sea
Indian Ocean
Southern Ocean
Tasman Sea
200 m
200 m
W.A.
N.T.
Qld
S.A.
N.S.W.
A.C.T.
Vic.
Tas.
a
b
c
d
e
f
g
h
i
j
k
m
1
2
3
4
5
6
7
8
9
10
11
Letter and number codes are as follows:
a Great Barrier Reef
b NE coast
c SE coast
d Bass Strait
e Tas. coast
f S coast
g Great Australian Bight
h SW coast
i W coast
j NW Shelf
k N coast
m Gulf of Carpentaria
1 NE coastal
2 SE coastal
3 Murray–Darling basin
4 Bulloo River basin
5 S Gulfs
6 Lake Eyre Basin
7 W Plateau
8 SW coastal
9 NW coastal
10 N coastal
11 N Gulf

EDITORIAL PREFACE

INTRODUCTION

An objective of the Australian Biological Resources Study is to stimulate research and publications on the taxonomy and distribution of Australian fauna and flora. Consistent with this aim, the *Zoological Catalogue of Australia* was conceived as a concise, computer-based data bank consisting of current taxonomic and biological knowledge of the Australian fauna, accessible to all interested in such information. As the project developed, the advantages of publication of this information were recognised.

Data for inclusion in the *Catalogue* are assembled in four separate files: a genus taxonomic arrangement file, a species taxonomic arrangement file, a genus available name file and a species available name file. The contents of appropriate files are then integrated by computer. This methodology yields a standard format which will be maintained throughout the volumes of the *Catalogue* and provides consistency in the data. The format and style of presentation are, therefore, the responsibility of the Bureau of Flora and Fauna. The authors are responsible only for the information content.

Each volume of the *Catalogue*, treating specific taxa, will cite by name and original reference all species known from Australia. The species are arranged taxonomically by family and genus. Information for each species includes synonymy, literature citation, location/status of the type material and type locality for each available name in synonymy, a brief summary of geographical distribution and ecological attributes, and important references on various aspects of the biology. It is designed to serve primarily as a bibliographic directory to the most comprehensive and recent information available on each species.

This data base is intended to provide a substantial assessment of current knowledge and to stimulate and provide a starting point for future investigations. It is estimated that the Australian fauna exceeds 150 000 species of which about half have yet to be recognised and described. As knowledge of the Australian fauna advances, the data base will be updated and expanded.

TAXONOMIC INFORMATION

Nomenclature in the *Catalogue* adheres to the provisions established in the International Code of Zoological Nomenclature. The author and date of all names appearing in the *Catalogue* are presented so that the user may understand the nature and relationships of the names and all names appear in their legitimate form, not as they appeared in their original presentation. The valid genus and species group names and their allocation to families are determined by the contributors. No new genus or species group names are introduced in the *Catalogue* although new combinations may be established. Synonymies do not include new combinations. Treatment of family group names is not included.

ECOLOGY AND DISTRIBUTION

Information on ecology and distribution is given with each valid species. The ecological descriptors are general terms derived from a list prepared by the Bureau of Flora and Fauna. These descriptors act as computer search terms for use with the data base.

Distribution data are based on a standardised list of computer search terms established by the Bureau of Flora and Fauna. Both political and geographical region descriptors are included (see map). Political areas include the adjacent waters. Terrestrial geographical terms are based on the drainage systems of continental Australia, while marine terms are self explanatory except as follows: the boundary between the coastal and the oceanic zones is the 200 m contour; the Arafura Sea extends from Cape York to 124°E longitude; and the boundary between the Tasman and Coral Seas is considered to be the latitude of Fraser Island, also regarded as the southern terminus of the Great Barrier Reef. Ecological or distributional terms in parentheses imply that the information is unconfirmed but, in the opinion of the contributor, likely to be correct. Terms for terrestrial habitat or vegetation type follow Specht, R.L. (1970). Vegetation. pp 44–67 *in* Leeper, G.W. (ed.) *The Australian Environment.* 4th edn. Melbourne : CSIRO-Melbourne Univ. Press.

BIBLIOGRAPHIC INFORMATION

Where possible, selected references are provided as an introduction to the biology of a species. Literature citations throughout the *Catalogue* are given in full. Older works, with extended subtitles, in some cases have been shortened but only if their identity is preserved. Serial titles are abbreviated in a manner designed to facilitate library research. The number and variety of sources for serial abbreviations employed by workers of different nationalities and among the various taxa precluded use of a standard guide. References or titles originally issued in a script other than Roman and lacking a Romanised translation are transliterated with the original language shown in brackets. Common abbreviations are listed in Appendix I. Acronyms of museums or collections, given as part of the Type data, are defined in Appendix II.

ACKNOWLEDGEMENTS

Within the Bureau of Flora and Fauna, Janet Godsell, Keith Houston, Richard Longmore and Barry Richardson assisted in the editing of the volume. David Berman and Chris Curtis managed the database and Wendy Riley, Cindy Wolter, Cindy Rombouts, Ha Diep, Gloria Downing and Victoria Ochiltree the entry and revisions of the data.

To all those involved, grateful acknowledgement is extended.

D. W. Walton

MYGALOMORPHAE

Barbara York Main

INTRODUCTION

The Mygalomorphae, commonly called "trapdoor" and "funnel-web" spiders, are predominantly terrestrial burrowing spiders, which occasionally make tubular silk nests in bark or moss on tree trunks. They are readily distinguished from the Araneomorphae (or "true spiders") by (1) the (par-axial) arrangement of the chelicerae, which are hinged to the prosoma (cephalothorax) such that the fangs strike downward and parallel to one another and (2) by the presence of two pairs of lungbooks, the opercula and spiracles of which are clearly visible on the ventral surface of the opisthosoma (abdomen). Most species are large, upwards of 15 mm, while a few are much smaller, *viz.* some American species are less than three millimetres. The "bird-eating" spiders or tarantulas of the family Theraphosidae attain a very large size, with a body length of at least 50 mm and leg span of at least 150 mm.

The front part of the body has to rise for the spider to strike with its fangs because of the horizontal alignment of the first segment of the chelicerae. Thus, when aroused the spiders assume an aggressive, threatening pose, with anterior legs and palps held aloft and the fangs spread open. Although only one species of the genus *Atrax* ("funnel-web" spiders) has caused human deaths, species of other genera can cause discomfort and some, such as *Selenocosmia* and *Missulena*, are also potentially dangerous. Reported bites of *Missulena* have not caused severe reactions but the venom is apparently potent (Sutherland, 1980).

MORPHOLOGY

Terminology used in the following text is standard and explanations may be found in any general text or guide to families of spiders *e.g.* Comstock revised by Gertsch (1948); Forster (1967); Hickman (1967); Kaston (1953,1981); Locket & Millidge (1951); Main (1964); Millot (1949).

BIOLOGY

Life Cycle

Unlike most araneomorphs, mygalomorphs are long-lived and take several years to mature. In his studies of American tarantulas (Theraphosidae), Baerg (1963) noted that two females and two males of certain species took 10 and 11 and 10 and 13 years, respectively, to mature and that specimens in captivity lived for 20 to 22 years. Baerg & Peck (1970) noted a specimen which probably reached 24 years. Non-burrowing theraphosids have been kept alive in various European museums for 20 or more years. Possibly the only study on a field population in which age at maturation and longevity have been noted is that of Main (1978 and unpublished data). Although Main (1978) estimated that males of *Anidiops villosus* took at least six years to mature, further data (unpublished) indicate that males mature at a minimum of eight years and females at nine or ten years. Females can live at least 20 years though the upper age limit is not known.

The life cycle of mygalomorphs generally conforms to the following pattern. Young disperse from the female's nest, establish a nest (burrow or silk tube) and enlarge and repair it as needed. A female lives in one nest for its entire life. In regions which experience seasonal drought, spiderlings disperse (generally at night) during or following rainy weather. They establish a burrow (with door, if door-building species) in a few hours and may continue to deepen the burrow over several days or nights. Burrowing is facilitated in rain-softened soil. A few species, notably *Conothele* and *Missulena*, disperse on silk threads.

Although adult female spiders moult once a year, juvenile spiders probably moult two or more times annually during the first few years. Some species seal the nest prior to moulting. Baerg & Peck (1970) documented the moulting patterns of some American tarantulas.

Sexual dimorphism is pronounced in adults but the sex of immature specimens is indistinguishable externally until the penultimate or sub-penultimate instars. Males are generally recognizable at this stage by the swollen tarsus of the pedipalp in which the intromittent organ is developing. After the final moult, males are markedly different from both females and immature males. They have longer legs, are generally more spinose and may have spined apophyses on the first and/or second legs, and the pedipalp bears the embolus on the tarsus. The palp tibia of the male frequently has characteristic groups of spines or apophyses. Females have no external sexual characters *e.g.* there is no epigynum. The marked sexual dimorphism of mature males and females and the frequent difficulty of associating conspecific males and females poses difficulties for the systematist.

When mature, males leave the burrow or tube and wander in search of a female. Wandering of males generally occurs during humid or rainy weather. Reproductive behaviour of many species is seasonal, *i.e.* some species mate during autumn/winter, others during the summer; the behaviour of some is associated with the onset of summer thunderstorms. Certainly a combination of temperature and humidity is significant for particular genera and possibly some species. Males of most species wander at night. The exception is *Missulena* in which the males wander during the day; this behaviour accounts for the large number of male *Missulena* specimens in many museum collections. Walker (see Hickman 1964) trapped male funnel-web spiders (*Atrax infensus*) by placing female "baits" in gauze-sealed jars in pitfall traps into which the males had to actively crawl, rather than fall haphazardly. Species-specific trapping of males suggests that females probably attract males by a pheromone. While males of some species enter the females' nests, there is some evidence (author unpublished) that some species mate, or courtship occurs, outside burrows. Males probably mate with more than one female but their adult lives are generally short. Females deposit the eggs in silk cocoons which may be suspended in the burrow (general for Ctenizidae, Migidae, Actinopodidae, some Dipluridae); others (especially Barychelidae) "nurse" the cocoon or leave it loose at the bottom of the burrow. Most spiders seal their nests prior to egg laying; some plug the entrances with soil wads or "mud" plugs. Females may fast for at least twelve months in those species which plug their nests prior to egg laying. Brood young generally do not feed before dispersal. Juveniles generally disperse about a year after the female has mated. There is some evidence that dispersal occurs about eighteen months after mating in certain summer-breeding species (especially Diplurinae).

The most vulnerable periods of the spider's life cycle to both predation and exposure to physical elements (*e.g.* desiccation) are (a) during juvenile dispersal and establishment of a nest and (b) when males are wandering. Thus the advantage of a close association between favourable seasonal weather conditions and reproductive behaviour and dispersal is obvious.

Foraging methods

Spiders generally forage at the entrance of their burrows or tubes, lying in wait for chance prey. Most door-building species pounce on prey by only partially emerging from the nest while many tube dwellers run out of the nest and attack prey in the open. A few species range freely away from the nest. A unique way of increasing the foraging area has evolved in several ctenizid genera, particularly in the Aganippini. Such spiders attach fans of twigs or plant debris with silk to the rim of the burrow entrance. These twig-lines function as sensory, prey-detection lines thereby increasing the area over which the spiders can capture prey (Main 1957b,1976,1982b). Prey consists predominantly of insects and other arthropods, earthworms, snails and occasionally small vertebrates. Many species store the rejectamenta in the nests by closing it off in pockets along the walls or depositing it at the bottom of the nest. A few spiders, *e.g.* *Conothele* flick food remains away from the burrow entrance (Main 1957a). Few studies have been made of the food of Australian mygalomorphs in the wild. Main (1956a) identified the prey items in twelve nests of *Missulena insignis*, and authors, such as Hickman and Raven, have mentioned food items of species in their taxonomic studies. Although spiders in wetter, forest habitats appear to be catholic in their feeding habits, species in arid and semi-arid habitats, where indeed the dietary opportunities are less diverse, feed mainly on ants. It appears that spiders are generally opportunistic rather than selective feeders. There are several records of Australian mygalomorphs feeding on vertebrates, particularly frogs, for example *Atrax formidabilis* (McKeown 1963); *Aganippe raphiduca* (Butler & Main 1959); *Selenocosmia stirlingi* (*nec crassipes*) (Main & Main 1956). In addition, McKeown (1963) recorded that *Selenotypus plumipes* catches and feeds on chickens and I have observed *Atrax validus* eating a lizard.

Burrows

The burrow or tube serves as a life-time shelter and retreat for the spider from predators, a cushion against unfavourable seasonal conditions (*e.g.* summer drought or seasonal flooding), an incubation chamber for eggs and spiderlings and an ambush site for predatory attacks. The diversity of burrow structures, convergences and specializations associated with distinctive adaptive behaviours are manifold.

Predators and Parasites

Juveniles, while dispersing and establishing burrows, are particularly vulnerable to predation by birds, small mammals, lizards, frogs and arthropods such as centipedes, scorpions and other spiders. Centipedes enter burrows and feed on adult spiders (Main 1957b) and one genus of scorpion, *Isometroides*, is a specialist predator of burrowing spiders, mainly mygalomorphs (Main 1956b). Goannas (*Varanus*, especially *V. gouldii*) dig out the burrows and feed on spiders (author unpublished observations); similar predatory behaviour is exhibited by bandicoots. Pompilid wasps prey on spider species which have burrows with open entrances and several wasp genera (tribe Ctenoceratini) are specialized ectoparasites of door-building species (Evans 1972).

For a general account of the natural history, burrowing behaviour and adaptations of mygalomorphs see Main (1976).

SYSTEMATICS

Of the thirteen or more mygalomorph families currently recognized, seven occur in Australia. In a recent zoogeographic review of the Australian Mygalomorphae, Main (1981) recognized six families. Raven (1980a) separated certain genera from the Dipluridae by elevating the Hexathelinae to family rank, the Hexathelidae; there are thus a total of seven families represented in Australia. Of these, the Ctenizidae and Dipluridae are taxonomically the most complex, and systematically and behaviourally, the most diverse. The Actinopodidae is represented by only one genus, the Migidae by two. Undoubtedly, there are more genera and many species which require description. The cutoff date for the inclusion of new names in this catalogue was 1 Apr. 1984.

Historical

The earliest described (and still identifiable) species of Australian mygalomorph was *Missulena occatoria*, described by Walckenaer in 1805. Only a few more genera and species were described prior to those by Koch (1873,1874). Thereafter many more genera and species were described before the turn of the century, mainly by various European workers (Ausserer, Koch, Karsch, Thorell and others), by Cambridge and Pocock in England and by Rainbow in Australia. Most of the described specimens were collected by individuals on expeditions sponsored by European museums, by naturalists (or surgeons) on exploration ships, or by officials in the various British colonial centres. Most specimens were deposited in European museums, a few in the British Museum (Natural History) and some in the Hope Museum, Oxford. In the early part of the twentieth century, Hogg, who had been in Australia, described many genera and species, and lodged at least a few syntypes in Australian museums. Kulczynski (1908) continued the taxonomic work on specimens in the National Hungarian Museum; his specimens were ultimately deposited in Warsaw. Apart from odd specimens, it was not until 1914, with Rainbow's work in the Australian Museum on material collected by Australian workers, that collections were finally deposited on home ground in Australian museums. The deposition of early collections in overseas museums, some of which were damaged by war, greatly inhibited taxonomic research by Australian workers until 1958 when I was able to locate and examine most of the extant overseas material. In the meantime, Womersley (1943) had attempted the difficult task of revising *Missulena* without access to early types. Rainbow (1914), Rainbow & Pulleine (1918) and Hickman (1927, 1928,1929,1930,1933,1936,1944,1964) described many new species. The systematic status of many of the species described prior to these papers has remained in doubt until recently. Some of the synonyms proposed in this catalogue derive directly from my examination (in 1958 and later) of extant early types. It should be noted that until recently no loans of type material were permitted from the British Museum (Natural History) or the Australian Museum. Taxonomists, therefore, had to visit these institutions in order to examine such specimens. Taxonomic studies by Rainbow and Hickman have been followed by those of Main (beginning in 1952) and Raven (1976 and onwards). Type localities shown in the *Catalogue* refer to primary types unless otherwise stated.

Early studies on Australian Mygalomorphae were primarily taxonomic. Rainbow (1914) and Rainbow & Pulleine (1918) presented a few natural history notes. Hickman, in his many papers, gave detailed accounts of nest structures, egg cocoons and habitat characteristics for many of the Tasmanian species he described. McKeown (1952 and later editions) gave summary accounts in his popular book *Australian Spiders*, but often without precise identification of the species discussed. Main (1952, 1956a,1957a,1957b,1962,1969,1972) recorded biological observations on *Idiosoma, Missulena, Conothele* and the Aganippini, *Cataxia* and *Stanwellia.* Generalized accounts of the comparative biology and adaptations of various mygalomorph genera are included in Main (1964,1976,1981) and a preliminary account of a demographic study of *Anidiops villosus* was presented by Main (1978). Mascord (1970,1981) also presented general natural history notes and Main & Mascord (1974) recorded the unusual aerial tube of the ctenizid *Dyarcyops robertsi.* Main (1982b) gives a summary account of arid-adapted mygalomorphs. Behavioural studies undertaken by graduate students are recorded in unpublished theses, notably Gray (1968), Moran (1980), Raven (1981b) and Rowell (1980). Raven, at the Queensland Museum, has recently studied the Hexathelidae (Raven 1976,1978,1980a) and Dipluridae (Raven 1980b,1981a). Research into the venom of *Atrax* was undertaken by Kaire (1963) and continued by Sutherland (1972a,b,c;1980; Sutherland *et al.* 1981).

Geographic Distribution

Despite the number of species and genera from Australia described over a period of a hundred and fifty years, no generalizations were made concerning the distributions of the mygalomorph fauna, apart from Pocock's early work (Pocock 1903), until recent accounts by Main on the Aganippini, *Cataxia* and *Stanwellia* (Main 1957b,1962,1969,1972) and the Mygalomorphae as a group within Australia and affinities elsewhere (Main 1976,1981,1982). Raven (1980a) provided a comprehensive biogeographical account of the Hexathelidae and a discussion on the zoogeographic distribution of the genus *Masteria* (Raven 1979).

Briefly, the broad patterns of distribution of families of Mygalomorphae and their affinities on a global scale, as interpreted largely by Main (1981,1982a), are as follows:

1. Ancient families (Gondwanaland relicts) with southern continent distribution *e.g.* (a) Actinopodidae, one genus in Australia - *Missulena*, two in South and/or Central America - *Neocteniza* and *Actinopus*, (b) Migidae, two genera in Tasmania and eastern Australia, one of which, *Migas*, is also represented in Madagascar, New Zealand, New Caledonia and South America, the other, *Heteromigas* , is possibly endemic, (c) Hexathelidae, Hexathelinae, one genus in South America, one in New Zealand and several in Australia.

2. Circumtemperate and/or tropical families (or subfamilies) with large, diverse endemic elements in Australia *e.g.* Ctenizidae, all genera endemic except *Conothele*, *Arbanitis* and *Misgolas* and Dipluridae, Diplurinae of which all genera except *Stanwellia* are endemic, some with no close affinity elsewhere.

3 Neotropical and/or South Pacific and/or southeast Asian affinities *e.g. Masteria* of the Masteriinae, Dipluridae (all regions) (Raven 1979,1981c); *Stenygrocercus*, New Caledonia. Both the foregoing groups occur only in the northeast or east of Australia.

4. Recent immigrants of the circumtropical family Theraphosidae, several genera only in Australia and the island of New Guinea (with Indo-Malayan affinities), are confined to tropical and inland Australia.

5. A combination of possibly old southern elements (with South American affinities) and recent northern entrants (with Indo-Malayan and South Pacific and, possibly, Neotropical affinities) represented in the Barychelidae.

Within Australia there are many complex patterns but some of the more obvious patterns are:

1. The Ctenizidae are abundant in southern and northeastern Australia but rare in the northwest and N.T. The greatest diversity is represented by the Aganippini (west of the Great Dividing Range); *Dyarcyops* (=*Misgolas*) occurs east of the Flinders Ranges to southeast Qld., including Tas.; *Arbanitis* occurs in southwestern and eastern Australia excluding S.A.; *Blakistonia* is the only ctenizid genus with a central Australian radiation and is more or less associated with the Flinders Ranges block, spreading into Eyre Peninsula and across the Nullarbor and into western Vic. and southwestern N.S.W. Other genera have more restricted ranges: *Idiosoma* is confined to a narrow strip in southwestern W.A. and *Cataxia* to eastern Qld.

2. Within the Dipluridae, Diplurinae, one genus at least *i.e. Aname* occurs throughout mainland Australia and Tasmania; *Stanwellia* occurs in southeastern and south coastal mainland Australia and Tas.; *Ixamatus* and *Namea* occur in mid- and northeastern Australian rainforests; *Kiama* is restricted to mid-coastal N.S.W.; *Teyl* to southwestern Western Australia. In the Ischnothelinae, *Cethegus* is widespread except in southeastern mainland Australia and Tas.; other genera are confined to eastern Australia (Raven 1984).

Distribution patterns within Australia can only be interpreted through an understanding of past climates and the ecological needs of respective genera and species. Main (1981) gave a summary of the broad distributions of ctenizid genera in relation to the generalized habitats occupied. In a very general sense, the less tolerant mygalomorph genera are confined to humid and forested regions and few genera inhabit extreme habitats such as alpine or desert habitats. There are, however, a large number of genera and species adapted to semi-arid habitats *i.e.* the periphery of the great central desert region of Australia and many species also persist in favourable minor habitats within the general arid region. One of the few inhabitants of the central desert proper is the theraphosid *Selenocosmia stirlingi*, a large spider with a very deep burrow. Main (1982b) has suggested that it is a recent coloniser and that ecologically it fills the role of a small mammal or reptile.

ACKNOWLEDGEMENTS

I am grateful to the following colleagues who either assisted in the search for types or loaned or made available such types as are listed in the accompanying catalogue: AM, M.R. Gray, the late A. Musgrave; BMNH, G.O. Evans, K.H. Hyatt, F. Wanless, P. Hillyard and the late D.J. Clark; MCZ, H.W. Levi; MNH, J.I. Balogh; MNHP, M. Vachon, M. Hubert; MCG, E. Tortonese; NHMW, E. Kritscher; OUM, H.E. Varley, E. Taylor; QM, E. Dahms, V.E. Davies; QVM, R.H. Green; SAMA, D.C. Lee, G. Grosse, H. Womersley; SMF, O. Kraus; SMNS, H. Janus; WAM, L.E. Koch, the late L. Glauert; ZMB, W. Crome; ZMH, H. Weidner, G. Rack; IZW, J. Proszynski.

I am grateful for the two following grants which enabled me to examine otherwise unavailable type collections: a University of Western Australia Research Grant (1954) which supported my initial visit to the Australian Museum, National Museum of Victoria (now Museum of Victoria), South Australian Museum and Queensland Museum; an Alice Hamilton Fellowship (1958) from the International Federation of University Women which supported a visit to overseas museums. I acknowledge with many thanks the company and assistance in the field over many years of all those colleagues who thereby facilitated the search for, discovery and collection of new topotypic material. The Zoology Department, University of Western Australia is thanked for providing laboratory space and facilities. The Western Australian Museum has also provided laboratory space, facilities and access to collections. Preparation of the Catalogue was undertaken during tenure of a Consultancy with the Bureau of Flora and Fauna.

B.Y.M.

References

Baerg, W.J. (1963). Tarantula life history records. *J. N.Y. Entomol. Soc.* **71**: 233–238

Baerg, W.J. & Peck, W.B. (1970). A note on the longevity and moult cycle of two tropical Theraphosids. *Bull. Br. Arachnol. Soc.* **11**: 107–108

Butler, W.H. & Main, B.Y. (1959). Predation on vertebrates by mygalomorph spiders. *West. Aust. Nat.* **7**: 52

Comstock, J.H. (1948). *The Spider Book.* New York : Doubleday, Doran & Co. 729 pp. (revised by Gertsch, W.J.)

Evans, H.E. (1972). The tribe Ctenoceratini in Australia (Hymenoptera : Pompilidae). *J. Aust. Entomol. Soc.* **11**: 244–252

Forster, R.R. (1967). The spiders of New Zealand. Part I. *Otago Mus. Bull. No. 1* 124 pp.

Gray, M.R. (1968). *Comparison of three genera of trapdoor spiders (Ctenizidae, Aganippini) with respect to survival under arid conditions.* Unpubl. M.Sc. Thesis. Nedlands : Univ. Western Australia 76pp.

Hickman, V.V. (1927). Studies in Tasmanian spiders. Part I. *Pap. Proc. R. Soc. Tasm.* **1926**: 52–86

Hickman, V.V. (1928). Studies in Tasmanian spiders. Part II. *Pap. Proc. R. Soc. Tasm.* **1927**: 158–175

Hickman, V.V. (1929). Studies in Tasmanian spiders. Part III. *Pap. Proc. R. Soc. Tasm.* **1928**: 96–118

Hickman, V.V. (1930). Studies in Tasmanian spiders. Part IV. *Pap. Proc. R. Soc. Tasm.* **1929**: 87–122

Hickman, V.V. (1933). A new ctenizid spider from New South Wales. *Ann. Mag. Nat. Hist. 10* **12**: 210–216 figs 1–4

Hickman, V.V. (1936). A new mygalomorph spider belonging to the genus *Hexathele* Ausserer. *Pap. Proc. R. Soc. Tasm.* **1935**: 131–137 fig 1A–F

Hickman, V.V. (1944). The Simpson Desert Expedition, 1939 - Scientific Reports — No. 1. Biology — Scorpions and spiders. *Trans. R. Soc. S. Aust.* **68**(1): 18–48 pls 1–3

Hickman, V.V. (1964). On *Atrax infensus* sp. n. (Araneida : Dipluridae) its habits and a method of trapping the males. *Pap. Proc. R. Soc. Tasm.* **1964**: 107–112 text figs 1–5, pl 1 figs 1–2, pl 2 figs 3–4

Hickman, V.V. (1967). *Some Common Spiders of Tasmania.* Hobart : Tasmanian Museum and Art Gallery 112 pp.

Hogg, H.R. (1901). On Australian and New Zealand spiders of the suborder Mygalomorphae. *Proc. Zool. Soc. Lond.* **1901(2)**: 218–279

Hogg, H.R. (1902). On some additions to the Australian spiders of the suborder Mygalomorphae. *Proc. Zool. Soc. Lond.* **1902(2)**: 121–142

Kaire, G.H. (1963). Observations on some funnel-web spiders (*Atrax* species) and their venoms, with particular reference to *Atrax robustus. Med. J. Aust.* **1963**: 307–311

Kaston, B.J. (1948). Spiders of Connecticut. *Geol. & Nat. Hist. Surv. Conn. Bull.* **70**: 1–874 144 pls (revised 1981)

Kaston, B.J. (1953). *How to Know the Spiders.* Dubuque, Iowa : W.C. Brown Co. 220 pp.

Koch, L. (1873). *Die Arachniden Australiens.* Lf. 9 pp. 409–472 pls 33–36 Nürnberg : Bauer & Raspe

Koch, L. (1874). *Die Arachniden Australiens.* Lf. 10 pp. 473–528 pls 37–40 Nürnberg : Bauer & Raspe

Kulczynski, W. (1908). Araneae musei nationalis Hungarici in regionibus Indica et Australica A Ludovici Biro collectae. *Ann. Hist. Nat. Mus. Natl. Hung.* **6**: 428–494 pl 9

Locket, G.H. & Millidge, A.F. (1951). *British Spiders.* London : The Ray Society Vol. I 310 pp.

Main, B.Y. (1952). Notes on the genus *Idiosoma*, a supposedly rare Western Australian trap-door spider. *West. Aust. Nat.* **3**: 130–137

Main, B.Y. (1956a). Observations on the burrow and natural history of the trap-door spider *Missulena* (Ctenizidae). *West. Aust. Nat.* **5**: 73–80

Main, B.Y. (1956b). Taxonomy and biology of the genus *Isometroides* Keyserling (Scorpionida). *Aust. J. Zool.* **4**: 158–164

Main, B.Y. (1957a). Occurrence of the trap-door spider *Conothele malayana* (Doleschall) in Australia (Mygalomorphae : Ctenizidae). *West. Aust. Nat.* **5**: 209–216

Main, B.Y. (1957b). Biology of aganippine trapdoor spiders (Mygalomorphae : Ctenizidae). *Aust. J. Zool.* **5**: 402–473

Main, B.Y. (1962). Adaptive responses and speciation in the spider genus *Aganippe* Cambridge. pp. 359–369 *in* Leeper, G.W. (ed.) *The Evolution of Living Organisms.* Melbourne : R. Soc. Vict.

Main, B.Y. (1964). *Spiders of Australia.* Brisbane : Jacaranda 124 pp. (reprinted 1981, Adelaide : Axiom)

Main, B.Y. (1969). The trap-door spider genus *Cataxia* Rainbow (Mygalomorphae : Ctenizidae) – Taxonomy and natural history. *J. Aust. Entomol. Soc.* **8**: 192–209

Main, B.Y. (1972). The mygalomorph spider genus *Stanwellia* Rainbow and Pulleine (Dipluridae) and its relationship to *Aname* Koch and certain other diplurine genera. *J. R. Soc. West. Aust.* **55**: 100–114

Main B.Y. (1976). *Spiders.* Sydney : Collins 296 pp.

Main, B.Y. (1978). Biology of the arid-adapted Australian trap-door spider *Anidiops villosus* (Rainbow). *Bull. Br. Arachnol. Soc.* **4**: 161–175

Main, B.Y. (1981). Eco-evolutionary radiation of mygalomorph spiders in Australia. pp. 855–872 *in* Keast, A. (ed.) *Ecological Biogeography of Australia.* The Hague : Junk

Main, B.Y. (1982a). Some zoogeographic considerations of families of spiders occurring in New Guinea. pp. 583–602 *in* Gressitt, J.L. (ed.) *Biogeography and Ecology in New Guinea.* The Hague : Junk

Main, B.Y. (1982b). Adaptations to arid habitats by mygalomorph spiders. pp. 273–283 *in* Barker, W.R. & Greenslade, P.J.M. (eds.) *Evolution of the Flora and Fauna of Arid Australia.* Frewville, S.A. : Peacock Publications

Main, B.Y. & Main, A.R. (1956). Spider predator on a vertebrate. *West. Aust. Nat.* **5**: 139

Main, B.Y. & Mascord, R.M. (1974). Description and natural history of a tube-building" species of *Dyarcyops* from New South Wales and Queensland (Mygalomorphae : Ctenizidae). *J. Entomol. Soc. Aust. (N.S.W.)* **8**: 15–21

Mascord, R. (1970). *Australian Spiders in Color.* Sydney : Reed 112 pp.

Moran, R. (1980). *The ecology and water relations of* Atrax *and* Hadronyche *(Araneae : Dipluridae).* Unpubl. Hons. Thesis. Canberra : Australian National Univ.

McKeown, K. (1952 & 1963). *Australian Spiders.* Sydney : Angus & Robertson and Sirius Books 274 & 284 pp. (revised by Roberts, N., 1963)

Millot, J. (1949). Ordre des Araneides. pp. 589–743 *in* Grassé, P. (ed.) *Traité de Zoologie.* Tom. VI Paris : Masson

Pocock, R.J. (1903). On the geographic distribution of the spiders of the order Mygalomorphae. *Proc. Zool. Soc. Lond.* **1903(1)**: 340–386

Rainbow, W.J. (1914). Studies in Australian Araneidea, No. 6. The Terretelariae. *Rec. Aust. Mus.* **10**: 187–210

Rainbow, W.J. & Pulleine, R.H. (1918). Australian trapdoor spiders. *Rec. Aust. Mus.* **12**: 81–169

Raven, R.J. (1976). A new spider of the genus *Hexathele* Ausserer (Dipluridae : Mygalomorphae) from Australia. *Proc. R. Soc. Qd.* **87**: 53–61

Raven, R.J. (1978). Systematics of the spider subfamily Hexathelinae (Dipluridae : Mygalomorphae : Arachnida). *Aust. J. Zool. Suppl. Ser.* **65**: 1–75

Raven, R.J. (1979). Systematics of the mygalomorph spider genus *Masteria* (Masteriinae : Dipluridae : Arachnida). *Aust. J. Zool.* **27**: 623–626

Raven, R.J. (1980a). The evolution and biogeography of the mygalomorph spider family Hexathelidae (Araneae, Chelicerata). *J. Arachnol.* **8**: 251–266

Raven, R.J. (1980b). The Australian mygalomorph spider genus *Ixamatus* Simon (Dipluridae : Diplurinae) and its affinities. *Bull. Br. Arachnol. Soc.* **5**: 43–49

Raven, R.J. (1981a). A review of the Australian genera of the mygalomorph spider subfamily Diplurinae (Dipluridae : Chelicerata). *Aust. J. Zool.* **29**: 321–363

Raven, R.J. (1981b). *Taxonomy, cladistics and biogeography of Australian Dipluridae and Hexathelidae (Araneae, Mygalomorphae).* Unpubl. Ph.D. Thesis. St. Lucia : Univ. Queensland 725pp.

Raven, R.J. (1981c). Three new mygalomorph spiders (Dipluridae, Masteriinae) from Colombia. *Bull. Am. Mus. Nat. Hist.* **170**: 57–63

Raven, R.J. (1984). Systematics of the Australian curtain web spiders (Ischnothelinae : Dipluridae : Chelicerata). *Aust. J. Zool. Suppl. Ser.* **93**: 1–102

Rowell, D. (1980). *Morphology and biosystematics of the genera* Atrax *and* Hadronyche *with notes on* Bymainiella brindabella *(Mygalomorphae, Dipluridae).* Unpubl. Hons. Thesis. Canberra : Australian National Univ.

Sutherland, S.K. (1972a). The Sydney funnel-web spider (*Atrax robustus*). 1: A review of published studies on the crude venom. *Med. J. Aust.* **2**: 528–530

Sutherland, S.K. (1972b). The Sydney funnel-web spider (*Atrax robustus*). 2: Fractionation of the female venom into five different components. *Med. J. Aust.* **2**: 593–596

Sutherland, S.K. (1972c). The Sydney funnel-web spider (*Atrax robustus*). 3: A review of some clinical records of human envenomation. *Med. J. Aust.* **2**: 643–647

Sutherland, S.K. (1980). Recent advances in the knowledge of Australian venoms. *Med. J. Aust.* **2**: 248–250

Sutherland, S.K., Tibballs, J. & Duncan, A.W. (1981). Funnel-web spider *Atrax robustus* antivenom. 1. Preparation and laboratory testing. *Med. J. Aust.* **2**: 522–524

Womersley, H. (1943). A revision of the spiders of the genus *Missulena* Walkenaer, 1805. *Rec. S. Aust. Mus.* **7**: 249–269

ACTINOPODIDAE

INTRODUCTION

Broad high caput with vertical declivity of fovea. Eyes small and spread almost completely across caput. Cephalothorax and abdomen both squat and almost square in outline; legs relatively short, scopulae absent. Labium conical with anterior cluster of finely pointed cuspules; labium in one piece with sternum; posterior sigilla of sternum large. Rastellum on a process, tapering, finely pointed spine-like teeth. Spiders usually dark coloured to black, occasionally with white or yellow patch on dorsum of abdomen; males often with red chelicerae and blue abdomen. Males without modification of anterior legs. Two pairs of spinnerets; posterior lateral pair with short, rounded terminal segment. Burrow always with two doors at entrance and sideshaft closed by a vertically hinged door. Juveniles disperse on gossamer. Occurs only in Australia (and possibly New Guinea), South and Central America. One genus only in Australia, *Missulena*, which is widespread across the continent (absent from Tas.) but never abundant; generally absent from tropical rainforests.

Missulena Walckenaer, 1805

Missulena Walckenaer, C.A. (1805). *Tableau des Araneides* ou Caractères Essentiels des Tribus, Genres, Familles et Races que Renferme le Genre Aranea de Linné, avec la Designation des Espèces Comprises dans Chacune de ces Divisions. Paris : Dentu xii 88 pp. 9 pls [8]. Type species *Missulena occatoria* Walckenaer, 1805 by monotypy.

Eriodon Latreille, P.A. (1806). *Genera Crustaceorum et Insectorum.* Paris Vol. 1 pp. 82–127 [86] [*Eriodon* as used by Latreille, P.A. (1804). Tableau des méthodique insectes. *N. Dict. Hist. Nat.* **24**: 131–136 is a *nom. nud.*]. Type species *Eriodon occatorius* Latreille, 1806 by monotypy. Compiled from secondary source: Bonnet, P. (1956–1957). *Bibliographia Araneorum.* Toulouse : Douladoure Vol. 2 Pt. 2, Pt. 3 [1782,2931].

Closterochilus Ausserer, A. (1871). Beiträge zur Kenntniss der Arachniden-Familie der Territelariae Thorell (Mygalidae Autor.). *Verh. Zool.-Bot. Ges. Wien* **21**: 117–224 [141 pl 1]. Type species *Pachyloscelis nigripes* Lucas, 1834 by monotypy.

Theragretes Ausserer, A. (1871). Beiträge zur Kenntniss der Archniden-Familie der Territelariae Thorell (Mygalidae Autor.). *Verh. Zool.-Bot. Ges. Wien* **21**: 117–224 [142 pl 1]. Type species *Actinopus walckenaerii* Lucas, 1837 by monotypy.

Taxonomic decision of Bonnet, P. (1957). *Bibliographia Araneorum.* Toulouse : Douladoure Vol. 2 Pt. 3 [2931].

This group may occur in Papua New Guinea, see Rainbow, W.J. (1920). Trapdoor spiders of the *Chevert* Expedition. *Rec. Aust. Mus.* **13**: 77–85 [78]; Main, B.Y. (1981). Eco-evolutionary radiation of Mygalomorph spiders in Australia. pp. 853–872 *in* Keast, A. (ed.) *Ecological Biogeography of Australia.* The Hague : Junk Vol. 2 [858]; Main, B.Y. (1982). Some zoogeographic considerations of families of spiders occurring in New Guinea. pp. 583–602 *in* Gressitt, J.L. (ed.) *Biogeography and Ecology in New Guinea.* The Hague : Junk [587].

Missulena bradleyi Rainbow, 1914

Missulena bradleyi Rainbow, W.J. (1914). Studies in Australian Araneidae. No. 6. The Terretelariae. *Rec. Aust. Mus.* **10**: 187–270 [15 Aug. 1914] [268]. Type data: holotype, AM K36151, KS6402 M. adult, paratype AM K57495, KS6401 F. adult, from North Sydney, N.S.W.

Distribution: SE coastal, N.S.W.; Sydney. Ecology: terrestrial, predator.

Missulena granulosa (Cambridge, 1869)

Eriodon granulosum Cambridge, O.P. (1869). Descriptions of a new genus and six new species of spiders. *J. Linn. Soc. Zool.* **10**: 264–276 [8 Jan. 1869] [268 pl 9]. Type data: holotype, OUM M. adult missing, from Swan River, W.A.

Eriodon crassum Cambridge, O.P. (1869). Descriptions of a new genus and six new species of spiders. *J. Linn. Soc. Zool.* **10**: 264–276 [8 Jan. 1869] [269 pl 9]. Type data: holotype, OUM F. adult missing, from Swan River, W.A.

Eriodon incertum Cambridge, O.P. (1877). On some new genera and species of Araneidea. *Ann. Mag. Nat. Hist. (4)* **19**: 26–39 pls 6–7 [Jan. 1877] [30]. Type data: holotype, BMNH 43/14 M. adult, from Swan River, W.A.

Taxonomic decision of Womersley, H. (1943). A revision of the spiders of the genus *Missulena* Walckenaer, 1805. *Rec. S. Aust. Mus.* **7**: 249–269.

Distribution: SW coastal, W.A.; Perth environs and Darling Ranges. Ecology: open forest, burrowing, terrestrial, predator; found in open *Casuarina* / eucalypt forest, sandy soil.

Missulena hoggi Womersley, 1943

Missulena hoggi Womersley, H. (1943). A revision of the spiders of the genus *Missulena* Walckenaer, 1805. *Rec. S. Aust. Mus.* **7**: 249–269 [30 May 1943] [262]. Type data: holotype (probable), BMNH 43/14 M. adult from Swan River, W.A.; Hogg (1901) redescribed *Eriodon incertum* Cambridge, 1877 from a specimen in the BMNH. Womersley (1943) noted that this description did not agree with Cambridge's original description (Cambridge, 1877) of *E. incertum* but that it did agree with a second specimen referred to by Cambridge in the same paper. Womersley therefore proposed for this specimen the new name *Missulena hoggi*. He could not have seen Cambridge's and Hogg's specimen and gave a description based on another specimen (not specified). He listed five specimens in the WAM from W.A. localities but did not state on which he based the description. I have interpreted Hogg's (Cambridge's) specimen (which I have seen) as the holotype. F. Wanless (Curator at BMNH), at my suggestion, has relocated and labelled the specimen as "Probable Holotype" of *M. hoggi* Womersley, 1943.

Distribution: SW coastal, W.A.; Darling Ranges, parts of coastal plain, Rottnest Is. and as far inland as about Kellerberrin. Ecology: open forest, woodland, burrowing, terrestrial, predator; found in sandy and clay soils.

Missulena insignis (Cambridge, 1877)

Eriodon insigne Cambridge, O.P. (1877). On some new genera and species of Araneidea. *Ann. Mag. Nat. Hist. (4)* **19**: 26–39 pls 6–7 [Jan. 1877] [29]. Type data: holotype, BMNH 50/7 M. adult, from Swan River, W.A.

Eriodon semicoccineum Simon, E. (1896). Liste der Arachniden der Semon'schen Sammlung in Australien und dem Malayischen Archipel. *in* Semon, R. *Zoologische Forschungreisen in Australien und dem Malayischen Archipel.* Fünfter Band. Systematik, Tiergeographie, Anatomie, Wirbelloser Tiere. *Denkschr. Med. Naturw. Ges. Jena* **8**: 343–351 [343]. Type data: holotype, ?MNHP M. adult missing, from Burnett River, Qld.

Actinopus formosus Rainbow, W.J. (1896). Descriptions of some new Araneidae of New South Wales. No. 6. *Proc. Linn. Soc. N.S.W.* **21**: 320–344 pls 18–20 [328]. Type data: holotype, AM K11920, KS7676 M. adult, from Menindee, N.S.W.

Taxonomic decision of Main, B.Y., this work; a specimen which Rainbow, W.J. (1903). Studies in Australian Araneidae. No. 1. *Rec. Aust. Mus.* **5**: 62–66, described as a female of *Eriodon rubrocapitatum* is believed to be *M. insignis*.

Distribution: NW coastal, SW coastal, W plateau, S Gulfs, Murray-Darling basin, W.A., S.A., Vic., N.S.W.; only eastern part of SW coastal, southern W plateau and western Murray-Darling basin. Ecology: open woodland, burrowing, terrestrial, predator; in clay soils. Biological references: Main, B.Y. (1956). Observations on the burrow and natural history of the trap-door spider *Missulena* (Ctenizidae). *West. Aust. Nat.* **5**: 73–80 (distribution, biology, feeding).

Missulena nigripes (Lucas, 1834)

Pachyloscelis nigripes Lucas, H. (1834). Mémoire sur un nouveau-genre d'Araneide de l'ordre des Pulmonaires. *Ann. Soc. Entomol. Fr.* **3**: 359–365 pl 7 [364]. Type data: holotype, MNHP M. adult missing, from South America.

Actinopus walckenaeri Lucas, H. (1837). Observations sur les Araneides du genre *Pachyloscelis*, et synonymie de ce genre. *Ann. Soc. Entomol. Fr.* **6**: 369–392 [377 pl 13] [*nom. nov.* for male of *Sphodros abboti* Walckenaer, 1835]. Type is same specimen as holotype of *Pachyloscelis nigripes* Lucas, 1834 according to Simon, E. (1892). *Histoire Naturelles des Araignées.* Paris : Roret Vol. 1 [81]; Bonnet, P. (1957). *Bibliographia Araneorum.* Toulouse : Douladoure Vol. 2 Pt. 3 [2933].

Taxonomic decision of Bonnet, P. (1957). *Bibliographia Araneorum.* Toulouse : Douladoure Vol. 2 Pt. 3 [2933].

Distribution: type locality of the objective synonyms recorded as South America but Simon, E. (1892). *Histoire Naturelle des Araignées.* Paris : Roret Vol. 2 [81] suggested that this was an error and that the specimen came from Australia, locality unknown; description of specimen suggests similarity with *Missulena pruinosa* Levitt-Gregg; because of uncertainty, absence of definite type locality and loss of type of *nigripes* recommend retention of name *pruinosa* for readily identifiable specimens; *M. nigripes* is thus not a recognizable species.

Missulena occatoria Walckenaer, 1805

Missulena occatoria Walckenaer, C.A. (1805). *Tableau des Araneides* ou Caractères Essentiels des Tribus, Genres, Familles et Races que Renferme le Genre Aranea de Linné, avec la Designation des Espèces Comprises dans Chacune de ces Divisions. Paris : Dentu xii 88 pp. 9 pls [8]. Type data: holotype, F. adult missing, from Australia (as New Holland).

Eriodon occatorius Latreille, P.A. (1806). *Genera Crustaceorum et Insectorum.* Araneides. I. Paris [86]. Type data: holotype, ?MHNP F. adult missing probably lost, from N.S.W.?, Australia.

Eriodon formidabile Cambridge, O.P. (1869). Descriptions of a new genus and six new species of spiders. *J. Linn. Soc. Zool.* **10**: 264–276 [8 Jan. 1869] [267 pl 9]. Type data: holotype, OUM M. adult missing, from Australia (as New Holland).

Eriodon rubrocapitatum Keyserling, E. (1875). *in* Ausserer, A. (1875). Zweiter beiträge zur Kenntniss der

Arachniden-Familie der Territelariae Thorell (Mygalidae Autor.). *Verh. Zool.-Bot. Ges. Wien* **25**: 125–206 pls 5–7 [140]. Type data: holotype, Keyserling collection ZMH? M. adult missing, from Australia.

Eriodon rugosum Ausserer, A. (1875). Zweiter beiträge zur Kenntniss der Arachniden-Familie der Territelariae Thorell. (Mygalidae Autor.). *Verh. Zool.-Bot. Ges. Wien* **25**: 125–206 pls 5–7 [140]. Type data: holotype, Keyserling collection ZMH? M. adult missing, from Australia.

Taxonomic decision of Womersley, H. (1943). A revision of the spiders of the genus *Missulena* Walckenaer, 1805. *Rec. S. Aust. Mus.* **7**: 249–269 [251]; Bonnet, P. (1957). *Bibliographia Araneorum.* Toulouse : Douladoure Vol. 2 Pt. 3 [2933]; Main, B.Y., this work.

Distribution: SE coastal, Murray-Darling basin, N.S.W., Vic.; Womersley (1943) appears to have confused (in part) *occatoria* and *insignis*. *M. occatoria* is here considered to occur only in eastern Australia, doubtful occurrence in Papua New Guinea. Ecology: open forest, woodland, burrowing, terrestrial, predator.

Missulena pruinosa Levitt-Gregg, 1966

Missulena pruinosa Levitt-Gregg, V. (1966). A new species of *Missulena* (Ctenizidae : Araneae). *Aust. Zool.* **13**: 391–393 [14 Dec. 1966] [391]. Type data: holotype, AM K68593, KS6403 M. adult, from Groote Eylandt, Gulf of Carpentaria, N.T.

Distribution: N coastal, N Gulf, W.A.; Wyndham, W.A., Darwin and Groote Eylandt, N.T. Ecology: terrestrial, predator.

Missulena reflexa Rainbow and Pulleine, 1918

Missulena reflexa Rainbow, W.J. & Pulleine, R.H. (1918). Australian trap-door spiders. *Rec. Aust. Mus.* **12**: 81–169 pls 12–24 [24 Dec. 1918] [87]. Type data: holotype, AM K40832, KS6404 M. adult, from Keith, S.A.

Distribution: S Gulfs, S.A.; southeastern portion. Ecology: terrestrial, predator.

BARYCHELIDAE

INTRODUCTION

Medium-sized spiders, usually hairy. Most Australian genera with eyes in three distinct rows of 2,2,4 (except some unrecorded forms and *Idioctis* Koch); eyes on a tubercle or with low profile. Chelicerae with or without a rastellum. Two tarsal claws and pronounced claw tufts; dense scopulae on at least palps and anterior pairs of legs; legs not heavily spinose. Spinnerets short, with rounded terminal segments; one or two pairs of spinnerets. Burrows always with a door; some forms with two doors at the surface, some with door(s) below the surface; generally terrestrial, some arboreal. At present *Idiommata* is the only readily identifiable genus in Australia. As well as the named genera and species there are many unnamed species and several genera not recorded in the literature as occurring in Australia. Australia-wide in distribution except the extreme southwest and southeast mainland and Tas. Widely distributed in South America, Africa, Madagascar, India and New Guinea. The family includes the only mygalomorph genera recorded from oceanic islands (South Pacific).

Encyocrypta Simon, 1889

Encyocrypta Simon, E. (1889). Études arachnologiques. 21e Mém. XXXII. Descriptions d'espèces et de genres nouveaux de Nouvelle Calédonie. *Ann. Soc. Entomol. Fr. (6)* **8**: 237–247 [247]. Type species *Encyocrypta meleagris* Simon, 1889 by monotypy.

This group is also found in Papua New Guinea(?) and New Caledonia, see Bonnet, P. (1957). *Bibliographia Araneorum.* Toulouse : Douladoure Vol. 2 Pt. 2 [1655], Pt. 3 [2287] (under *Idiommata*).

Encyocrypta reticulata (L. Koch, 1874)

Idiommata reticulata Koch, L. (1874). *Die Arachniden Australiens.* Lf. 10 pp. 473–528 pls 37–40 Nürnberg : Bauer & Raspe [474]. Type data: lectotype, ZMH (MGH 8094) M. adult from Sydney, N.S.W. (description of one F. adult from Port MacKay, probably of a different species, preceeds that of the M. from Sydney), paralectotype ZMH (MGH 8094) F. adult (Port MacKay), type status doubtful, ZMH (MGH 8094) F. adult (no locality), designation by Main, B.Y., this work; Rack, G. (1961). Die entomologischen Sammlungen des Zoologischen Staatsinstitut und Zoologischen Museums Hamburg. II. Teil: Chelicerata II: Araneae. *Mitt. Hamb. Zool. Mus. Inst.* **59**: 1–60 listed the Sydney male specimen as "Holotype", the female from Port MacKay as paratype, and did not mention the second female. Koch described the female and male as above, did not mention the second female. However, in 1958 I saw three specimens all with the MGH number as above and consisting of the male from Sydney, the female from Port MacKay and a female with no locality data. All three specimens were labelled "TYPE". Clearly one specimen should be selected as a lectotype, not holotype, and Rack's designation is corrected accordingly. There is also in the BMNH a female specimen "*Idiommata reticulata* L.K." from Bowen and part of the Koch collection (BMNH 19.9.18. 5693). It is also a possible syntype although not mentioned by Koch.

Taxonomic decision of Pocock, R.I. (1895). Notes on the identity of some of the types of Mygalomorphae in the collection of the British Museum. *Ann. Mag. Nat. Hist. (6)* **16**: 223–230 [225].

Distribution: SE coastal, N.S.W. Ecology: terrestrial, predator. Biological references: Bonnet, P. (1957). *Bibliographia Araneorum.* Toulouse : Douladoure Vol. 2 Pt. 3 [2288] (synonymy, listed under *Idiommata*).

Idioctis L. Koch, 1874

Idioctis Koch, L. (1874). *Die Arachniden Australiens.* Lf. 10 pp. 473–528 pls 37–40 Nürnberg : Bauer & Raspe [484]. Type species *Idioctis helva* L. Koch, 1874 by monotypy.

This group is also found in Singapore, Fiji and Samoa, see Roewer, C. Fr. (1942). *Katalog der Araneae.* Bremen : Kommissions - Verl. von "NATURA" Vol. 1 [214]; Marples, B.J. (1951). Mygalomorph spider in Samoa. *Nature* **168**: 300–301; Main, B.Y. (1982). Some geographic considerations of families of spiders occurring in New Guinea. pp. 583–602 *in* Gressitt, J.L. (ed.) *Biogeography and Ecology of New Guinea.* The Hague : Junk. Species now not known to occur in Australia: *Idioctis helva* L. Koch, 1874, see Hogg,

H.R. (1901). *Proc. Zool. Soc. Lond.* **1901(2)**: 242, but predictably may occur in Qld. and Kimberleys, W.A.

Idioctis helva L. Koch, 1874

Idioctis helva Koch, L. (1874). *Die Arachniden Australiens.* Lf. 10 pp. 473–528 pls 37–40 Nürnberg : Bauer & Raspe [484]. Type data: syntypes, ZMH (MGH 8097) F. adult, juv., from Ovalau, Fiji.

Distribution: Ovalau, Fiji and Samoa, though said to be in Australia no definite examples known. Ecology: littoral, burrowing, terrestrial, predator; this species is found on ocean beaches, above the tide level. Biological references: Marples, B.J. (1951). Mygalomorph spider in Samoa. *Nature* **168**: 300–301 (biology, biogeography); Marples, B.J. (1955). Spiders from Western Samoa. *J. Linn. Soc. Zool.* **42**: 453–504 (biology, biogeography).

Idiommata Ausserer, 1871

Idiommata Ausserer, A. (1871). Beiträge zur Kenntniss der Arachniden-Familie der Territelariae Thorell (Mygalidae Autor.). *Verh. Zool.-Bot. Ges. Wien* **21**: 117–224 [183]. Type species *Idiops blackwalli* Cambridge, 1870 by monotypy.

Lampropodus Rainbow, W.J. & Pulleine R.H. (1918). Australian trap-door spiders. *Rec. Aust. Mus.* **12**: 81–169 [24 Dec. 1918] [135]. Type species *Lampropodus scintillans* Rainbow and Pulleine, 1918 by subsequent designation, see Petrunkevitch, A. (1928). Systema Aranearum. *Trans. Conn. Acad. Arts Sci.* **29**: 1–270 [Jan. 1928] [75].

Taxonomic decision of Main, B.Y., this work.

This group is also found in Malaysia and Papua New Guinea, see Bonnet, P. (1957). *Bibliographia Araneorum.* Toulouse : Douladoure Vol. 2 Pt. 3 [2286–2288].

Idiommata blackwalli (Cambridge, 1870)

Idiops blackwalli Cambridge, O.P. (1870). Supplementary notice on the genus *Idiops. Proc. Zool. Soc. Lond.* **1870**: 152–157 pl 8 (10 Mar. 1870) [154]. Type data: holotype, BMNH 43.14 M. adult, from Swan River, W.A.

Distribution: NW coastal, SW coastal, (S.A.), W.A.; offshore islands W.A. *e.g.* Rottnest, Garden, Abrolhos, possibly to Flinders and Mt. Lofty Ranges, S.A. Ecology: terrestrial, burrowing, predator, low open shrubland, open forest, woodland; found in coastal dunes, open forest, woodland, grassy and shrubby open scrub, burrow made with tough parchment-like silk closed with door of toughened silk. Biological references: Ausserer, A. (1871). Beiträge zur Kenntniss der Arachniden-Familie der Territelariae Thorell (Mygalidae Autor.). *Verh. Zool.-Bot. Ges. Wien* **21**: 117–224; Bonnet, P. (1957). *Bibliographia Araneorum.* Vol. 2 Pt. 3 pp. 1927–3026 (taxonomy); Main, B.Y. (1976). *Spiders.* Sydney : Collins 296 pp. (natural history).

Idiommata fusca L. Koch, 1874

Idiommata fusca Koch, L. (1874). *Die Arachniden Australiens.* Lf. 10 pp. 473–528 pls 37–40 Nürnberg : Bauer & Raspe [478]. Type data: holotype, ZMH (MGH 8096) F. adult, from Rockhampton, Qld.

Distribution: NE coastal, Qld.; Rockhampton. Ecology: terrestrial, predator.

Idiommata iridescens (Rainbow and Pulleine, 1918)

Lampropodus iridescens Rainbow, W.J. & Pulleine, R.H. (1918). Australian trap-door spiders. *Rec. Aust. Mus.* **12**: 81–169 pls 12–24 [24 Dec. 1918] [137 pls 19,23]. Type data: holotype, AM K40950, KS5190 M. adult, from Eidsvold, Qld.

Taxonomic decision of Main, B.Y., this work (*nov. comb.*).

Distribution: NE coastal, Qld.; Eidsvold. Ecology: open forest, burrowing, terrestrial, predator.

Idiommata scintillans (Rainbow and Pulleine, 1918)

Lampropodus scintillans Rainbow, W.J. & Pulleine, R.H. (1918). Australian trap-door spiders. *Rec. Aust. Mus.* **12**: 81–169 pls 12–24 [24 Dec. 1918] [135 pls 17–18,23]. Type data: lectotype, from Black Hill, Mt. Lofty Range, S.A., AM KS5191 M. adult, paralectotype AM KS5191 F. adult, K40921, KS5192 F. immature (the latter lost), designation by Main, B.Y., this work; female paralectotype probably not same species as male lectotype; male has maxillary stridulating organ which female lacks.

Taxonomic decision of Main, B.Y., this work (*nov. comb.*).

Distribution: S Gulfs, S.A.; Mt. Lofty Ranges, possibly Flinders Range. Ecology: woodland, open scrub, burrowing, terrestrial, predator. Biological references: Rainbow, W.J. & Pulleine, R.H. (1918). Australian trap-door spiders. *Rec. Aust. Mus.* **12**: 81–169 (biology); Main, B.Y. (1976). *Spiders.* Sydney : Collins 296 pp. (natural history).

Synothele Simon, 1908

Synothele Simon, E. (1908). Araneae. Pt. I *in* Michaelsen, W. & Hartmeyer, R. (eds.) *Die Fauna Südwest-Australiens.* Bd. 1 Lf. 12 pp. 359–446 Jena : Fischer [364]. Type species *Synothele michaelseni* Simon, 1908 by monotypy.

Synothele michaelseni Simon, 1908

Synothele michaelseni Simon, E. (1908). Araneae. Pt. 1 *in* Michaelsen, W. & Hartmeyer, R. (eds.) *Die Fauna*

Südwest-Australiens. Bd. 1 Lf. 12 pp. 359–446 Jena : Fischer [364]. Type data: holotype, ZMB M. adult, from Mt. Helena W.A. (as Lion Mill).

Distribution: SW coastal, W.A.; Mt. Helena (as Lion Mill), possibly throughout Darling Range. Ecology: open forest, burrowing, terrestrial, predator.

Synothele parifusca (Main, 1954)

Encyocrypta parifusca Main, B.Y. (1954). Spiders and Opiliones. pp. 37–53 *Australian Geographical Society Reports.* No. 1. The Archipelago of the Recherche. Pt. 6 Melbourne : Australian Geographical Society [37]. Type data: holotype, WAM 82/109 F. adult, paratype WAM 82/110 F. adult, from Mondrain Is., (Recherche Archipelago), W.A., paratype from Twin Peak Is. (Recherche Archipelago), W.A.

Taxonomic decision of Main, B.Y., this work (*nov. comb.*).

Distribution: SW coastal, W.A.; Mondrain and Twin Peaks Ils., Recherche Archipelago (and probably adjacent areas of the south coast). Ecology: burrowing, terrestrial, predator.

Trittame L. Koch, 1874

Trittame Koch, L. (1874). *Die Arachniden Australiens.* Lf. 10 pp. 473–528 pls 37–40 Nürnberg : Bauer & Raspe [482]. Type species *Trittame gracilis* L. Koch, 1874 by monotypy.

Trittame gracilis L. Koch, 1874

Trittame gracilis Koch, L. (1874). *Die Arachniden Australiens.* Lf. 10 pp. 473–528 pls 37–40 Nürnberg : Bauer & Raspe [482 pl 37 fig 2]. Type data: holotype, ZMH (MGH 8095) M. adult, from Bowen, Qld.

Distribution: NE coastal, Qld.; Bowen. Ecology: terrestrial, predator.

Incertae sedis

Idiommata fuliginea Thorell, T. (1881). Studi sui ragni Malesi e Papuani. Pt. 3. Ragni dell'Austro-Malesi e del A Capo York, conservati del Museo Civico di Storia Naturale di Genova. *Ann. Mus. Civ. Stor. Nat. Genova* **17**: vii–xxvii 1–720 [243] [generic status doubtful, not *Idiommata*]. Type data: holotype, MCG M. adult, from Somerset, Cape York, Qld.

Aganippe bancrofti Rainbow, W.J. & Pulleine, R.H. (1918). Australian trap-door spiders. *Rec. Aust. Mus.* **12**: 81–169 pls 13–24 [24 Dec. 1918] [95] [generic placement of this species incorrect; this name is a secondary homonym of *Aganippe bancrofti* (Rainbow, 1914)]. Type data: lectotype, AM K40848, KS6151 F. adult, from Eidsvold, Qld., paralectotype AM KS6151 F. adult, *nec* paralectotype AM KS6152 F. adult (=Dipluridae), designation by Main, B.Y., this work.

CTENIZIDAE

INTRODUCTION

Medium to large spiders, usually not very hairy. Typical "trapdoor spiders". Eyes in compact group of two rows or wide spaced with three rows of 2,2,4; if in compact group may be on a low tubercle. Chelicerae usually with a rastellum. Legs with or without scopulae; may be heavily or sparsely spinose. Three tarsal claws; all Australian forms with single row of teeth on paired claws and smooth median claw. Spinnerets short, usually two pairs; lateral pair with short, rounded terminal segments. Well represented in Australia, occurring all over the continent and Tas.; greatest diversity of genera south of the tropics. Sedentary habits, burrowers, occasionally arboreal *e.g. Conothele* and *Homogona* (in part); burrow usually closed with a door. Occur abundantly on all continents with greatest diversity in subtropical and temperate regions. At present Australian genera are grouped in two subfamilies, the Ctenizinae (most genera) and the Homogoninae (*Homogona* and an undescribed genus).

Aganippe Cambridge, 1877

Aganippe Cambridge, O.P. (1877). On some genera and species of Araneidae. *Ann. Mag. Nat. Hist. (4)* **19**: 26–39 pls 6–7 [Jan. 1877] [28]. Type species *Aganippe subtristis* Cambridge, 1877 by subsequent designation, see Simon, E. (1892). *Histoire Naturelle des Araignées.* Paris : Roret [106].

Aganippe bancrofti (Rainbow, 1914)

Blakistonia bancrofti Rainbow, W.J. (1914). Studies in Australian Araneidae. No. 6. The Terretelariae. *Rec. Aust. Mus.* **10**: 187–270 [15 Aug. 1914] [203]. Type data: holotype, AM K33359, KS7177 F. adult, from Eidsvold, Qld.

Taxonomic decision of Main, B.Y., this work (*nov. comb.*), this becomes senior secondary homonym to *Aganippe bancrofti* Rainbow and Pulleine, 1918.

Distribution: NE coastal, (Murray-Darling basin), Qld.; found only in the region of Eidsvold. Ecology: open forest, predator, terrestrial, burrowing; found in giant brigalow and vine forest in gilgais, burrow with door.

Aganippe berlandi Rainbow, 1914

Aganippe berlandi Rainbow, W.J. (1914). Studies in Australian Araneidae. No. 6. The Terretelariae. *Rec. Aust. Mus.* **10**: 187–270 [15 Aug. 1914] [199]. Type data: holotype, AM K34948, KS1668 M. adult, from Narrabri, N.S.W.

Distribution: Murray-Darling basin, N.S.W.; Narrabri district. Ecology: terrestrial, burrowing, predator. Biological references: Main, B.Y. (1957). Biology of aganippine trapdoor spiders (Mygalomorphae : Ctenizidae). *Aust. J. Zool.* **5**: 402–473 pls 1–6 (tentatively synonymized with *Aganippe smeatoni* Hogg, 1902).

Aganippe cupulifex Main, 1957

Aganippe cupulifex Main, B.Y. (1957). Biology of aganippine trapdoor spiders (Mygalomorphae : Ctenizidae). *Aust. J. Zool.* **5**: 402–473 pls 1–6 [Dec. 1957] [436]. Type data: holotype, WAM 64–61 M. adult, paratype WAM F. adult, from Chittering Lake, W.A., see Koch, L.E. (1980). The primary types of Arachnida, Chilopoda, Diplopoda, Insecta, Onycophora and Pycnogonida in the Western Australian Museum. *Rec. West. Aust. Mus.* **8**: 295–326 [298].

Distribution: SW coastal, W.A.; Avon Valley, Brockman River-Chittering Lakes area, northeastward onto Victoria plains. Ecology: open forest, open woodland, terrestrial, burrowing, predator; burrow with door, several related, undescribed species confused with *A. cupulifex*.

Aganippe modesta Rainbow and Pulleine, 1918

Aganippe modesta Rainbow, W.J. & Pulleine, R.H. (1918). Australian trap-door spiders. *Rec. Aust. Mus.* **12**: 81–169 pls 12–24 [24 Dec. 1918] [98]. Type data: holotype, AM K40853, KS6153 F. adult, from Black Hill, Mt. Lofty, S.A.

Distribution: S Gulfs, S.A.; E of St. Vincent's Gulf, Mt. Lofty Ranges (Black Hill). Ecology: burrowing, terrestrial, predator. Biological references: Main, B.Y. (1957). Biology of aganippine trapdoor spiders (Mygalomorphae : Ctenizidae). *Aust. J. Zool.* **5**: 402–473 (tentatively synonymized with *Aganippe smeatoni* Hogg, 1902; here kept separate from *A. smeatoni,* but species status doubtful).

Aganippe occidentalis Hogg, 1903

Aganippe occidentalis Hogg, H.R. (1903). Two new Australian spiders of the family Ctenizidae. *Ann. Mag. Nat. Hist. (7)* **11**: 308–312 [Mar. 1903] [309]. Type data: holotype, BMNH 03.2.11.6 M. adult, from Roebourne, W.A.

Distribution: NW coastal, W.A.; Roebourne district. Ecology: burrowing, terrestrial, predator. Biological references: Main, B.Y. (1957). Biology of aganippine trapdoor spiders (Mygalomorphae : Ctenizidae). *Aust. J. Zool.* **5**: 402–473 (species confused with undescribed species of W plateau and eastern part of SW coastal region).

Aganippe pelochroa Rainbow and Pulleine, 1918

Aganippe pelochroa Rainbow, W.J. & Pulleine, R.H. (1918). Australian trap-dooor spiders. *Rec. Aust. Mus.* **12**: 81–169 pls 12–24 [24 Dec. 1918] [100]. Type data: holotype, AM K41248, KS6252 F. adult, from Mt. Lofty, S.A.

Distribution: S Gulfs, S.A.; E of St. Vincent's Gulf, Mt. Lofty Ranges. Ecology: terrestrial, burrowing, predator. Biological references: Main, B.Y. (1957). Biology of aganippine trapdoor spiders (Mygalomorphae : Ctenizidae). *Aust. J. Zool.* **5**: 402–473 pls 1–6 (tentatively synonymized with *Aganippe smeatoni* Hogg, 1902; here kept separate from *A. smeatoni*, but species status doubtful).

Aganippe raphiduca Rainbow and Pulleine, 1918

Aganippe raphiduca Rainbow, W.J. & Pulleine, R.H. (1918). Australian trap-door spiders. *Rec. Aust. Mus.* **12**: 81–169 pls 12–24 [24 Dec. 1918] [93]. Type data: syntypes, AM K40845, KS6253 2M. adult., AM K40847, KS6255 2F. adult, AM K40846, KS6254 F. adult, from Kings Park, Perth, W.A.

Distribution: SW coastal, W.A.; valleys of Darling scarp S of Avon-Swan to about Bunbury and SE to about Albany. Ecology: open forest, terrestrial, burrowing, predator; in wet creek banks, flood flats of water courses, open patches of ground, clay soil, (may occur inland into western wheatbelt), burrow with trapdoor. Biological references: Main, B.Y. (1957). Biology of aganippine trapdoor spiders (Mygalomorphae : Ctenizidae). *Aust. J. Zool.* **5**: 402–473 pls 1–6 (distribution, biology; "type race" only, other races probably distinct undescribed species).

Aganippe robusta Rainbow and Pulleine, 1918

Aganippe robusta Rainbow, W.J. & Pulleine, R.H. (1918). Australian trap-door spiders. *Rec. Aust. Mus.* **12**: 81–169 pls 12–24 [24 Dec. 1918] [97]. Type data: lectotype, AM K40852, KS6155 F. adult, from reed beds, Mellor Is., S.A., designation by Main, B.Y., this work.

Distribution: Murray-Darling Basin, S.A.; coastal only, such as Mellor Is. Ecology: terrestrial, burrowing, predator. Biological references: Main, B.Y. (1957). Biology of aganippine trapdoor spiders (Mygalomorphae : Ctenizidae). *Aust. J. Zool.* **5**: 402–473 pls 1–6 (synonymized with *Aganippe subtristis* Cambridge, 1877 but reinstated here).

Aganippe simpsoni Hickman, 1944

Aganippe simpsoni Hickman, V.V. (1944). The Simpson Desert Expedition, 1939. Scientific Reports. No. 1. Biology — Scorpions and spiders. *Trans. R. Soc. S. Aust.* **68**: 18–48 pls 1–3 [28 July 1944] [22]. Type data: holotype, AM K67694, KS6256 F. adult, from 17.7 km E Hale River, Simpson Desert, S.A.

Distribution: Lake Eyre basin, S.A. Ecology: desert, terrestrial, burrowing, predator. Biological references: Main, B.Y. (1957). Biology of aganippine trapdoor spiders (Mygalomorphae : Ctenizidae). *Aust. J. Zool.* **5**: 402–473 pls 1–6 (tentatively synonymized with *Aganippe occidentalis* Hogg, 1903 but reinstated here).

Aganippe smeatoni Hogg, 1902

Aganippe smeatoni Hogg, H.R. (1902). On some additions to the Australian spiders of the suborder Mygalomorphae. *Proc. Zool. Soc. Lond.* **1902(2)**: 121–142 [3 June 1902] [126 pl 13]. Type data: syntypes, 4M. adult, probably originally BMNH and SAMA, SAMA 3M. adult probably syntypes, 1M. adult lost, from S.A.; Hogg cited four males sent by Mr. T.D. Smeaton of Blakiston, but without a locality for the specimens; Blakiston, in Mt. Lofty Ranges, S.A. probably correct locality.

Distribution: S Gulfs, S.A. Ecology: terrestrial, burrowing, predator, open forest.

Aganippe subtristis Cambridge, 1877

Aganippe subtristis Cambridge, O.P. (1877). On some new genera and species of Araneidea. *Ann. Mag. Nat. Hist. (4)* **19**: 26–39 pls 6–7 [Jan. 1877] [28] [AM K40835 M. adult, not part of type series, labelled allotype and described by Rainbow, W.J. & Pulleine, R.H. (1918). Australian trapdoor spiders. *Rec. Aust. Mus.* **12**: 81–169 pls 12–24]. Type data: holotype, BMNH 59–52 F. adult, from Adelaide, S.A.

Idiommata schomburgki Karsch, F. (1878). Exotisch-Araneologishes. II. *Zeit. Ges. Naturw.* **51**: 771–826 pl 9 [820]. Type data: holotype, ZMB 2106 M. adult, from Adelaide, S.A.

Aganippe pulleinei Hogg, H.R. (1902). On some additions to the Australian spiders of the suborder Mygalomorphae. *Proc. Zool. Soc. Lond.* **1902(2)**: 121–142 [3 June 1902] [128 pl 13]. Type data: lectotype, BMNH 1924.III.1.5–6 F78 F. adult, from Blakiston, S.A., designation by Main, B.Y., this work, paralectotypes BMNH 1924.III.1.5–6 F78 M. adult, BMNH 1924.1.7 M. adult, SAMA N198157 M. adult possibly part of type series.

Taxonomic decision of Main, B.Y. (1957). Biology of Aganippine trapdoor spiders (Mygalomorphae : Ctenizidae). *Aust. J. Zool.* **5**: 402–473 [428] (in part) and Main, B.Y., this work.

Distribution: S Gulfs, S.A.; Mt. Lofty and Flinders Ranges. Ecology: open forest, woodland, open scrub, burrowing, terrestrial, predator.

Anidiops Pocock, 1897

Anidiops Pocock, R.I. (1897). On some trapdoor spiders of the family Ctenizidae from South and West Australia, contained in the collection of the British Museum. *Ann. Mag. Nat. Hist. (6)* **19**: 109–116 [Jan. 1897] [114]. Type species *Anidiops manstridgei* Pocock, 1897 by monotypy.

Gaius Rainbow, W.J. (1914). Studies in Australian Araneidae. No.6. The Terretelariae. *Rec. Aust. Mus.* **10**: 187–270 [15 Aug. 1914] [195]. Type species *Gaius villosus* Rainbow, 1914 by monotypy.

Taxonomic decision of Main, B.Y. (1957). Biology of aganippine trapdoor spiders (Mygalomorphae : Ctenizidae). *Aust. J. Zool.* **5**: 402–473 pls 1–6 [424].

Anidiops manstridgei Pocock, 1897

Anidiops manstridgei Pocock, R.I. (1897). On some trapdoor spiders of the family Ctenizidae from South and West Australia, contained in the collection of the British Museum. *Ann. Mag. Nat. Hist. (6)* **19**: 109–116 [Jan. 1897] [114] [AM K40874 M. adult, not part of type series, designated allotype by Rainbow, W.J. & Pulleine, R.H. (1918). Australian trap-door spiders. *Rec. Aust. Mus.* **12**: 81–169 pls 12–24]. Type data: holotype, BMNH 96.11.4.2 F. adult, from Lawlers, W.A.

Aganippe whitei Rainbow, W.J. (1915). Scientific notes on an expedition into the north-western regions of South Australia. (h) Arachnida. *Trans. Proc. R. Soc. S. Aust.* **39**: 772–793 pls 67–68 [Dec. 1915] [774]. Type data: holotype, SAMA N198158 F. adult, from between Musgrave and Everard Ranges, S.A.

Gaius hirsutus Rainbow, W.J. & Pulleine R.H. (1918). Australian trap-door spiders. *Rec. Aust. Mus.* **12**: 81–169 pls 12–24 (24 Dec. 1918) [102]. Type data: holotype, AM K40895, KS6257 F. adult, from Port Augusta, S.A.

Taxonomic decision of Main, B.Y. (1957). Biology of aganippine trapdoor spiders (Mygalomorphae : Ctenizidae). *Aust. J. Zool.* **5**: 402–473 pls 1–6 [426].

Distribution: W plateau, S Gulfs, Lake Eyre Basin, W.A., S.A. Ecology: low shrubland, open scrub, woodland, burrowing, terrestrial, predator; in saltbush clay pans and mulga and eucalypt litter.

Anidiops villosus (Rainbow, 1914)

Gaius villosus Rainbow, W.J. (1914). Studies in Australian Araneidae. No. 6. The Terretelariae. *Rec. Aust. Mus.* **10**: 187–270 [15 Aug. 1914] [195] [WAM 40–203 M. adult, not a part of the type series, was designated allotype by Main, B.Y. (1957). Biology of aganippine trapdoor spiders (Mygalomorphae : Ctenizidae). *Aust. J. Zool.* **5**: 402–473 pls 1–6]. Type data: holotype, AM K34393, KS6259 F. adult, paratype WAM 1913–6944 F. adult, from Minnivale, W.A.

Distribution: SW coastal, W plateau, W.A. Ecology: open scrub, burrowing, terrestrial, predator; burrows in litter in mulga and acacia groves of woodlands. Biological references: Main, B.Y. (1957). Biology of aganippine trapdoor spiders (Mygalomorphae : Ctenizidae). *Aust. J. Zool.* **5**: 402–473 pls 1–6 (biology, evolution, taxonomy); Main, B.Y. (1978). Biology of the arid adapted Australian trapdoor spider *Anidiops villosus* (Rainbow). *Bull. Brit. Arachnol. Soc.* **4**: 161–175 (biology, population structure); Main, B.Y. (1979). An unusual method of soil disposal during burrow excavation by the trapdoor spider *Anidiops villosus* (Rainbow). *West. Aust. Nat.* **14**: 115–117 (burrowing behaviour).

Arbanitis L. Koch, 1874

Pholeuon Koch, L. (1873). *Die Arachniden Australiens.* Lf. 9 pp. 409–472 pls 33–36 Nürnberg : Bauer & Raspe [471] [*non Pholeuon* Hampe, 1856]. Type species *Pholeuon longipes* L. Koch, 1873 by monotypy.

Arbanitis Koch, L. (1874). *Die Arachniden Australiens.* Lf. 10 pp. 473–528 pls 37–40 Nürnberg : Bauer & Raspe [491] [*nom. nov.* for *Pholeuon* L. Koch, 1873].

Maoriana Hogg, H.R. (1901). On Australian and New Zealand spiders of the suborder Mygalomorphae. *Proc. Zool. Soc. Lond.* **1901(2)**: 218–279 [4 June 1901] [236] [*non Maoriana* Suter, 1891]. Type species *Maoriana dendyi* Hogg, 1901 by original designation.

Cantuaria Hogg, H.R. (1902). On some additions to the Australian spiders of the suborder Mygalomorphae. *Proc. Zool. Soc. Lond.* **1902(2)**: 121–142 pl 13 [3 June 1902] [123] [*nom. nov.* for *Maoriana* Hogg, 1901].

Evoplos Rainbow, W.J. (1914). Studies in Australian Araneidae. No.6. The Terretelariae. *Rec. Aust. Mus.* **10**: 187–270 [15 Aug. 1914] [217] [*Euoplos* (original spelling) emended to *Evoplos* by Bonnet, P. (1956). *Bibliographia Araneorum.* Toulouse : Douladoure Vol. 2 (1892)]. Type species *Evoplos spinnipes* Rainbow, 1914 by monotypy.

Tambouriniana Rainbow, W.J. & Pulleine, R.H. (1918). Australian trap-door spiders. *Rec. Aust. Mus.* **12**: 81–169 pls 12–24 [120]. Type species *Tambouriniana variabilis* Rainbow and Pulleine, 1918 by monotypy.

Albaniana Rainbow, W.J. & Pulleine, R.H. (1918). Australian trap-door spiders. *Rec. Aust. Mus.* **12**: 81–169 pls 12–24 [24 Dec. 1918] [122]. Type species *Albaniana inornata* Rainbow and Pulleine, 1918 by subsequent designation, see Petrunkevitch, A. (1928). Systema Aranearum. *Trans. Conn. Acad. Arts Sci.* **29**: 1–270 [63].

Bancroftiana Rainbow, W.J. & Pulleine, R.H. (1918). Australian trap-door spiders. *Rec. Aust. Mus.* **12**: 81–169 pls 12–24 [24 Dec. 1918] [127]. Type species *Bancroftiana speciosa* Rainbow and Pulleine, 1918 by monotypy.

Armadalia Rainbow, W.J. & Pulleine, R.H. (1918). Australian trap-door spiders. *Rec. Aust. Mus.* **12**: 81–169 pls 12–24 [24 Dec. 1918] [129]. Type species *Armadalia ornata* Rainbow and Pulleine, 1918 by subsequent designation, see Petrunkevitch, A. (1928). Systema Aranearum. *Trans. Conn. Acad. Arts Sci.* **29**: 1–270 [63].

Korua Todd, V. (1945). Systematic and biological account of the New Zealand Mygalomorphae (Arachnida). *Trans. R. Soc. N.Z.* **74**(4): 375–407 pls 55–58 [Mar. 1945] [389]. Type species *Korua wanganuiensis* Todd, 1945 by original designation.

Taxonomic decision of Main, B.Y., this work; Main, B.Y. (1964) *Spiders of Australia*. Brisbane : Jacaranda 124 pp.

This group is also found in New Zealand, where species currently placed in *Cantuaria* Hogg, 1902.

Arbanitis annulipes (C.L. Koch, 1842)

Mygale annulipes Koch, C.L. (1842). *Die Arachniden*. Nürnberg Bd. 9 pp. 1–108 pls CCLXXXIX–CCCXXIV [52]. Type data: holotype, missing, from Tas. (as Van Diemensland). Compiled from secondary source: Bonnet, P. (1957). *Bibliographia Araneorum*. Toulouse : Douladoure Vol. 2 Pt. 3 [2989]; Ausserer, A. (1871). Beiträge zur Kenntniss der Arachniden-Familie der Territelariae Thorell (Mygalidae Autor.). *Verh. Zool.-Bot. Ges. Wien* **21**: 117–224 pl 1 [217].

Macrothele aculeata Urquhart, A.T. (1893). On new species of Tasmanian Araneae. *Pap. Proc. R. Soc. Tasm.* **1892**: 94–130 [June 1893] [94]. Type data: holotype, M. adult whereabouts unknown, from Tas.

Arbanitis maculipes Hogg, H.R. (1903). Two new Australian spiders of the family Ctenizidae. *Ann. Mag. Nat. Hist. (7)* **11**: 308–312 [Mar. 1903] [311]. Type data: holotype, BMNH 1901.6.4.13 F. adult, from Tas.

Taxonomic decision of Musgrave, A. (1948). A catalogue of Tasmanian spiders. *Rec. Queen Victoria Mus.* **2**: 75–91 [76]; Main, B.Y., this work adding *Macrothele aculeata* Urquhart, 1893.

Distribution: Tas. Ecology: closed forest, burrowing, terrestrial, predator; burrow without trapdoor. Biological references: Hickman, V.V. (1967). *Some Common Spiders of Tasmania*. Hobart : Tas. Mus. Art Gallery 112 pp. (description of male, biology).

Arbanitis festivus Rainbow and Pulleine, 1918

Arbanitis festivus Rainbow, W.J. & Pulleine, R.H. (1918). Australian trap-door spiders. *Rec. Aust. Mus.* **12**: 81–169 pls 12–24 [24 Dec. 1918] [111]. Type data: holotype, AM K41255 F. adult, paratype WAM 64–67 F. adult, from Nannup, Blackwood River, W.A.

Distribution: SW coastal, W.A.; found only in the extreme SW. Ecology: tall open forest, burrowing, terrestrial, predator; in karri forest (*Eucalyptus diversicolor*), creek banks and open slopes (not heavily littered floor), burrow with a door.

Arbanitis inornatus (Rainbow and Pulleine, 1918)

Albaniana inornata Rainbow, W.J. & Pulleine, R.H. (1918). Australian trap-door spiders. *Rec. Aust. Mus.* **12**: 81–169 pls 12–24 [24 Dec. 1918] [122]. Type data: holotype, AM K40912, KS1621 F. adult, from Narrogin, Albany Road, W.A. emended to Armadale, W.A., by Main, B.Y., this work.

Albaniana flavomaculata Rainbow, W.J. & Pulleine, R.H. (1918). Australian trap-door spiders. *Rec. Aust. Mus.* **12**: 81–169 pls 12–24 [24 Dec. 1918] [125]. Type data: holotype, AM K40915, KS1622 F. adult, from Narrogin, Albany Road, W.A., emended to Armadale, W.A., by Main, B.Y., this work.

Armadalia setosa Rainbow, W.J. & Pulleine, R.H. (1918). Australian trap-door spiders. *Rec. Aust. Mus.* **12**: 81–169 pls 12–24 [24 Dec. 1918] [130]. Type data: holotype, AM K40918, KS7174 F. juv., from Armadale, W.A.

Taxonomic decision of Main, B.Y., this work (*nov. comb.*); Main, B.Y. (1976) *Spiders*. Sydney : Collins 296 pp. (as *Arbanitis flavomaculata*).

Distribution: SW coastal, W.A.; valleys of Darling scarp. Ecology: open forest, terrestrial, burrowing, predator; in jarrah and redgum forests (*Eucalyptus marginata* and *E. callophylla*), burrow with thick, plug-like trapdoor.

Arbanitis hoggi (Simon, 1908)

Cantuaria hoggi Simon, E. (1908). Araneae. Pt. 1 pp. 359–446 *in* Michaelsen, W. & Hartmeyer, R. (eds.) *Die Fauna Süd-west Australiens*. Jena : Fischer Vol. 1 Pt. 12 [361]. Type data: holotype, ZMB M. adult, from Eradu, W.A.

Distribution: NW coastal, SW coastal, W plateau, W.A.; along boundaries where these three regions meet. Ecology: woodland, terrestrial, burrowing, predator; in creek banks and flood flats of salt lakes, burrow with thick, plug-like door. Biological references: Main, B.Y. (1964,1981). *Spiders of Australia*. Brisbane : Jacaranda & S.A. : Axiom 124 pp.; Main, B.Y. (1981). Eco-evolutionary radiation of Mygalomorph spiders in Australia. pp. 853–872 *in* Keast, A. (ed.) *Ecological Biogeography of Australia*. The Hague : Junk Vol. 2.

Arbanitis longipes (L. Koch, 1873)

Pholeuon longipes Koch, L. (1873). *Die Arachniden Australiens*. Lf. 9 pp. 409–472 pls 33–36 Nürnberg : Bauer & Raspe [472]. Type data: holotype, ZMH (MGH) M. adult lost, from Bowen, Qld.

Distribution: type locality given as Bowen, Qld., no later specimens attributable to this species have been collected. Ecology: terrestrial. Biological references: Koch, L. (1874). *Die Arachniden Australiens*. Lf. 10 pp. 473–528 pls 37–40 Nürnberg : Bauer & Raspe (taxonomy); Rainbow, W.J. (1914). Studies in Australian Araneidae. No. 6. The Terretelariae. *Rec. Aust. Mus.* **10**: 187–270 (doubtful description of the female).

Arbanitis ornatus (Rainbow and Pulleine, 1918)

Albaniana ornata Rainbow, W.J. & Pulleine, R.H. (1918). Australian trap-door spiders. *Rec. Aust. Mus.* **12**: 81–169 pls 12–24 [24 Dec. 1918] [123]. Type data: holotype, AM K40913, KS1623 F. adult, from Eidsvold, Qld.

Bancroftiana speciosa Rainbow, W.J. & Pulleine, R.H. (1918). Australian trap-door spiders. *Rec. Aust. Mus.* **12**: 81–169 pls 12–24 [24 Dec. 1918] [127]. Type data: holotype, AM K40974, KS7176 M. adult, from Eidsvold, Qld.

Armadalia ornata Rainbow, W.J. & Pulleine, R.H. (1918). Australian trap-door spiders. *Rec. Aust. Mus.* **12**: 81–169 pls 12–24 [24 Dec. 1918] [129]. Type data: holotype, AM K40916, KS6268 F. adult, paratypes SAMA F. adult, immature M., from Eidsvold, Qld.

Taxonomic decision of Main, B.Y., this work (*nov. comb.*).

Distribution: NE coastal, Qld.; Eidsvold. Ecology: open forest, terrestrial, burrowing, predator; in giant brigalow and dry-vine forest, gilgais, burrow with trapdoor.

Arbanitis similaris Rainbow and Pulleine, 1918

Arbanitis similaris Rainbow, W.J. & Pulleine, R.H. (1918). Australian trap-door spiders. *Rec. Aust. Mus.* **12**: 81–169 pls 12–24 [24 Dec. 1918] [112]. Type data: holotype, AM K40880, KS6267 F. adult, from Kedron Brook, Brisbane, Qld.

Distribution: NE coastal, Qld. Ecology: open forest, terrestrial, burrrowing, predator; in creek banks, burrow with thick plug-like trapdoor.

Arbanitis spinnipes (Rainbow, 1914)

Evoplos spinnipes Rainbow, W.J. (1914). Studies in Australian Araneidae. No. 6. The Terretelariae. *Rec. Aust. Mus.* **10**: 187–270 [15 Aug. 1914] [219]. Type data: holotype, AM K34426, KS1612 F. adult, from Upper Burnett River, Qld.

Taxonomic decision of Main, B.Y., this work (*nov. comb.*).

Distribution: NE coastal, Qld. Ecology: low open forest, woodland, terrestrial, burrowing, predator; heavy clay soil, creek banks, burrow with thick plug-like trapdoor. Biological references: Main, B.Y. (1981). Eco-evolutionary radiation of Mygalomorph spiders in Australia. pp. 853–872 *in* Keast, A. (ed.) *Ecological Biogeography of Australia.* The Hague : Junk Vol. 2 (distribution, habitat, under *Evoplos*).

Arbanitis tasmanica (Hickman, 1928)

Aganippe tasmanica Hickman, V.V. (1928). Studies in Tasmanian spiders. Pt. II. *Pap. Proc. R. Soc. Tasm.* **1927**: 158–175 pls 21–24 [29 Feb. 1928] [58]. Type data: holotype, QVM 1957/13/21 *F. adult, from Prince of Wales Bay, Derwent Park, Tas.

Distribution: Tas.; Prince of Wales Bay, Derwent Park, probably northern regions only. Ecology: supralittoral, terrestrial, burrowing, predator; banks above high tide, burrow with trapdoor. Biological references: Main, B.Y. (1957). Biology of aganippine trapdoor spiders (Mygalomorphae : Ctenizidae). *Aust. J. Zool.* **5**: 402–473 pls 1–6 (taxonomy); Main, B.Y. (1981). Eco-evolutionary radiation of Mygalomorph spiders in Australia. pp. 853–872 *in* Keast, A. (ed.) *Ecological Biogeography of Australia.* The Hague : Junk Vol. 2 (discussed under *Cantuaria*).

Arbanitis variabilis (Rainbow and Pulleine, 1918)

Tambouriniana variabilis Rainbow, W.J. & Pulleine, R.H. (1918). Australian trap-door spiders. *Rec. Aust. Mus.* **12**: 81–169 pls 12–24 (24 Dec. 1918) [121]. Type data: holotype, AM K40899, KS6409 F. adult, paratype SAMA F. adult, from Tamborine Mt., Qld.

Tambouriniana variabilis flavomaculata Rainbow, W.J. & Pulleine, R.H. (1918). Australian trap-door spiders. *Rec. Aust. Mus.* **12**: 81–169 pls 12–24 (24 Dec. 1918) [122]. Type data: holotype, AM K40907, KS6138 F. adult, from Tamborine Mt., Qld.

Albaniana villosa Rainbow, W.J. & Pulleine, R.H. (1918). Australian trap-door spiders. *Rec. Aust. Mus.* **12**: 81–169 pls 12–24 [24 Dec. 1918] [124]. Type data: holotype, AM K40914, KS1620 F. adult, from Tamborine Mt., Qld.

Taxonomic decision of Main, B.Y. (1964). *Spiders of Australia.* Brisbane : Jacaranda 124 pp. [28]; Main, B.Y., this work.

Distribution: NE coastal, SE coastal, Qld., N.S.W.; mainly in MacPherson Ranges. Ecology: closed forest, tall open forest, terrestrial, burrowing, predator; bare soil patches in forest floor especially banks and slopes, burrow with very thick, plug-like trapdoor.

Arbanitis zorodes (Rainbow and Pulleine, 1918)

Armadalia zorodes Rainbow, W.J. & Pulleine, R.H. (1918). Australian trap-door spiders. *Rec. Aust. Mus.* **12**: 81–169 pls 12–24 [24 Dec. 1918] [131]. Type data: holotype, AM K41247, KS7175 F. adult, from Mt. Lofty, S.A., may be an error.

Taxonomic decision of Main, B.Y., this work (*nov. comb.*).

Distribution: S Gulfs, S.A.; in the Mt. Lofty Ranges. Ecology: open forest, terrestrial, burrowing, predator.

Blakistonia Hogg, 1902

Blakistonia Hogg, H.R. (1902). On some additions to the Australian spiders of the suborder Mygalomorphae. *Proc. Zool. Soc. Lond.* **1902(2)**: 121–142 [3 June 1902] [131 pl 13]. Type species *Blakistonia aurea* Hogg, 1902 by original designation.

Cantuarides Strand, E. (1907). Aviculariidae und Atypidae des Kgl. Naturalien-kabinetts in Stuttgart. *Jahr.*

Ges. Naturkd. Würt. **63**: 1–100 [8]. Type species *Cantuarides exsiccatus* Strand, 1907 by original designation.

Taxonomic decision of Main, B.Y., this work.

Blakistonia aurea Hogg, 1902

Blakistonia aurea Hogg, H.R. (1902). On some additions to the Australian spiders of the suborder Mygalomorphae. *Proc. Zool. Soc. Lond.* **1902(2)**: 121–142 [3 June 1902] [132 pl 13]. Type data: syntypes, BMNH and SAMA probably, but presumed lost M.,4F. adult, from Adelaide, S.A. (M. syntype) and Blakiston and Mt. Lofty Ranges, S.A. (F. syntypes).

Aganippe villosa Rainbow, W.J. & Pulleine, R.H. (1918). Australian trap-door spiders. *Rec. Aust. Mus.* **12**: 81–169 pls 12–24 [24 Dec. 1918] [96]. Type data: holotype, AM K40851, KS6156 F. adult, from Bridgewater, S.A.

Taxonomic decision of Main, B.Y., this work.

Distribution: S Gulfs, S.A.; in the Mt. Lofty Ranges and eastward, N into Flinders Ranges as far as Wilpena Pound, northern part Eyre Peninsula(?). Ecology: open forest, woodland, terrestrial, burrowing, predator; creek banks and clay pans, burrow in heavy clay soil, inset thick soil door with long hinge-line. Biological references: Main, B.Y. (1976). *Spiders.* Sydney : Collins 296 pp. (natural history, door); Main, B.Y. (1981). Eco-evolutionary radiation of Mygalomorph spiders in Australia. pp. 853–872 *in* Keast, A. (ed.) *Ecological Biogeography of Australia.* The Hague : Junk Vol. 2 (distribution, habitat).

Blakistonia exsiccatus (Strand, 1907)

Cantuarides exsiccatus Strand, E. (1907). Aviculariidae und Atypidae des Kgl. Naturalien-kabinetts in Stuttgart. *Jahr. Ges. Naturkd. Würt.* **63**: 1–100 [8]. Type data: syntypes, SMNS 4F. adult lost, from central Australia.

Taxonomic decision of Main, B.Y., this work (*nov. comb.*).

Distribution: Lake Eyre basin, S.A.; central Australia. Ecology: open woodland, terrestrial, burrowing, predator; rocky cliffs of creek gorges, thick trapdoor. Biological references: Main, B.Y. (1981). Eco-evolutionary radiation of Mygalomorph spiders in Australia. pp. 853–872 *in* Keast, A. (ed.) *Ecological Biogeography of Australia.* The Hague : Junk Vol. 2 (distribution, species included in *Arbanitis*).

Blakistonia rainbowi (Pulleine, 1919)

Aganippe rainbowi Pulleine, R.H. (1919). A new species of *Aganippe* from Kangaroo Island. *Trans. Proc. R. Soc. S. Aust.* **43**: 74–76 pl 12 [12 Dec. 1919] [74]. Type data: syntypes, AM M.,2F. adult lost, SAMA 2F. adult lost, from American River, Kangaroo Is., S.A.

Taxonomic decision of Main, B.Y., this work (*nov. comb.*).

Distribution: S coast, S.A.; American River, Kangaroo Is. Ecology: supralittoral, terrestrial, burrowing, predator; banks above high tide, burrow with trapdoor.

Cataxia Rainbow, 1914

Cataxia Rainbow, W.J. (1914) Studies in Australian Araneidae. No. 6. The Terretelariae. *Rec. Aust. Mus.* **10**: 187–270 [15 Aug. 1914] [222]. Type species *Cataxia maculata* Rainbow, 1914 by monotypy.

Cataxia babindaensis Main, 1969

Cataxia babindaensis Main, B.Y. (1969). The trap-door spider genus *Cataxia* Rainbow (Mygalomorphae : Ctenizidae): taxonomy and natural history. *J. Aust. Entomol. Soc.* **8**: 192–209 [31 Dec. 1969] [203]. Type data: holotype, QM W3125 F. adult, paratype AM juv., from The Boulders Natl. Park, Babinda, Qld.

Distribution: NE coastal, Qld.; known only from The Boulders Natl. Park, Babinda. Ecology: closed forest, terrestrial, burrowing, predator; wet, unlittered soil, burrow with soft door.

Cataxia eungellaensis Main, 1969

Cataxia eungellaensis Main, B.Y. (1969). The trap-door spider genus *Cataxia* Rainbow (Mygalomorphae : Ctenizidae): taxonomy and natural history. *J. Aust. Entomol. Soc.* **8**: 192–209 [31 Dec. 1969] [205]. Type data: holotype, QM W3126 F. adult, paratype QM W3127, AM F. adult, from Eungella Natl. Park, Clarke Range (creek crossing on road between Valley Look-out and Broken River), Qld.

Distribution: NE coastal, Qld.; Clark Range, Eungella Natl. Park. Ecology: closed forest, tall open forest, terrestrial, burrowing, predator; burrow with thin trapdoor.

Cataxia maculata Rainbow, 1914

Cataxia maculata Rainbow, W.J. (1914). Studies in Australian Araneidae. No. 6. The Terretelariae. *Rec. Aust. Mus.* **10**: 187–270 [15 Aug. 1914] [223]. Type data: syntypes, AM K34977, KS6335 3F. immature (Rainbow stated description of M. but appeared to be that of an immature F.), from Upper Burnett River, Qld.

Arbanitis inornatus Rainbow, W.J. & Pulleine, R.H. (1918). Australian trap-door spiders. *Rec. Aust. Mus.* **12**: 81–169 pls 12–24 [24 Dec. 1918] [119]. Type data: syntypes, AM K40890, KS6264 F. adult, M. immature, from Eidsvold, Qld.

Cataxia tetrica Rainbow, W.J. & Pulleine, R.H. (1918). Australian trapdoor spiders. *Rec. Aust. Mus.* **12**: 81–169 pls 12–24 [24 Dec. 1918] [133]. Type data: holotype, AM K41250, KS1630 F. adult, from Eidsvold, Qld.

Taxonomic decision of Main, B.Y. (1969). The trap-door spider genus *Cataxia* Rainbow (Mygalomorphae : Ctenizidae): taxonomy and natural history. *J. Aust. Entomol. Soc.* **8**: 192–209 [193].

Distribution: NE coastal, Qld.; Eidsvold-Proserpine region. Ecology: open forest, terrestrial, burrowing, predator; giant brigalow and dry-vine forest, burrow with wafer-like trapdoor.

Cataxia pallida (Rainbow and Pulleine, 1918)

Armadalia pallida Rainbow, W.J. & Pulleine, R.H. (1918). Australian trap-door spiders. *Rec. Aust. Mus.* **12**: 81–169 pls 12–24 [24 Dec. 1918] [132]. Type data: holotype, AM K40878, KS6269 M. juv., from Eidsvold, Qld.

Taxonomic decision of Main, B.Y., this work (*nov. comb.*).

Distribution: NE coastal, Qld.; Eidsvold. Ecology: open forest, terrestrial, burrowing, predator.

Cataxia spinipectoris Main, 1969

Cataxia spinipectoris Main, B.Y. (1969). The trap-door spider genus *Cataxia* Rainbow (Mygalomorphae : Ctenizidae): taxonomy and natural history. *J. Aust. Entomol. Soc.* **8**: 192–209 [31 Dec. 1969] [201]. Type data: holotype, QM W2876 F. adult, paratype QM W2877 M. adult, from Toowoomba, Qld.

Distribution: NE coastal, Qld. Ecology: tall open forest, terrestrial, burrowing, predator; dry vine forest.

Conothele Thorell, 1878

Conothele Thorell, T. (1878). Studi sui ragni Malesi e Papuani. Pt. 2. Ragni di Amboina raccolti dal Prof. O. Beccari. *Ann. Mus. Civ. Stor. Nat. Genova* **13**: 1–317 [303]. Type species *Cteniza malayana* Doleschall, 1859 by original designation.

This group is also found on the island of New Guinea, adjacent islands *e.g.* Bismarck Archipelago, Solomon Ils., Vanuatu, Malaysia and Micronesia, see Bonnet, P. (1956). *Bibliographia Araneorum.* Toulouse : Douladoure Vol. 2 [1197]; Roewer, C.Fr. (1963). Insects of Micronesia Araneina; Orthognatha, Labidognatha. *Insects of Micronesia* **3**(4): 105–132 1 map. Species now known not to occur in Australia: *Conothele malayana* (Doleschall, 1859), Main, B.Y., this work.

Conothele arboricola Pocock, 1898

Conothele arboricola Pocock, R.I. (1898). Scorpions, Pedipalpi and spiders collected by Dr. Willey in New Britain, the Solomon Islands, Loyalty Islands, etc. pp. 95–120 *in* Willey, A. *Zoological Results Based on Material from New Britain, New Guinea, Loyalty Islands and Elsewhere.* Pt. I Cambridge : Cambridge Univ. Press [101 pl 10 fig 3]. Type data: holotype, BMNH 1898.12.5.9 F. adult, from New Britain.

Distribution: Gulf of Carpentaria, Qld.; Thursday Is., Qld., New Britain and New Guinea(?). Ecology: closed forest, tall open forest, predator; trapdoor nest in bark of trees. Biological references: Crome, W. (1962). Die Wohnröhre von *Conothele arboricola* Pocock nebst Bemerkungen Über die Lebenweise dieser Falltürspinne. *Zool. Anz.* **168**: 450–459 (nest structure).

Conothele doleschalli Thorell, 1881

Conothele doleschalli Thorell, T. (1881). Studi sui ragni Malesi e Papuani. Pt. III. Ragni dell'Austro-Malesiae del Capo York, conservati del Museo Civico di Storia Naturale di Genova. *Ann. Mus. Civ. Stor. Nat. Genova* **17**: 1–720 pls 7–27 (Territelariae, pp. 236–254) [6 May 1881] [237]. Type data: syntypes, MCG F. juv., F. ?lost, from Somerset, Cape York, Qld. (Fly River specimen lost?).

Distribution: NE coastal, N Gulf, Qld.; Somerset, Cape York, also SE Papua New Guinea. Ecology: woodland, terrestrial, burrowing, predator; burrow with trapdoor.

Eucyrtops Pocock, 1897

Eucyrtops Pocock, R.I. (1897). On some trapdoor spiders of the family Ctenizidae from South and West Australia, contained in the collection of the British Museum. *Ann. Mag. Nat. Hist. (6)* **19**: 109–116 [Jan. 1897] [113]. Type species *Aganippe latior* Cambridge, 1877 by original designation.

Eucyrtops eremaea Main, 1957

Eucyrtops eremaea Main, B.Y. (1957). Biology of aganippine trapdoor spiders (Mygalomorphae : Ctenizidae). *Aust. J. Zool.* **5**: 402–473 pls 1–6 [Dec. 1957] [421]. Type data: holotype, WAM 64–60 M. adult, paratype WAM F. adult, from Cardinia Creek, Malcolm, W.A.

Distribution: W plateau, W.A.; Malcolm-Laverton area only. Ecology: open shrubland, open scrub, terrestrial, burrowing, predator; mulga predominant cover, clay loams of braided watercourses, burrow with trapdoor.

Eucyrtops latior (Cambridge, 1877)

Aganippe latior Cambridge, O.P. (1877). On some new genera and species of Araneidae. *Ann. Mag. Nat. Hist. (6)* **19**: 26–39 pls 6–7 [29] [for description of WAM male designated allotype, see Main, B.Y. (1957). Biology of aganippine trapdoor spiders (Mygalomorphae : Ctenizidae). *Aust. J. Zool.* **5**: 402–473 pls 1–6 (416)]. Type data: holotype, BMNH 59.52.47.109 F. adult, from W.A.

Distribution: SW coastal, W.A.; Darling Ranges, Perth coastal plain. Ecology: open forest, terrestrial, burrowing, predator; in heavily littered sandy and loam soils, burrow with thin trapdoor. Biological references: Pocock, R.I. (1897). On some trapdoor spiders of the family Ctenizidae from South and West Australia, contained in the collection of the British Museum. *Ann. Mag. Nat. Hist. (6)* **19**: 109–116 (taxonomy); Main, B.Y. (1957). Biology of aganippine trapdoor spiders

(Mygalomorphae : Ctenizidae). *Aust. J. Zool.* **5**: 402–473 pls 1–6 (natural history, habitat, burrow, feeding, seasonal life; distribution may refer in part to undescribed species).

Eucyrtops riparia Main, 1957

Eucyrtops riparia Main, B.Y. (1957). Biology of aganippine trapdoor spiders (Mygalomorphae : Ctenizidae). *Aust. J. Zool.* **5**: 402–473 pls 1–6 [Dec. 1957] [419]. Type data: holotype, WAM 64–59 F. adult, paratype WAM M. adult, from 1.6 km W of Mt. Misery, Moora, W.A.

Distribution: SW coastal, W.A.; from west of Moora to Northhampton and Abrolhos Ils. Ecology: woodland, low woodland, terrestrial, burrowing, predator; creek banks, friable sedimentary soils, burrow with stiff, flap-like trapdoor. Biological references: Main, B.Y. (1957). Biology of aganippine trapdoor spiders (Mygalomorphae : Ctenizidae). *Aust. J. Zool.* **5**: 402–473 pls 1–6 (natural history, habitat, burrow, feeding, seasonal life; refers in part to several undescribed species included under *E. riparia*).

Homogona Rainbow, 1914

Homogona Rainbow, W.J. (1914). Studies in Australian Araneidae. No. 6. The Terretelariae. *Rec. Aust. Mus.* **10**: 187–270 [15 Aug. 1914] [189]. Type species *Homogona pulleinei* Rainbow, 1914 by monotypy.

Homogona cunicularius Main, 1983

Homogona cunicularius Main, B.Y. (1983). Systematics of the trapdoor spider genus *Homogona* Rainbow (Mygalomorphae : Ctenizidae : Homogoninae). *J. Aust. Entomol. Soc.* **22**: 81–92 figs 1–29 [89]. Type data: holotype, QM W2472 F. adult, paratype QM W3654 M. adult, from North Cedar Creek (between Atherton and Ravenshoe), Qld.

Distribution: NE coastal, Qld; northern mountainous areas only. Ecology: closed forest, burrowing, terrestrial, predator; burrow with collar-like entrance and no door. Biological references: Main, B.Y. (1983). Systematics of the trapdoor spider genus *Homogona* Rainbow (Mygalomorphae : Ctenizidae : Homogoninae). *J. Aust. Entomol. Soc.* **22**: 81–92 figs 1–29.

Homogona pulleinei Rainbow, 1914

Homogona pulleinei Rainbow, W.J. (1914). Studies in Australian Araneidae. No. 6. The Terretelariae. *Rec. Aust. Mus.* **10**: 187–270 [15 Aug. 1914] [189]. Type data: holotype, AM K18299, KS6394 F. adult, from Lismore, N.S.W.

Distribution: NE coastal, SE coastal, Murray-Darling basin, Qld., N.S.W.; in southeastern Qld.-northeastern N.S.W. border area, *e.g.* MacPherson Ranges. Ecology: closed forest, terrestrial, burrowing, predator; burrow with thin flap-like trapdoor in soil, logs and tree butts. Biological references: Main, B.Y. (1976). *Spiders.* Sydney : Collins 296 pp. (natural history, habitat).

Idiosoma Ausserer, 1871

Idiosoma Ausserer, A. (1871). Beiträge zur Kenntniss der Arachniden-Familie der Territelariae Thorell (Mygalidae Autor.). *Verh. Zool.-Bot. Ges. Wien* **21**: 117–224 [150]. Type species *Idiops sigillatus* Cambridge, 1870 by monotypy.

Idiosoma hirsutum Main, 1952

Idiosoma hirsutum Main, B.Y. (1952). Notes on the genus *Idiosoma*, a supposedly rare Western Australian trap-door spider. *West. Aust. Nat.* **3**: 130–137 (15 Sept. 1952) [131]. Type data: holotype, WAM 29–1060 F. adult, paratype WAM 1914–1069 F. adult, from Victoria Park, W.A.

Distribution: SW coastal, W.A.; Swan River coastal plain, edge of Darling scarp. Ecology: low open forest, shrubland, terrestrial, burrowing, predator; burrow with flap door and twiglines in *Casuarina* and *Jacksonia* litter. Biological references: Main, B.Y. (1957). Biology of aganippine trapdoor spiders (Mygalomorphae : Ctenizidae). *Aust. J. Zool.* **5**: 407–473 pls 1–6 (considered as a hybrid between *Idiosoma sigillatum* and *Aganippe raphiduca* but here regarded as a valid species with restricted distribution).

Idiosoma nigrum Main, 1952

Idiosoma nigrum Main, B.Y. (1952). Notes on the genus *Idiosoma*, a supposedly rare Western Australian trap-door spider. *West. Aust. Nat.* **3**: 130–137 (15 Sept. 1952) [133] [WAM 40–1070 M. adult designated allotype by Main, B.Y. (1957). Biology of aganippine trapdoor spiders (Mygalomorphae : Ctenizidae). *Aust. J. Zool.* **5**: 402–473 pls 1–6]. Type data: holotype, WAM 64–58 F. adult, paratype AM K67724, KS6392 F. adult, from The Wongan Hills, W.A.

Distribution: SW coastal, W.A.; from eastern edge of Darling Ranges to Kellerberrin in the east, Murchison River in the north (western part of wheatbelt). Ecology: low open woodland, terrestrial, burrowing, predator; in York gum (*Eucalyptus loxophleba*) and jam (*Acacia acuminata*) woodlands, burrow with thin trapdoor and fan of twig-lines in litter, loamy-clay soils. Biological references: Main, B.Y. (1957). Biology of aganippine trapdoor spiders (Mygalomorphae : Ctenizidae). *Aust. J. Zool.* **5**: 402–473 pls 1–6 (habitat, burrow, feeding, seasonal behaviour).

Idiosoma sigillatum (Cambridge, 1870)

Idiops sigillatus Cambridge, O.P. (1870). Monograph of the genus *Idiops*, including descriptions of several species new to science. *Proc. Zool. Soc. Lond.* **1870**: 101–108 pl 8 [10 Feb. 1870] [105] [BMNH 93.7.4.47 F. adult described as *Idiosoma sigillatum* (Cambridge) by Pocock,

R.I. (1897). On some trapdoor spiders of the family Ctenizidae from South and West Australia, contained in the collection of the British Museum. *Ann. Mag. Nat. Hist. (6)* **19**: 109–116]. Type data: holotype, OUM M. adult, from Swan River, W.A.

Distribution: SW coastal, W coast, W.A.; Darling Ranges (northern part only), Swan River coastal plain, Rottnest Is. Ecology: open forest, terrestrial, burrowing, predator; in *Casuarina* groves in jarrah forest, burrow in litter on sandy-gravel soils, with thin trapdoor and fan of twig-lines. Biological references: Ausserer, A. (1871). Beiträge zur Kenntniss der Arachniden-Familie der Territelariae Thorell (Mygalidae Autor.). *Verh. Zool.-Bot. Ges. Wien* **21**: 117–224 (taxonomy); Main, B.Y. (1957). Biology of aganippine trapdoor spiders (Mygalomorphae : Ctenizidae). *Aust. J. Zool.* **5**: 402–473 pls 1–6 (distribution, habitat, burrow, seasonal life).

Misgolas Karsch, 1878

Misgolas Karsch, F. (1878). Exotisch-Araneologisches. II. *Zeit. Ges. Naturw.* **51**: 771–826 pl 9 [821]. Type species *Misgolas rapax* Karsch, 1878 by original designation.

Hermeas Karsch, F. (1878). Exotisch-Araneologisches. II. *Zeit. Ges. Naturw.* **51**: 771–826 pl 9 [823]. Type species *Hermeas crispus* Karsch, 1878 by original designation.

Dyarcyops Hogg, H.R. (1902). On some additions to the Australian spiders of the suborder Mygalomorphae. *Proc. Zool. Soc. Lond.* **1902(2)**: 121–142 [3 June 1902] [130 pl 13]. Type species *Dyarcyops andrewsi* Hogg, 1902 by original designation.

Megalosara Rainbow, W.J. (1914). Studies in Australian Araneidae. No.6. The Terretelariae. *Rec. Aust. Mus.* **10**: 187–270 [15 Aug. 1914] [205]. Type species *Megalosara villosa* Rainbow, 1914 by monotypy.

Taxonomic decision of Main, B.Y., this work.

This group is also found in Papua New Guinea(?) (in *Dyarcyops*) and New Zealand (in *Cantuaria*), see Main, B.Y. (1981). Eco-evolutionary radiation of mygalomorph spiders in Australia. pp. 853–872 *in* Keast, A. (ed.) *Ecological Biogeography of Australia.* The Hague : Junk Vol. 2 [856]. Two species, *Arbanitis bradleyi* Rainbow and *A. villosus* Rainbow [Rainbow, W.J. (1920). Trapdoor spiders of the *Chevert* Expedition. *Rec. Aust. Mus.* **13**: 77–85], allegedly from Papua New Guinea, transferred to *Dyarcyops* Hogg, 1902 by Main, B.Y. (1977). Preliminary notes towards a revision of the mygalomorph spider genus *Dyarcyops* (Ctenizidae). *Aust. Entomol. Mag.* **4**: 69–72, are here placed in *Misgolas* Karsch, 1878.

Misgolas andrewsi (Hogg, 1902)

Dyarcyops andrewsi Hogg, H.R. (1902). On some additions to the Australian spiders of the sub-order Mygalomorphae. *Proc. Zool. Soc. Lond.* **1902(2)**: 121–142 pl 13 [3 June 1902] [130] [lectotype originally a cotype; erroneously referred to as a holotype by Main, B.Y. (1977). Preliminary notes towards a revision of the mygalomorph spider genus *Dyarcyops* (Ctenizidae). *Aust. Entomol. Mag.* **4**(4): 69–72]. Type data: lectotype, BMNH 1924.III.1.19–20 F. adult, paralectotype BMNH 1924.III.1.19–20 juv., probable paralectotype SAMA F. adult, from Mt. Compass, S.A., designation by Main, B.Y., this work.

Taxonomic decision of Main, B.Y., this work (*nov. comb.*).

Distribution: S Gulfs, S.A.; Mt. Lofty Ranges, Mt. Compass only. Ecology: open forest, terrestrial, burrowing, predator; burrow with open entrance (no trapdoor). Biological references: Main, B.Y. (1977). Preliminary notes towards a revision of the mygalomorph spider genus *Dyarcyops* (Ctenizidae). *Aust. Entomol. Mag.* **4**: 69–72 (taxonomy, as *Dyarcyops andrewsi*).

Misgolas biroi (Kulczynski, 1908)

Dyarcyops biroi Kulczynski, W. (1908). Araneae Musei Nationalis Hungarici in regionibus Indici et Australica a Ludovici Biro collectae. *Ann. Hist.-Nat. Mus. Natl. Hung.* **6**: 428–494 pl 9 (15 Dec. 1908) [435]. Type data: holotype, IZW F 1509 F. ?immature, from Mt. Victoria, Blue Mts., N.S.W.

Taxonomic decision of Main, B.Y., this work (*nov. comb*).

Distribution: SE coastal, N.S.W.; Mt. Victoria, Blue Mts. Ecology: tall open forest, terrestrial, burrowing, predator; probably open burrow in dense understorey of gullies, possibly in montane heath.

Misgolas crispus (Karsch, 1878)

Hermeas crispus Karsch, F. (1878). Exotisch-Araneologisches. II. *Zeit. Ges. Naturw.* **51**: 771–826 pl 9 [823]. Type data: holotype, ZMB 2093 F. adult, from Tas. (as Van Diemensland).

Arbanitis scaurus Hickman, V.V. (1927). Studies in Tasmanian spiders. Pt. I. *Pap. Proc. R. Soc. Tasm.* **1926**: 52–86 pls 4–11 [24 Feb. 1927] [57]. Type data: syntypes, QVM 1957/13/12 *M. adult, QVM 1957/13/13 *F. adult, from Westmoreland Falls, Mole Creek, Tas.

Taxonomic decision of Main, B.Y., this work (*nov. comb.*).

Distribution: Tas. Ecology: tall open forest, terrestrial, burrowing, predator; burrow without a trapdoor. Biological references: Hitchman, V.V. (1927). Studies in Tasmanian spiders. Pt. 1. *Pap. Proc. R. Soc. Tasm.* **1926**: 52–86, pls 4–11 (as *Arbantis scaurus*).

Misgolas elegans (Rainbow and Pulleine, 1918)

Arbanitis elegans Rainbow, W.J. & Pulleine, R.H. (1918). Australian trap-door spiders. *Rec. Aust. Mus.* **12**:

81–169 pls 12–24 [24 Dec. 1918] [115]. Type data: holotype, AM K40889, KS6261 F. adult, from Kaianga, Narooma, N.S.W.

Taxonomic decision of Main, B.Y., this work (*nov. comb.*).

Distribution: SE coastal, N.S.W.; known only from Kaianga, Narooma. Ecology: terrestrial, burrowing, predator.

Misgolas gracilis (Rainbow and Pulleine, 1918)

Arbanitis gracilis Rainbow, W.J. & Pulleine, R.H. (1918). Australian trap-door spiders. *Rec. Aust. Mus.* **12**: 81–169 pls 12–24 [24 Dec. 1918] [110]. Type data: holotype, AM K40879, KS6262 F. adult, paratypes AM K41444 3F. adult lost, from The Domain, Sydney, N.S.W.

Taxonomic decision of Main, B.Y., this work, *nov. comb.*

Distribution: SE coastal, N.S.W.; Sydney & environs, W into Blue Mts. and northward (probably confined mainly to Hawksbury sandstones). Ecology: open forest, terrestrial, burrowing, predator; burrow with thin trapdoor in friable soil. Biological references: Main, B.Y. (1976). *Spiders.* Sydney : Collins 296 pp. (burrow, as *Dyarcyops gracilis*).

Misgolas hirsutus (Rainbow and Pulleine, 1918)

Arbanitis hirsutus Rainbow, W.J. & Pulleine, R.H. (1918). Australian trap-door spiders. *Rec. Aust. Mus.* **12**: 81–169 pls 12–24 [24 Dec. 1918] [116]. Type data: holotype, AM K40886, KS6263 F. adult, from Kedron Brook, Brisbane, Qld.

Taxonomic decision of Main, B.Y., this work (*nov. comb.*).

Distribution: SE coastal, NE coastal, Qld., N.S.W.; known only from MacPherson Ranges *e.g.* Tamborine Mt., Lamington Plateau, Tooloom Scrub, Brisbane area and Mt. Glorious. Ecology: closed forest, tall open forest, terrestrial, burrowing, predator; burrow with no trapdoor. Biological references: Main, B.Y. (1976). *Spiders.* Sydney : Collins 296 pp. (natural history, habitat, burrow structure as *Dyarcyops hirsutus*); Main, B.Y. (1977). Preliminary notes towards a revision of the mygalomorph spider genus *Dyarcyops* (Ctenizidae). *Aust. Entomol. Mag.* **4**: 69–72 (taxonomy).

Misgolas maculosus (Rainbow and Pulleine, 1918)

Dyarcyops maculosus Rainbow, W.J. & Pulleine, R.H. (1918). Australian trap-door spiders. *Rec. Aust. Mus.* **12**: 81–169 pls 12–24 (24 Dec. 1918) [108]. Type data: syntypes, AM K41614 F. adult lost, AM K41615 3F. adult lost, from La Perouse, Botany, N.S.W.

Taxonomic decision of Main, B.Y., this work (*nov. comb.*).

Distribution: SE coastal, N.S.W.; Sydney environs only. Ecology: open forest, terrestrial, burrowing, predator.

Misgolas mestoni (Hickman, 1928)

Arbanitis mestoni Hickman, V.V. (1928). Studies in Tasmanian spiders. Pt. II. *Pap. Proc. R. Soc. Tasm.* **1927**: 158–175 pls 21–26 [29 Feb. 1928] [162]. Type data: syntypes, QVM 1957/13/14 *M. adult, QVM 1957/13/15 *F. adult, from Woodsdale, Tas.

Taxonomic decision of Main, B.Y., this work (*nov. comb.*).

Distribution: Tas. Ecology: terrestrial, burrowing, predator; burrow without trapdoor in grassy sites.

Misgolas ornata (Rainbow, 1914)

Idioctis ornata Rainbow, W.J. (1914). Studies in Australian Araneidae. No. 6. The Terretelariae. *Rec. Aust. Mus.* **10**: 187–270 [15 Aug. 1914] [230]. Type data: syntypes, AM K34491, KS1370 4M.,F. adult, from Upper Burnett River, Qld. but Eidsvold, Qld. on specimen label.

Aganippe ornata Rainbow, W.J. & Pulleine, R.H. (1918). Australian trap-door spiders. *Rec. Aust. Mus.* **12**: 81–169 pls 12–24 [24 Dec. 1918] [99]. Type data: syntypes, AM K40850, KS6154 2F. adult, from Eidsvold, Qld.

Taxonomic decision of Main, B.Y., this work (*nov. comb.*).

Distribution: Murray-Darling basin, NE coastal, Qld.; Eidsvold, Cracow, Upper Burnett River area. Ecology: open forest, terrestrial, burrowing, predator; burrow Y-shaped, sometimes in dry vine forests, clay and alluvial soils.

Misgolas papilliosus (Rainbow and Pulleine, 1918)

Arbanitis papilliosus Rainbow, W.J. & Pulleine, R.H. (1918). Australian trap-door spiders. *Rec. Aust. Mus.* **12**: 81–169 pls 12–24 [24 Dec. 1918] [118]. Type data: holotype, AM K40887, KS6266 F. adult, from Tamborine Mt., Qld.

Taxonomic decision of Main, B.Y., this work (*nov. comb.*).

Distribution: NE coastal, Qld.; Tamborine Mt. only. Ecology: closed forest, terrestrial, burrowing, predator.

Misgolas pulchellus (Rainbow and Pulleine, 1918)

Arbanitis pulchra Rainbow, W.J. & Pulleine, R.H. (1918). Australian trapdoor spiders. *Rec. Aust. Mus.* **12**: 81–169 pls 12–24 [24 Dec. 1918] [86] [*nom. nud.*; see *Arbanitis pulchellus* Rainbow and Pulleine, 1918].

Arbanitis pulchellus Rainbow, W.J. & Pulleine, R.H. (1918). Australian trap-door spiders. *Rec. Aust. Mus.* **12**: 81–169 pls 12–24 [24 Dec. 1918] [114] [*nom. nov.* for *Arbanitis pulchra* Rainbow and Pulleine, 1918]. Type data: holotype?, probably AM K40881, KS6327 F. adult (mislabelled *Arbanitis pulchra* by Rainbow), from Tamborine Mt., Qld.

Aname pulchra Rainbow, W.J. & Pulleine, R.H. (1918). Australian trap-door spiders. *Rec. Aust. Mus.* **12**: 81–169 pls 12–24 [24 Dec. 1918] [153]. Type data: holotype, AM K40933, KS8375 M. adult, from Tamborine Mt., Qld.

Taxonomic decision of Main, B.Y., this work (*nov. comb.*).

Distribution: NE coastal, SE coastal, Qld., N.S.W.; MacPherson Ranges. Ecology: closed forest, tall open forest, terrestrial, burrowing, predator; burrow with "folding-door" collar in bare patches of soil. Biological references: Main, B.Y. (1976). *Spiders.* Sydney : Collins 296 pp. (habitat, nest structure, ecological associations, as *Dyarcyops pulchellus*); there is a type specimen of *Arbanitis pulchra* but no description, whereas there is a description of *Arbanitis pulchellus* but no type specimen labelled as such. Rainbow probably mislabelled the type as *pulchra* and Rainbow and Pulleine discussed the species under this name in their introduction, see Rainbow, W.J. & Pulleine, R.H. (1918). Australian trap-door spiders. *Rec. Aust. Mus.* **12**: 81–169, pls 12–24.

Misgolas rapax Karsch, 1878

Misgolas rapax Karsch, F. (1878). Exotisch-Araneologisches. II. *Zeit. Ges. Naturw.* **51**: 771–826 pl 9 [823]. Type data: holotype, ZMB 2094 F. adult, from N.S.W.

Megalosara villosa Rainbow, W.J. (1914). Studies in Australian Araneidae. No. 6. The Terretelariae. *Rec. Aust. Mus.* **10**: 187–270 (15 Aug. 1914) [206]. Type data: holotype, AM K16199, KS7178 M. adult, from Enfield, Sydney, N.S.W.

Arbanitis fuscipes Rainbow, W.J. (1914). Studies in Australian Araneidae. No. 6. The Terretelariae. *Rec. Aust. Mus.* **10**: 187–270 [15 Aug. 1914] [210]. Type data: holotype, AM K30279, KS6336 F. adult, from Willoughby, N.S.W.

Dyarcyops melancholicus Rainbow, W.J. & Pulleine, R.H. (1918). Australian trap-door spiders. *Rec. Aust. Mus.* **12**: 81–169 pls 12–24 (24 Dec. 1918) [106]. Type data: syntypes, AM K40885, KS1631 F.,M. adult, SAMA F. adult, from Clifton Gardens, Sydney, N.S.W.

Dyarcyops ionthus Rainbow, W.J. & Pulleine, R.H. (1918). Australian trap-door spiders. *Rec. Aust. Mus.* **12**: 81–169 pls 12–24 (24 Dec. 1918) [107]. Type data: holotype, AM K41284, KS6337 F. adult, from Burwood, N.S.W.

Arbanitis montanus Rainbow, W.J. & Pulleine, R.H. (1918). Australian trap-door spiders. *Rec. Aust. Mus.* **12**: 81–169 pls 12–24 [24 Dec. 1918] [116]. Type data: syntypes, AM K40888, KS6265 M.,F. adult, from Jenolan Caves, N.S.W.

Arbanitis chisholmi Hickman, V.V. (1933). A new ctenizid spider from New South Wales. *Ann. Mag. Nat. Hist. (10)* **12**: 210–216 [Aug. 1933] [212]. Type data: holotype, AM K66900, KS6260 F. adult, from Comboyne, N.S.W.

Taxonomic decision of Main, B.Y., this work.

Distribution: SE coastal, N.S.W.; Sydney environs, Blue Mts., coastal area to north, mainly sandstone country. Ecology: open forest, terrestrial, burrowing, predator; burrow with open (no trapdoor) collar-like entrance especially in undergrowth and grassy areas. Biological references: Musgrave, A. (1949). Spiders harmful to man. *Aust. Mus. Leaflet* No. 16: 1–13 (record of bite, as *Arbanitis fuscipes*); Main, B.Y. (1964). *Spiders of Australia.* Brisbane : Jacaranda 124 pp. (natural history, as *Dyarcyops fuscipes*); Mascord, R. (1970). *Australian Spiders in Colour.* Sydney : Reed 112 pp. (natural history, as *Dyarcyops fuscipes*); Main, B.Y. (1976). *Spiders.* Sydney : Collins 296 pp. (natural history, as *Dyarcyops fuscipes*).

Misgolas robertsi (Main and Mascord, 1974)

Dyarcyops robertsi Main, B.Y. & Mascord, R. (1974). Description and natural history of a tube-building" species of *Dyarcyops* from New South Wales and Queensland (Mygalomorphae : Ctenizidae). *J. Entomol. Soc. Aust. (N.S.W.)* **8**: 15–21 (31 Jan. 1974) [17]. Type data: holotype, AM KS12 F. adult, paratypes AM KS13 F. adult, AM KS16 M. adult, from Minnamurra Falls, Kiama, N.S.W.

Taxonomic decision of Main, B.Y., this work (*nov. comb.*).

Distribution: NE coastal, SE coastal, Qld., N.S.W.; mountainous areas only from Minnamurra Falls, N.S.W. to about Yarraman, Qld. Ecology: closed forest, terrestrial, burrowing, predator; burrow with aerial tubes up to about 30 cm.

Misgolas villosus (Rainbow, 1920)

Arbanitis villosus Rainbow, W.J. (1920). Trap-door spiders of the *Chevert* Expedition. *Rec. Aust. Mus.* **13**: 77–85 [4 Dec. 1920] [78] [the syntypes are allegedly from Papua New Guinea, but may have come from Australia]. Type data: syntypes, AM K41699 F.,M. adult, AM K41700 F. adult, from Papua, Papua New Guinea?.

Arbanitis bradleyi Rainbow, W.J. (1920). Trapdoor spiders of the *Chevert* Expedition. *Rec. Aust. Mus.* **13**: 77–85 [4 Dec. 1920] [80] [this species is allegedly from Papua New Guinea, but may have come from Australia]. Type data: syntypes, AM K41701 M. adult, F. juv., AM K41702 M. adult, from Papua, Papua New Guinea?.

Idioctis papuensis Rainbow, W.J. (1920). Trap-door spiders of the *Chevert* Expedition. *Rec. Aust. Mus.* **13**: 77–85 [82] [the type is allegedly from New Guinea, but may have come from Australia]. Type data: holotype, AM K41704, KS1369 F. adult, from Papua, Papua New Guinea?.

Taxonomic decision of Main, B.Y., this work (*nov. comb.*), species status doubtful, may be synonymous with an Australian species.

Distribution: if from Australia, unknown.

DIPLURIDAE

INTRODUCTION

Eyes in compact group of two rows; sometimes reduced or absent. Labium separated from sternum by a pair of sigilla which may be fused into a single broad sigillate area; rarely with cuspules, if present then very few. Chelicerae with or without rastellum (or pseudorastellum), never a process. Cheliceral furrow usually with a single row of macroteeth (promarginal) and with basal group of small teeth (microteeth) on retromargin, the latter occasionally extend about halfway along retromargin. Two pairs of spinnerets; posterior lateral pair with tapering terminal segment. Paired tarsal claws with single or double row of teeth. Burrowers, usually terrestrial (rarely arboreal); burrow rarely with door(s), sometimes with expansive silk sheet or curtain-like silk threads over entrance. Tropical to sub-temperate on all continents extending to Polynesia, very abundant throughout mainland Australia and Tas. Three subfamilies: the Masteriinae (*Masteria*); the Ischnothelinae (*Cethegus, Stenygrocercus, Australothele, Carrai* and *Namirea*); the Diplurinae (*Aname, Chenistonia, Ixamatus, Kiama, Kwonkan, Merredinia, Namea, Stanwellia, Teyl, Troglodiplura* and *Xamiatus*).

Aname L. Koch, 1873

Aname Koch, L. (1873). *Die Arachniden Australiens.* Lf. 9 pp. 409–472 pls 33–36 Nürnberg : Bauer & Raspe [465]. Type species *Aname pallida* L. Koch, 1873 by monotypy.

Dekana Hogg, H.R. (1902). On some additions to the Australian spiders of the suborder Mygalomorphae. *Proc. Zool. Soc. Lond.* **1902(2)**: 121–142 pl 13 [3 June 1902] [138]. Type species *Dekana diversicolor* Hogg, 1902 by original designation.

Sungenia Rainbow, W.J. & Pulleine, R.H. (1918). Australian trap-door spiders. *Rec. Aust. Mus.* **12**: 81–169 pls 12–24 [24 Dec. 1918] [162]. Type species *Chenistonia (Dekana) atra* Strand, 1913 by monotypy.

Dolichosternum Rainbow, W.J. & Pulleine, R.H. (1918). Australian trap-door spiders. *Rec. Aust. Mus.* **12**: 81–169 pls 12–24 [24 Dec. 1918] [168]. Type species *Dolichosternum attenuatum* Rainbow and Pulleine, 1918 by monotypy.

Taxonomic decision of Main, B.Y. (1982). Notes on the revised taxonomic position of the black wishbone spider *Dekana diversicolor* Hogg (Mygalomorphae : Dipluridae). *J. R. Soc. West. Aust.* **65**: 25–29; Raven, R.J. (1981). A review of the Australian genera of the mygalomorph spider subfamily Diplurinae (Dipluridae, Chelicerata). *Aust. J. Zool.* **29**: 321–363 [310].

Aname armigera Rainbow and Pulleine, 1918

Aname armigera Rainbow, W.J. & Pulleine, R.H. (1918). Australian trap-door spiders. *Rec. Aust. Mus.* **12**: 81–169 pls 12–24 [24 Dec. 1918] [150]. Type data: holotype, AM K40922, KS8193 F. adult, from Mullewa, W.A.

Distribution: SW coastal, W.A.; NE wheatbelt and goldfields. Ecology: low woodland, terrestrial, burrowing, predator; in red sandy-loam soils.

Aname aurea Rainbow and Pulleine, 1918

Aname aurea Rainbow, W.J. & Pulleine, R.H. (1918). Australian trap-door spiders. *Rec. Aust. Mus.* **12**: 81–169 pls 12–24 [24 Dec. 1918] [145]. Type data: holotype, AM K40938, KS8204 F. adult, paratype SAMA N198162 F. adult, from Broken Hill, N.S.W.

Distribution: Murray-Darling basin, N.S.W.; Broken Hill. Ecology: terrestrial, burrowing, predator; associated with areas dominated by salt bush.

Aname coenosa Rainbow and Pulleine, 1918

Aname coenosa Rainbow, W.J. & Pulleine, R.H. (1918). Australian trap-door spiders. *Rec. Aust. Mus.* **12**: 81–169 pls 12–24 [24 Dec. 1918] [152]. Type data: holotype, AM K40941, KS8187 F. adult, from Little Kalkabury, Yorke Peninsula, S.A.

Distribution: S Gulfs, S.A.; Little Kalkabury, Yorke Peninsula. Ecology: terrestrial, burrowing, predator.

Aname distincta (Rainbow, 1914)

Ixamatus distinctus Rainbow, W.J. (1914). Studies in Australian Araneidae. No. 6. The Terretelariae. *Rec. Aust. Mus.* **10**: 187–270 [15 Aug. 1914] [237]. Type data: syntypes, AM K34490, KS8188 3F. adult, from Eidsvold, Qld.

Aname villosa Rainbow, W.J. & Pulleine, R.H. (1918). Australian trap-door spiders. *Rec. Aust. Mus.* **12**: 81–169 pls 12–24 [24 Dec. 1918] [141]. Type data: syntypes, AM K40935, KS1395 2F. adult, from Eidsvold, Qld.; Raven, R.J. (1984). A new Diplurid genus from eastern Australia and a related *Aname* species (Dipluridae : Diplurinae : Araneae). *Aust. J. Zool. Suppl. Ser.* **96**: 1–51 [35] infers that a dry specimen from Tamborine Mt. mentioned by Rainbow and Pulleine is a syntype, but this specimen (AM K41450, KS1399) is not a syntype and is not the same species as the two syntypes with the same original registration number (AM K40935) which are both from Eidsvold.

Dolichosternum attenuatum Rainbow, W.J. & Pulleine, R.H. (1918). Australian trap-door spiders. Rec. Aust. Mus. **12**: 81–169 pls 12–24 [24 Dec. 1918] [169]. Type data: holotype, AM K41249, KS8213 F. adult, from Eidsvold, Qld.

Taxonomic decision of Raven, R.J. (1981). A review of the Australian genera of the mygalomorph spider subfamily Diplurinae (Dipluridae : Chelicerata). *Aust. J. Zool.* **29**: 321–363 [338].

Distribution: NE coastal, Qld.; Eidsvold. Ecology: terrestrial, predator.

Aname diversicolor (Hogg, 1902)

Dekana diversicolor Hogg, H.R. (1902). On some additions to the Australian spiders of the suborder Mygalomorphae. *Proc. Zool. Soc. Lond.* **1902(2)**: 121–142 pl 13 [3 June 1902] [138]. Type data: neotype, SAMA N1980196 M. adult, from 26 km NW Elliston, S.A. (original localities of Hogg untraceable, given as Deka Station, Blackhall probably Colona Station, Blackhill), designation by Main, B.Y. (1982). Notes on the revised taxonomic position of the black wishbone spider *Dekana diversicolor* Hogg (Mygalomorphae : Dipluridae). *J. R. Soc. West. Aust.* **65**: 25–29.

Chenistonia (Dekana) atra Strand, E. (1913). Über einige Australische Spinnen des Senckenbergischen Museum. *Zool. Jahrb.* **35**: 599–624 [601]. Type data: holotype, SMF 2664 M. adult, from central Australia.

Aname hirsuta Rainbow, W.J. & Pulleine, R.H. (1918). Australian trap-door spiders. *Rec. Aust. Mus.* **12**: 81–169 pls 12–24 [24 Dec. 1918] [142]. Type data: holotype, AM K40943, KS8199 F. adult, from Mallala, S.A.

Aname comosa Rainbow, W.J. & Pulleine, R.H. (1918). Australian trap-door spiders. *Rec. Aust. Mus.* **12**: 81–169 pls 12–24 [24 Dec. 1918] [143]. Type data: holotype, AM K40939, KS8195 F. adult, from Pichi Richi, S.A.

Taxonomic decision of Main, B.Y. (1982). Notes on the revised taxonomic position of the black wishbone spider *Dekana diversicolor* Hogg (Mygalomorphae : Dipluridae). *J. R. Soc. West. Aust.* **65**: 25–29 [29]; Raven, R.J., (1981) A review of the Australian genera of the mygalomorph spider subfamily Diplurinae (Dipluridae : Chelicerta). *Aust. J. Zool.* **29**: 321–363 [338, *nov. comb. Aname diversicolor* (Hogg)]; Main, B.Y. (1972). The mygalomorph spider genus *Stanwellia* Rainbow and Pulleine (Dipluridae) and its relationship to *Aname* Koch and certain other diplurine genera. *J. R. Soc. West. Aust.* **55**: 100–114 [100].

Distribution: SW coastal, NW coastal, W plateau, S Gulfs, Lake Eyre basin, W.A., S.A. Ecology: low woodland, woodland, open forest, (open heath), terrestrial, burrowing, predator; Y-shaped burrow in sandy and loamy soils. Biological references: Main, B.Y. (1976). *Spiders.* Sydney : Collins 296 pp. (biology, burrow, under *Dekana diversicolor* Hogg).

Aname fuscocincta Rainbow and Pulleine, 1918

Aname fuscocincta Rainbow, W.J. & Pulleine, R.H. (1918). Australian trap-door spiders. *Rec. Aust. Mus.* **12**: 81–169 pls 12–24 [24 Dec. 1918] [153]. [status doubtful, may be synonym of *Aname diversicolor* (Hogg, 1902)] Type data: holotype, AM K40932, KS8205 F. adult (examined in AM in 1954, since lost), from Kalamunda, W.A.

Distribution: SW coastal, W.A.; Kalamunda. Ecology: open forest, terrestrial, burrowing, predator. Biological references: Main, B.Y. (1976). *Spiders.* Sydney : Collins 296 pp.; Main, B.Y. (1981). Eco-evolutionary radiation of mygalomorph spiders. pp. 853–872 *in* Keast, A. (ed.) *Ecological Biogeography of Australia.* The Hague : Junk Vol. 2 [biology, name *Ixamatus fuscocincta* (Rainbow and Pulleine) misapplied to an undescribed species].

Aname giraulti (Rainbow, 1914)

Chenistonia giraulti Rainbow, W.J. (1914). Studies in Australian Araneidae. No. 6. The Terretelariae. *Rec. Aust. Mus.* **10**: 187–270 [15 Aug. 1914] [243] [male syntype (=lectotype) only, or may be synonym of *Aname pallida* L. Koch, 1873; note: other syntypes at AM and QM belong in Theraphosidae, *incertae sedis* (Main, unpublished)]. Type data: lectotype, AM K41453, KS6391 M. adult, from Gordonvale, Qld. (as Nelson, Qld.), designation by Main, B.Y., this work, paralectotypes AM K41453 F. adult now lost but present in 1954, QM W22–23 2F. adult, other F. missing.

Distribution: NE coastal, Qld.; Nelson, but if synonymous with *Aname pallida* L. Koch, then with wider distribution. Ecology: terrestrial, burrowing, predator. Biological references: Raven, R.J. (1981). A review of the Australian genera of the mygalomorph spider subfamily Diplurinae (Dipluridae : Chelicerata). *Aust. J. Zool.* **29**: 321–363 (*nov. comb.*).

Aname grandis Rainbow and Pulleine, 1918

Aname grandis Rainbow, W.J. & Pulleine, R.H. (1918). Australian trap-door spiders. *Rec. Aust. Mus.* **12**: 81–169 pls 12–24 [24 Dec. 1918] [144]. Type data: holotype, AM K40946, KS8202 F. adult, paratype SAMA N198161 F. adult (Woolshed Flat, S.A.), from Pichi Richi, S.A.

Distribution: S Gulfs, S.A.; southern Flinders and Gawler Ranges. Ecology: low woodland, burrowing, terrestrial, predator; in sandy and loamy soils.

Aname kirrama Raven, 1984

Aname kirrama Raven, R.J. (1984). A new diplurid genus from eastern Australia and related *Aname* species (Diplurinae : Dipluridae : Araneae). *Aust. J. Zool. Supp. Ser.* **96**: 1–51 [26 Jan 1984] [47]. [generic status undecided, believe it is misplaced in *Aname*, Main, B.Y., this work] Type data: holotype, QM S1186 *M. adult, from Kirrama, Qld.

Distribution: NE coastal, Qld. Ecology: closed forest, terrestrial, (burrowing), predator.

Aname maculata (Rainbow and Pulleine, 1918

Ixamatus maculatus Rainbow, W.J. & Pulleine, R.H. (1918). Australian trap-door spiders. *Rec. Aust. Mus.* **12**: 81–169 pls 12–24 [24 Dec. 1918] [163]. Type data: lectotype (with paralectotype), AM K40956, KS7778 2F. adult, from Armadale, W.A., designation by Main, B.Y., this work.

Taxonomic decision of Main, B.Y. this work (*nov. comb.*).

Distribution: SW coastal, W.A.; Darling scarp and coastal plain near Perth. Ecology: low open forest, closed heath, burrowing, terrestrial, predator; in gullies of the scarp and swamp fringes on the coastal plain.

Aname minor Kulczynski, 1908

Aname minor Kulczynski, W. (1908). Araneae musei nationalis Hungarici in regionibus Indici et Australiica a Ludovici Biro collectae. *Ann. Hist.-Nat. Mus. Natl. Hung.* **6**: 428–494 pl 9 [15 Dec. 1908] [457]. Type data: holotype, IZW 1523 F. adult, from Mt. Victoria, Blue Mts., N.S.W.

Distribution: SE coastal, N.S.W.; Mt. Victoria. Ecology: terrestrial, burrowing, predator.

Aname pallida L. Koch, 1873

Aname pallida Koch, L. (1873). *Die Arachniden Australiens.* Lf. 9 pp. 409–472 pls 33–36 Nürnberg : Bauer & Raspe [465 pl 35 fig 8]. Type data: holotype, ZMH MGH 8104 M. immature, from Bowen, Qld.

Distribution: NE coastal, Qld.; mideastern Qld. Ecology: terrestrial, burrowing, predator; open sclerophyll forest and vine thickets in coastal areas. Biological references: Raven, R.J. (1981). A review of the Australian genera of the mygalomorph spider subfamily Diplurinae (Dipluridae : Chelicerata). *Aust. J. Zool.* **29**: 321–363 (distribution, habitat, description of adults).

Aname robusta Rainbow and Pulleine, 1918

Aname robusta Rainbow, W.J. & Pulleine, R.H. (1918). Australian trap-door spiders. *Rec. Aust. Mus.* **12**: 81–169 pls 12–24 [24 Dec. 1918] [154]. Type data: holotype, AM K40931, KS8203 F. adult, paratype SAMA N198160 F. adult, from Cross Roads, Mannum, S.A.

Distribution: S Gulfs, S.A.; Cross Roads, Mannum. Ecology: terrestrial, burrowing, predator.

Aname tasmanica Hogg, 1902

Aname tasmanica Hogg, H.R. (1902). On some additions to the Australian spiders of the suborder Mygalomorphae. *Proc. Zool. Soc. Lond.* **1902(2)**: 121–142 pl 13 [3 June 1902] [140]. Type data: holotype, BMNH 1924.III.1.25.F86 F. adult, from Table Cape, Tas.

Distribution: Tas.; Table Cape. Ecology: terrestrial, burrowing, predator.

Australothele Raven, 1984

Australothele Raven, R. J. (1984). Systematics of the Australian Curtain-web spiders (Ischnothelinae : Dipluridae : Chelicerata). *Aust. J. Zool. Supp. Ser.* **93**: 1–102 [5 Mar. 1984] [6]. Type species *Australothele maculata* Raven, 1984 by original designation.

Australothele bicuspidata Raven, 1984

Australothele bicuspidata Raven, R.J. (1984). Systematics of Australian Curtain-web Spiders (Ischnothelinae : Dipluridae : Chelicerata). *Aust. J. Zool. Supp. Ser.* **93**: 1–102 [5 Mar. 1984] [10]. Type data: holotype, QM S578 *F. adult, paratype AM KS4451 F. adult, from Point Lookout, New England Natl. Park, N.S.W.

Distribution: SE coastal, N.S.W.; New England Natl. Park. Ecology: closed forest, terrestrial, (burrowing), predator.

Australothele jamiesoni Raven, 1984

Australothele jamiesoni Raven, R.J. (1984). Systematics of Australian Curtain-web Spiders (Ischnothelinae : Dipluridae : Chelicerata). *Aust. J. Zool. Supp. Ser.* **93**: 1–102 [5 Mar. 1984] [11]. Type data: holotype, QM S550 *M. adult, paratypes QM S750 F. adult (described), QM S551 F. adult, QM S552 F. adult, QM S553 2M. adult, QM S554 2F. adult, QM S555 F. adult, QM S556 4F. adult, QM S557 3F. adult, EUQ M.,F. adult and others in QM & EUQ, from Little Yabba Creek, Conondale Range, SE Qld.

Distribution: NE coastal, SE coastal, Qld., N.S.W.; MacPherson Ranges and other mountains of SE Qld. and NE N.S.W. Ecology: closed forest, terrestrial, arboreal, predator; silk tubes under rocks, in logs and into tree trunks and creviced bark.

Australothele maculata Raven, 1984

Australothele maculata Raven, R.J. (1984). Systematics of Australian Curtain-web Spiders (Ischnothelinae : Dipluridae : Chelicerata). *Aust. J. Zool. Supp. Ser.* **93**:

1–102 [5 Mar. 1984] [7]. Type data: holotype, QM S541 *M. adult, paratypes QM S752 F. adult (described), QM S542 3F. adult, QM S543 F. adult, QM S544 F. adult, QM S545 F. adult, QM S546 F. adult, QM S547 M. adult, QM S548 F. adult, QM S549 F. adult, from Lamington Natl. Park, Qld.

Distribution: NE coastal, Qld. Ecology: terrestrial, predator; silk tubes in logs.

Australothele magna Raven, 1984

Australothele magna Raven, R.J. (1984). Systematics of Australian Curtain-web Spiders (Ischnothelinae : Dipluridae : Chelicerata). *Aust. J. Zool. Supp. Ser.* **93**: 1–102 [5 Mar. 1984] [15]. Type data: holotype, QM S572 *M. adult, paratypes QM S573 F. adult (described), QM S574 F. adult, QM S575 F. adult, QM S576 F. adult, QM S577 2F. adult, QM S754 F. adult, from Wratten's Camp, via Widgee, SE Qld.

Distribution: NE coastal, Qld.; Bunya Mts., Conondale and Blackbutt Ranges. Ecology: closed forest, terrestrial, predator; silk tubes under rocks in vine thickets.

Australothele montana Raven, 1984

Australothele montana Raven, R.J. (1984). Systematics of Australian Curtain-web Spiders (Ischnothelinae : Dipluridae : Chelicerata). *Aust. J. Zool. Supp. Ser.* **93**: 1–102 [5 Mar. 1984] [17]. Type data: holotype, QM S579 *F. adult, paratypes QM S580 M. immature, AM KS8226 F. adult, AM KS8271 F. adult, from New England Natl. Park, N.S.W.

Distribution: SE coastal, N.S.W.; New England Natl. Park. Ecology: closed forest, terrestrial, predator; tubular silk sheet under logs and rocks.

Australothele nambucca Raven, 1984

Australothele nambucca Raven, R.J. (1984). Systematics of Australian Curtain-web Spiders (Ischnothelinae : Dipluridae : Chelicerata). *Aust. J. Zool. Supp. Ser.* **93**: 1–102 [5 Mar. 1984] [19]. Type data: holotype, AM KS8215 *M. adult, paratypes AM KS8272 F. adult (described), AM KS9550 2F. adult, from Sawtell, N.S.W.

Distribution: SE coastal, N.S.W.; NE area only. Ecology: closed forest, terrestrial, predator.

Australothele nothofagi Raven, 1984

Australothele nothofagi Raven, R.J. (1984). Systematics of Australian Curtain-web Spiders (Ischnothelinae : Dipluridae : Chelicerata). *Aust. J. Zool. Supp. Ser.* **93**: 1–102 [5 Mar. 1984] [21]. Type data: holotype, QM S532 *M. adult, QM S534 F. adult (described), paratypes QM S533 M. adult, QM S751 F. adult, QM S535 F. adult, QM S536 3F. adult, QM S537 M. adult, QM S538 F. adult, QM S539 M.,F. adult, QM S540 F. adult, from Lamington Natl. Park, Qld.

Distribution: NE coastal, Qld.; Lamington Natl. Park and adjacent areas. Ecology: closed forest, terrestrial, arboreal, predator; silk sheets with tube in bark of *Nothofagus* butts and the ground.

Carrai Raven, 1984

Carrai Raven, R.J. (1984). Systematics of the Australian Curtain-web spiders (Ischnothelinae : Dipluridae : Chelicerata). *Aust. J. Zool. Supp. Ser.* **93**: 1–102 [5 Mar. 1984] [24]. Type species *Carrai afoveolata* Raven, 1984 by original designation.

Carrai afoveolata Raven, 1984

Carrai afoveolata Raven, R.J. (1984). Systematics of Australian Curtain-web Spiders (Ischnothelinae : Dipluridae : Chelicerata). *Aust. J. Zool. Supp. Ser.* **93**: 1–102 [5 Mar. 1984] [25]. Type data: holotype, QM S581 *M. adult, paratype AM KS4452 F. adult (described), QM S582 F. adult, from Carrai State Forest, N.S.W.

Distribution: SE coastal, N.S.W.; Carrai State Forest. Ecology: closed forest, terrestrial, predator; silk sheet and tube under logs.

Cethegus Thorell, 1881

Cethegus Thorell, T. (1881). Studi sui ragni Malesi e Papuani. Part III. Ragni dell'Austro-Malesi e del Capo York, conservati del Civico di Storia Naturale di Genova. *Ann. Mus. Civ. Stor. Nat. Genova* **17**: 1–720 [240]. Type species *Cethegus lugubris* Thorell, 1881 by original designation.

Palaevagrus Simon, E. (1908). Araneae. Pt. 1 pp. 359–446 *in* Michaelsen, W. & Hartmeyer, R. (eds.) *Die Fauna Südwest-Australiens.* Jena : Fischer Bd. 1 Lf. 12 [365]. Type species *Palaevagrus fugax* Simon, 1908 by monotypy.

Taxonomic decision of Main, B.Y. (1960). The genus *Cethegus* Thorell (Mygalomorphae : Macrothelinae). *J. R. Soc. West. Aust.* **43**: 30–34.

Cethegus barraba Raven, 1984

Cethegus barraba Raven, R.J. (1984). Systematics of Australian Curtain-web Spiders (Ischnothelinae : Dipluridae : Chelicerata). *Aust. J. Zool. Supp. Ser.* **93**: 1–102 [5 Mar. 1984] [31]. Type data: holotype, AM KS1532 *F. adult, paratype AM KS1380 F. adult, from Barraba, N.S.W.

Distribution: SE coastal, N.S.W.; Barraba and Mt. Kaputar Natl. Park. Ecology: open forest, terrestrial, predator; (silk sheet and tube).

Cethegus broomi (Hogg, 1901)

Stenygrocercus broomi Hogg, H.R. (1901). On Australian and New Zealand spiders of the suborder Mygalomorphae. *Proc. Zool. Soc. Lond.* **1901(2)**: 218–279 [4 June 1901] [270]. Type data: syntypes, BMNH F. adult, F. juv., from Hillgrove, N.S.W.

Taxonomic decision of Main, B.Y. (1960). The genus *Cethegus* Thorell (Mygalomorphae : Macrothelinae). *J. R. Soc. West. Aust.* **43**: 30–34.

Distribution: SE coastal, N.S.W.; Hillgrove. Ecology: open forest, terrestrial, burrowing, predator; web builder. Biological references: Raven, R.J. (1984). Systematics of the Australian

Curtain-web spiders (Ischnothelinae : Dipluridae : Chelicerata). *Aust. J. Zool. Supp. Ser.* **93**: 1–102 (burrow, web, phenology, habitat).

Cethegus colemani Raven, 1984

Cethegus colemani Raven, R.J. (1984). Systematics of Australian Curtain-web Spiders (Ischnothelinae : Dipluridae : Chelicerata). *Aust. J. Zool. Supp. Ser.* **93**: 1–102 [5 Mar. 1984] [35]. Type data: holotype, AM KS4447 *M. adult, paratype AM KS4448 F. adult (described), from Mareeba, Qld.

Distribution: NE coastal, Qld.; Mareeba and Cairns. Ecology: closed forest, terrestrial, predator; (silk sheet or web).

Cethegus daemeli Raven, 1984

Cethegus daemeli Raven, R.J. (1984). Systematics of Australian Curtain-web Spiders (Ischnothelinae : Dipluridae : Chelicerata). *Aust. J. Zool. Supp. Ser.* **93**: 1–102 [5 Mar. 1984] [37]. Type data: holotype, QM S456 *M. adult, paratypes QM S457 F. adult (described), QM S458–477 3M., 19F. adult, from Line Hill, near Lockhardt River Mission, Cape York, Qld.

Distribution: NE coastal, Qld.; Iron Range. Ecology: closed forest, terrestrial, arboreal, burrowing, predator; curtain-like silk webs with tubes leading into burrow, under logs, in tree buttresses and under bark.

Cethegus elegans Raven, 1984

Cethegus elegans Raven, R.J. (1984). Systematics of Australian Curtain-web Spiders (Ischnothelinae : Dipluridae : Chelicerata). *Aust. J. Zool. Supp. Ser.* **93**: 1–102 [5 Mar. 1984] [40]. Type data: holotype, QM S495 *M. adult, paratypes QM S496 F. adult (described), QM S755 2F. adult, from Forty Mile Scrub, SW of Mt. Garnet, Qld.

Distribution: NE coastal, Qld.; Forty Mile Scrub, and Mt. Garnet. Ecology: closed forest, terrestrial, burrowing, predator; in semi-evergreen vine thickets, multi-tubular webs under rocks.

Cethegus fugax (Simon, 1908)

Palaevagrus fugax Simon, E. (1908). Araneae. Pt. 1 pp. 359–446 *in* Michaelsen, W. & Hartmeyer, R. (eds.) *Die Fauna Südwest-Australiens.* Jena : Fischer Bd. 1 Lf. 12 [365] [Main, B.Y. (1960). The genus *Cethegus* Thorell (Mygalomorphae : Macrothelinae). *J. R. Soc. West. Aust.* **43**: 30–34 (31) described, figured and designated a male "allotype" WAM 82/664]. Type data: lectotype, MNHP 15106 *F. adult, paralectotypes ZMB juv., ZMH *juv., from Lion Mill, W.A. (now Mt. Helena), designation by Raven, R.J. (1984). Systematics of the Australian Curtain-web spiders (Ischnothelinae : Dipluridae : Chelicerata). *Aust. J. Zool. Suppl. Sec.* **93**: 1–102 [42,44].

Taxonomic decision of Main, B.Y. (1960). The genus *Cethegus* Thorell (Mygalomorphae : Macrothelinae). *J.R. Soc. West. Aust.* **43**: 30–34. Distribution: SW coastal, W plateau, S Gulfs, W.A., S.A. Ecology: open forest, woodland, low woodland, burrowing, terrestrial, predator; in sandy and gravel soils, with sheet web against stumps, logs, rocks. Biological references: Main, B.Y. (1960). The genus *Cethegus* Thorell (Mygalomorphae : Macrothelinae). *J. R. Soc. West. Aust.* **43**: 30–34 (taxonomy, natural history, web, burrow).

Cethegus hanni Raven, 1984

Cethegus hanni Raven, R.J. (1984). Systematics of Australian Curtain-web Spiders (Ischnothelinae : Dipluridae : Chelicerata). *Aust. J. Zool. Supp. Ser.* **93**: 1–102 [5 Mar. 1984] [45]. Type data: holotype, QM S483 *F. adult, from near Hann River, Qld.

Distribution: NE coastal, Qld.; near Hann River. Ecology: open forest, terrestrial, burrow, predator; funnel-like web and burrow.

Cethegus lugubris Thorell, 1881

Cethegus lugubris Thorell, T. (1881). Studi sui ragni Malesi e Papuani. Pt. 3. Ragni dell'Austro-Malesia e del Capo York, conservati del Museo Civico di Storia Naturale di Genova. *Ann. Mus. Civ. Stor. Nat. Genova* **17**: 1–720 (Territelariae pp. 236–254) [6 May 1881] [241]. Type data: syntypes, MCG F. adult, juv., M. immature, from Somerset, Cape York, Qld.

Distribution: NE coastal, N Gulf, Qld.; tip of Cape York. Ecology: closed forest, open forest, predator; walking, tubular webs. Biological references: Raven, R.J. (1984). Systematics of the Australian Curtain-web spiders (Ischnothelinae : Dipluridae : Chelicerata). *Aust. J. Zool. Supp. Ser.* **93**: 1–102 (habitat, webs); Main, B.Y. (1960). The genus *Cethegus* Thorell (Mygalomorphae : Macrothelinae). *J. R. Soc. West. Aust.* **43**: 30–34 [tentatively identified specimens from Macpherson Ranges and inland W.A. as *C. lugubris*; the inland W.A. specimens are of an undescribed species; those from Macpherson Ranges identified by Raven (1984), *op. cit.* as *Namirea planipes* Raven, 1984].

Cethegus multispinosus Raven, 1984

Cethegus multispinosus Raven, R.J. (1984). Systematics of Australian Curtain-web Spiders (Ischnothelinae : Dipluridae : Chelicerata). *Aust. J. Zool. Supp. Ser.* **93**: 1–102 [5 Mar. 1984] [46]. Type data: holotype, QM S481 F. adult, from 80km N Musgrave, Qld.

Distribution: NE coastal, Qld.; 80 km north of Musgrave. Ecology: (open forest), terrestrial, burrowing, predator; curtain-like web.

Cethegus pallipes Raven, 1984

Cethegus pallipes Raven, R.J. (1984). Systematics of Australian Curtain-web Spiders (Ischnothelinae : Dipluridae : Chelicerata). *Aust. J. Zool. Supp. Ser.* **93**: 1–102 [5 Mar. 1984] [48]. Type data: holotype, QM S484 *F. adult, paratypes QM S485 4F. adult, QM S486 F. adult, from Mt. Cook, Cooktown, Qld.

Distribution: NE coastal, Qld.; Cooktown. Ecology: closed forest, burrowing, terrestrial, predator; curtain-like webs under rocks, in tree buttresses and under bark.

Cethegus robustus Raven, 1984

Cethegus robustus Raven, R.J. (1984). Systematics of Australian Curtain-web Spiders (Ischnothelinae : Dipluridae : Chelicerata). *Aust. J. Zool. Supp. Ser.* **93**: 1–102 [5 Mar. 1984 [50]. Type data: holotype, QM S489 F. adult, paratypes QM S490 4F. adult, from Chillagoe, Qld.

Distribution: NE coastal, Qld.; Chillagoe, Wolfram, Herberton and Rosella Plains. Ecology: open forest, terrestrial, burrowing, predator; curtain webs over burrow.

Chenistonia Hogg, 1901

Chenistonia Hogg, H.R. (1901). On Australian and New Zealand spiders of the suborder Mygalomorphae. *Proc. Zool. Soc. Lond.* **1901(2)**: 218–279 [4 June 1901] [261]. Type species *Chenistonia maculata* Hogg, 1901 by original designation.

Proshermacha Simon, E. (1908). Araneae. Pt. 1 pp. 359–446 *in* Michaelsen, W. & Hartmeyer, R. (eds.) *Die Fauna Südwest-Australiens.* Jena : Fischer Bd. 1 Lf. 12 [363]. Type species *Proshermacha subarmata* Simon, 1908 by subsequent designation, see Rainbow, W.J. (1912). A census of Australian Araneidae. *Rec. Aust. Mus.* **10**: 107–319 [113].

Taxonomic decision of Main, B.Y. (1982). Further studies on the systematics of Australian Diplurinae (Araneae : Mygalomorphae : Dipluridae): the taxonomic status of *Proshermacha* Simon and *Chenistonia tepperi* Hogg. *Aust. Entomol. Mag.* **8**: 83–88 [84].

Chenistonia cuspidata Main, 1954

Chenistonia cuspidata Main, B.Y. (1954). Spiders and Opiliones. pp. 37–53 *Australian Geographical Society Reports. No. 1.* The Archipelago of the Recherche. Pt. 6 pl 1 figs 5–7, pl 2 fig 8 Melbourne : Australian Geographical Society [38]. Type data: holotype, WAM 82/111 F. adult, from Termination Is., Recherche Archipelago, W.A.

Distribution: SW coast, W.A.; Recherche Archipelago. Ecology: terrestrial, burrowing, predator.

Chenistonia maculata Hogg, 1901

Chenistonia maculata Hogg, H.R. (1901). On Australian and New Zealand spiders of the suborder Mygalomorphae. *Proc. Zool. Soc. Lond.* **1901(2)**: 218–279 [4 June 1901] [262]. Type data: syntypes, BMNH 03.2.10–9–13 M.,3F. adult, F. immature, from Macedon, Vic.

Distribution: SE coastal, Vic.; Macedon and Grampian Mts. Ecology: tall open forest, open forest, terrestrial, burrowing, predator; under logs. Biological references: Raven, R.J. (1981). A review of the Australian genera of the Mygalomorph spiders subfamily Diplurinae (Dipluridae : Chelicerata). *Aust. J. Zool.* **29**: 321–363 [synonymized *Chenistonia* (type species *maculata* Hogg) with *Aname*] retained as a distinct genus in Main, B.Y. (1982). Further studies on the sytematics of Australian Diplurinae (Araneae : Mygalomorphae : Dipluridae): the taxonomic status of *Proshermacha* Simon and *Chenistonia tepperi* Hogg. *Aust. Entomol. Mag.* **8**: 83–88 and in Main, B.Y. (1983). Further studies on the systematics of Australian Diplurinae (Chelicerata : Mygalomorphae : Dipluridae): two new genera from south Western Australia. *J. Nat. Hist.* **17**: 923–949.

Chenistonia tepperi Hogg, 1902

Chenistonia tepperi Hogg, H.R. (1902). On some additions to the Australian spiders of the suborder Mygalomorphae. *Proc. Zool. Soc. Lond.* **1902(2)**: 121–141 [3 June 1902] [137]. Type data: syntypes, BMNH 1924.III.1.23.24.F62 2F. adult, BMNH 1924.III.1.1507–8 F. adult, juv. (possibly syntypes), one syntype missing, from Ardrossan, S.A., four localities listed by Hogg, but only unequivocal syntypes are from Ardrossan, therefore designated type locality.

Proshermacha tigrina Simon, E. (1908). Araneae. Pt. 1 pp. 359–446 *in* Michaelsen, W. & Hartmeyer, R. (eds.) *Fauna Südwest Australiens.* Jena : Fischer Bd. 1 Lf. 12 [363]. Type data: lectotype, ZMB juv., F. adult? lost, from Jarrahdale, W.A.; only existing syntype is from Jarrahdale therefore designated as type locality, designation by Main, B.Y., this work.

Proshermacha subarmata Simon, E. (1908). Araneae. Pt. 1 pp. 359–472 *in* Michaelsen, W. & Hartmeyer, R. (eds.) *Fauna Südwest Australiens.* Fischer : Jena Bd. 1 Lf. 12 [364]. Type data: lectotype, ZMH F.? juv., paralectotypes ZMB 2 juv., WAM 4272 juv., from Harvey, W.A., designation by Main, B.Y., this work; see Main, B.Y. (1982). Further studies on the systematics of Australian Diplurinae (Araneae : Mygalomorphae : Dipluridae). The taxonomic status of *Proshermacha* Simon and *Chenistonia tepperi* Hogg. *Aust. Entomol. Mag.* **8**: 83–88.

Aname maculata Rainbow, W.J. & Pulleine, R.H. (1918). Australian trap-door spiders. *Rec. Aust. Mus.* **12**: 81–169 pls 12–24 [24 Dec. 1918] [151]. Type data: holotype, AM K40943, KS8190 F. adult, from Jarrahdale, W.A.; description and label on specimen state "Jarrahdale Road, Armadale" but introduction states collected "near Jarrahdale" further along the Armadale/Jarrahdale Road; type locality restricted here to Jarrahdale.

Aname intricata Rainbow, W.J. & Pulleine, R.H. (1918). Australian trap-door spiders. *Rec. Aust. Mus.* **12**: 81–169 pls 12–24 [24 Dec. 1918] [157]. Type data: holotype, AM K41256, KS8192 F. adult, from Nannup (Blackwood River), W.A.

Chenistonia auropilosa Rainbow, W.J. & Pulleine, R.H. (1918). Australian trap-door spiders. *Rec. Aust. Mus.* **12**: 81–169 pls 12–24 [24 Dec. 1918] [160]. Type data: holotype, AM K40967, KS8191 F. adult, from Armadale, W.A.

Taxonomic decision of Main, B.Y., this work; Main, B.Y. (1982). Further studies on the systematics of Australian Diplurinae (Araneae : Mygalomorphae : Dipluridae): The taxonomic status of *Proshermacha* Simon and *Chenistonia tepperi* Hogg. *Aust. Entomol. Mag.* **8**: 83–88.

Distribution: SW coastal, W plateau, S Gulfs, S.A., W.A. Ecology: open forest, woodland, low woodland, open scrub, open heath, burrowing, predator; in sandy and loamy soils with deep sinuous burrow. Biological references: Main, B.Y. (1981). *Spiders of Australia.* S.A. : Axiom 124 pp.; Main, B.Y. (1976). *Spiders.* Sydney : Collins 296 pp. (natural history).

Chenistonia trevallynia Hickman, 1926

Chenistonia trevallynia Hickman, V.V. (1926). Notes on Tasmanian Araneidae (with descriptions of new species). *Pap. Proc. R. Soc. Tasm.* **1925**: 171–186 pls 12–15 [5 Mar. 1926] [171]. Type data: lectotype, QVM 1957–13–7 M. adult, paralectotype QVM 1957–13–6 F. adult, from Trevallyn, Launceston, Tas., designation by Main B.Y., this work.

Distribution: Tas.; Trevallyn, Launceston. Ecology: terrestrial, burrowing, predator; silk tubes under logs and stones in wet situations.

Chenistonia villosa Rainbow and Pulleine, 1918

Chenistonia villosa Rainbow, W.J. & Pulleine, R.H. (1918). Australian trap-door spiders. *Rec. Aust. Mus.* **12**: 81–169 pls 12–24. [24 Dec. 1918] [161]. Type data: holotype, AM K41257, KS1394 F. adult, from Carlotta Brook, W.A.

Distribution: SW coastal, W.A. Ecology: tall open forest, terrestrial, arboreal, burrowing, predator; in karri (*Eucalyptus diversicolor*) forest, silk tubes or shallow burrows, sometimes arboreal under bark.

Ixamatus Simon, 1887

Ixalus Koch, L. (1873). *Die Arachniden Australiens.* Lf. 9 pp. 409–472 pls 33–36 Nürnberg : Bauer & Raspe [469] [*non Ixalus* Ogilby, 1837; *non Ixalus* Duméril and Bibron, 1841; *non Ixalus* Gistl., 1848]. Type species *Ixalus varius* L. Koch, 1873 by monotypy.

Ixamatus Simon, E. (1887). Quelques observations sur les Arachnides. *Bull. Soc. Entomol. Fr.* **1887**: 193–195 [195] [*nom. nov.* for *Ixalus* L. Koch, 1873].

Ixamatus barina Raven, 1982

Ixamatus barina Raven, R.J. (1982). Systematics of the Australian mygalomorph spider genus *Ixamatus* Simon (Diplurinae : Dipluridae : Chelicerata). *Aust. J. Zool.* **30**: 1035–1067 [1041]. Type data: holotype, QM S600 *M. adult, paratypes QM S602 F. adult (described), QM 3M.,13F. adult, from Baldy Mt., Atherton Tableland, NE Qld.

Distribution: NE coastal, Qld. Ecology: terrestrial, predator.

Ixamatus broomi Hogg, 1901

Ixamatus broomi Hogg, H.R. (1901). On Australian and New Zealand spiders of the suborder Mygalomorphae. *Proc. Zool. Soc. Lond.* **1901(2)**: 218–279 [4 June 1901] [260]. Type data: holotype, BMNH M. adult, from Hillgrove, N.S.W.

Distribution: NE coastal, SE coastal, Qld., N.S.W.; Hillgrove, MacPherson Ranges. Ecology: closed forest, tall open forest, terrestrial, predator; shallow burrows in humus or silk tubes in logs.

Ixamatus caldera Raven, 1982

Ixamatus caldera Raven, R.J. (1982). Systematics of the Australian mygalomorph spider genus *Ixamatus* Simon (Diplurinae : Dipluridae : Chelicerata). *Aust. J. Zool.* **30**: 1035–1067 [1049]. Type data: holotype, QM S626 M. adult, paratypes QM S627 F. adult (described), QM 9M.,4F. adult, from Bar Mt., Kyogle, N.S.W.

Distribution: SE coastal, N.S.W. Ecology: closed forest, terrestrial, predator.

Ixamatus candidus Raven, 1982

Ixamatus candidus Raven, R.J. (1982). Systematics of the Australian mygalomorph spider genus *Ixamatus* Simon (Diplurinae : Dipluridae : Chelicerata). *Aust. J. Zool.* **30**: 1035–1067 [1051]. Type data: holotype, QM S640 M. adult, paratypes QM S642 F. adult (described), QM 4M., from Poverty Point, Tenterfield, N.S.W.

Distribution: NE coastal, SE coastal, Qld., N.S.W.; SE Qld. and NE N.S.W. Ecology: closed forest, terrestrial, predator.

Ixamatus fischeri Raven, 1982

Ixamatus fischeri Raven, R.J. (1982). Systematics of the Australian mygalomorph spider genus *Ixamatus* Simon (Diplurinae : Dipluridae : Chelicerata). *Aust. J. Zool.* **30**: 1035–1067 [1055]. Type data: holotype, QM S727 *M. adult, paratypes QM S728 F. adult (described), AM KS8216, 8220, 8222 1M.,3F. adult, AM KS8219 M. adult, from Mt. Banda Banda Beech Reserve, N.S.W.

Distribution: SE coastal, N.S.W.; Mt. Boss - Mt. Banda Banda. Ecology: closed forest, terrestrial, predator; silk tubes in rotten logs.

Ixamatus musgravei Raven, 1982

Ixamatus musgravei Raven, R.J. (1982). Systematics of the Australian mygalomorph spider genus *Ixamatus* Simon (Diplurinae : Dipluridae : Chelicerata). *Aust. J. Zool.* **30**: 1035–1067 [1058]. Type data: holotype, QM S729 *M. adult, paratypes QM S730 F. adult (described), QM S730 M. adult, AM KS6067 M.,F. adult, from Point Lookout, New England Natl. Park, N.S.W.

Distribution: SE coastal, N.S.W. Ecology: closed forest, terrestrial, predator; silk tubes in logs.

Ixamatus varius (L. Koch, 1873)

Ixalus varius Koch, L. (1873). *Die Arachniden Australiens.* Lf. 9 pp. 409–472 pls 33–36 Nürnburg :

Bauer & Raspe [469]. Type data: syntypes, ZMH (MGH 9891) *M.,F. adult, BMNH 1919.9.18.5702–3 F. adult, juv., from Eungella Natl. Park, Qld. (Koch, 1873 referred to Bowen, Qld.), see Raven, R.J. (1981). A review of the Australian genera of the mygalomorph spider subfamily Diplurinae (Dipluridae, Chelicerata) *Aust. J. Zool.* **29**: 321–363 [339] (emended type locality).

Distribution: NE coastal, Qld.; Eungella Natl. Park and Credition. Ecology: closed forest, burrowing, terrestrial, predator. Biological references: Simon, E. (1887). Quelques observations sur les Arachnides. *Bull. Soc. Entomol. Fr.* **1887**: 193–195 (taxonomy); Raven, R.J. (1981). A review of the Australian genera of the mygalomorph spider subfamily Diplurinae (Dipluridae, Chelicerata). *Aust. J. Zool.* **29**: 321–363 (distribution).

Ixamatus webbae Raven, 1982

Ixamatus webbae Raven, R.J. (1982). Systematics of the mygalomorph spider genus *Ixamatus* Simon (Diplurinae : Dipluridae : Chelicerata). *Aust. J. Zool.* **30**: 1035–1067 [1062]. Type data: holotype, QM S731 *M. adult, paratypes QM S732 F. adult (described), QM 8M.,6F. adult, from Lamington Natl. Park, Qld.

Distribution: NE coastal, Qld. Ecology: closed forest, terrestrial, predator; silk tube under logs.

Kiama Main and Mascord, 1971

Kiama Main, B.Y. & Mascord, R. (1971). A new genus of diplurid spider (Araneae : Mygalomorphae) from New South Wales. *J. Entomol. Soc. Aust. (N.S.W.)* **6**: 24–30 [21 Jan. 1971] [24]. Type species *Kiama lachrymoides* Main and Mascord, 1971 by original designation.

Kiama lachrymoides Main and Mascord, 1971

Kiama lachrymoides Main, B.Y. & Mascord, R.M. (1971). A new genus of diplurid spider (Araneae : Mygalomorphae) from New South Wales. *J. Entomol. Soc. Aust. (N.S.W.)* **6**: 24–30 [21 Jan. 1971] [25]. Type data: holotype, AM K68776, KS6396 F. adult, paratypes AM K68777, KS6397 M. adult (described), AM K68941, KS6398 F. adult, AM K68942, KS6399 F. adult, AM K68943, KS6400 F. adult, from Kiama, S of Sydney, N.S.W.

Distribution: SE coastal, N.S.W.; Kiama (south of Sydney) and Royal Natl. Park. Ecology: closed forest, burrowing, terrestrial, predator. Biological references: Mascord, R. (pers. comm.) (distribution).

Kwonkan Main, 1983

Kwonkan Main, B.Y. (1983). Further studies on the systematics of Australian Diplurinae (Chelicerata : Mygalomorphae : Dipluridae): Two new genera from south Western Australia. *J. Nat. Hist.* **17**: 923–949 figs 1–70 [Dec. 1983] [925]. Type species *Dekana wonganensis* Main, 1977 by original designation.

Kwonkan anatolion Main, 1983

Kwonkan anatolion Main, B.Y. (1983). Further studies on the systematics of Australian Diplurinae (Chelicerata : Mygalomorphae : Dipluridae): Two new genera from south Western Australia. *J. Nat. Hist.* **17**: 923–949 figs 1–70 [929]. Type data: holotype, WAM 1982/1361, from 45 km W. of Penong, S.A.

Distribution: W plateau, S.A.; southeast area of S.A., Penong. Ecology: closed heath, burrowing, predator.

Kwonkan eboracum Main, 1983

Kwonkan eboracum Main, B.Y. (1983). Further studies on the systematics of Australian Diplurinae (Chelicerata : Mygalomorphae : Dipluridae): Two new genera from south Western Australia *J. Nat. Hist.* **17**: 923–949 figs 1–70 [929]. Type data: holotype, WAM 1982/1358 F. adult, paratype WAM 82–1359 M. adult (described), from Eboracum, 22 km N.E. of Tammin, W.A.

Distribution: SW coastal, W.A.; mideastern margin, Tammin environs. Ecology: open scrub, closed heath, burrowing, predator; cryptic burrows in litter.

Kwonkan goongarriensis Main, 1983

Kwonkan goongarriensis Main, B.Y. (1983). Further studies on the systematics of Australian Diplurinae (Chelicerata : Mygalomorphae : Dipluridae): Two new genera from south Western Australia. *J. Nat. Hist.* **17**: 923–949 figs 1–70 [926]. Type data: holotype, WAM 82–1352 M. adult, paratypes WAM 82–1353 F. adult (described), WAM 82–1354 F. adult, WAM 82–1355 juv., from Goongarrie (via Menzies 29°54'S, 121°10'E), W.A.

Distribution: SW coastal, W.A.; mideastern border with W plateau, Menzies environs. Ecology: open scrub, closed heath, burrowing, predator; cryptic burrows in litter.

Kwonkan moriartii Main, 1983

Kwonkan moriartii Main, B.Y. (1983). Further studies on the systematics of Australian Diplurinae (Chelicerata : Mygalomorphae : Dipluridae): Two new genera from south Western Australia. *J. Nat. Hist.* **17**: 923–949 figs 1–70 [930]. Type data: holotype, WAM 1982/1360 M. adult, from Kathleen Valley Station (via Wiluna), W.A.

Distribution: W plateau, W.A.; SW coastal border, Wiluna environs.

Kwonkan silvestre Main, 1983

Kwonkan silvestre Main, B.Y. (1983). Further studies on the systematics of Australian Diplurinae (Chelicerata : Mygalomorphae : Dipluridae): Two new genera from south Western Australia. *J. Nat. Hist.* **17**: 923–949 figs 1–70 [930]. Type data: holotype, WAM 82/1356 M. adult, paratypes WAM 82/1357 F. adult, BYM 53/514 M. adult, BYM 53/517 M. adult, from Juranda, W.A.

Distribution: W plateau, SW coastal, W.A.; Fraser Ranges and Juranda. Ecology: low open woodland, burrowing, predator.

Kwonkan wonganensis (Main, 1977)

Dekana wonganensis Main, B.Y. (1977). Spiders. pp. 101–107 *in* Kenneally, K.F. *The Natural History of the Wongan Hills.* Perth : West. Aust. Naturalist's Club [102]. Type data: holotype, WAM 78–596 F. adult, paratype WAM 1982–118 F. adult, from The Wongan Hills (S end of), W.A.

Distribution: SW coastal, W.A.; mid region, Wongan Hills. Ecology: open scrub, burrowing, predator; builds shallow burrows in gravelly loam, burrow Y-shaped, with turret of soil and pebbles at entrance.

Masteria L. Koch, 1873

Masteria Koch, L. (1873). *Die Arachniden Australiens.* Lf. 9 pp. 409–472 pls 33–36 Nürnberg : Bauer & Raspe [457]. Type species *Masteria hirsuta* L. Koch, 1873 by monotypy.

Accola Simon, E. (1889). Voyage de M.E. Simon au Vénézuéla (Décembre 1887-Avril 1888). 4e Mém. Arachnides. *Ann. Soc. Entomol. Fr. (6)* **9**: 169–220 pls 1–3 [191]. Type species *Accola lucifuga* Simon, 1889 by subsequent designation, see Simon, E. (1892). *Histoire Naturelle des Araignées.* Paris : Roret Vol. 1 [190].

Antrochares Rainbow, W.J. (1898). Contribution to a knowledge of the Arachniden fauna of British New Guinea. *Proc. Linn. Soc. N.S.W.* **23**: 328–356 pl 7 [328]. Type species *Antrochares macgregori* Rainbow, 1898 by monotypy.

Taxonomic decision of Raven, R.J. (1979). Systematics of the mygalomorph spider genus *Masteria* (Masteriinae : Dipluridae : Arachnida). *Aust. J. Zool.* **27**: 623–636 [624].

This group is also found in New Guinea, Fiji, Philippines, Venezuela, Chile and Panama, see Roewer, C.Fr. (1963). Insects of Micronesia Araneina : Orthognatha, Labidognatha. *Insects of Micronesia* **3**(4): 105–132; Raven, R.J. (1981). Three new mygalomorph spiders (Dipluridae, Masteriinae) from Colombia. *Bull. Am. Mus. Nat. Hist.* **170**: 57–63.

Masteria toddae Raven, 1979

Masteria toddae Raven, R.J. (1979). Systematics of the mygalomorph spider genus *Masteria* (Masteriinae : Dipluridae : Arachnida). *Aust. J. Zool.* **27**: 623–636 [Aug. 1979] [630]. from Home Rule, Qld.

Distribution: NE coastal, Qld.; northern coastal montane areas only. Ecology: closed forest, (burrowing), terrestrial, predator; in litter.

Merredinia Main, 1983

Merredinia Main, B.Y. (1983). Further studies on the systematics of Australian Diplurinae (Chelicerata : Mygalomorphae : Dipluridae): Two new genera from south Western Australia. *J. Nat. Hist.* **17**: 923–949 figs 1–70 [Dec. 1983] [931]. Type species *Merredinia damsonoides* Main, 1983 by original designation.

Merredinia damsonoides Main, 1983

Merredinia damsonoides Main, B.Y. (1983). Further studies on the systematics of Australian Diplurinae (Chelicerata : Mygalomorphae : Dipluridae): Two new genera from south Western Australia. *J. Nat. Hist.* **17**: 923–949 figs 1–70 [933]. Type data: holotype, WAM 82/1362 M. adult, paratypes WAM 82/1363 F. adult (described), WAM 82/1364, 1365 2F. adult, WAM 82/1366–1373 8M. adult, WAM 82/1374 F. adult and others, from Lake Cronin, 32°24′15″S, 119°41′10″E, W.A.

Distribution: SW coastal, W.A.; mid eastern area. Ecology: low open woodland, open heath, burrowing, predator; deep sinuous burrow, sparse silk, makes soil plug during summer.

Namea Raven, 1984

Namea Raven, R.J. (1984). A new diplurid genus from eastern Australia and a related *Aname* species (Diplurinae : Dipluridae : Araneae). *Aust. J. Zool. Supp. Ser.* **96**: 1–51 [26 Jan. 1984] [2]. Type species *Namea capricornia* Raven, 1984 by original designation.

Namea brisbanensis Raven, 1984

Namea brisbanensis Raven, R.J. (1984). A new diplurid genus from eastern Australia and related *Aname* species (Diplurinae : Dipluridae : Araneae). *Aust. J. Zool. Supp. Ser.* **96**: 1–51 [26 Jan. 1984] [12]. Type data: holotype, QM S767 M. adult, paratypes QM S768 F. adult (described), QM S769 M. adult, QM S770 M.,F. adult, QM S771 2M. adult, QM S772 M. adult, QM S773 M. adult, QM S774 M. adult, QM S775 F. adult, QM S776 F. adult, QM S777 M. adult, QM S778, QM S779 2F. adult, QM S210 F. adult, from Brookfield, Brisbane, Qld.

Distribution: NE coastal, Qld.; Moreton district. Ecology: closed forest, open forest, terrestrial, burrowing, predator.

Namea bunya Raven, 1984

Namea bunya Raven, R.J. (1984). A new diplurid genus from eastern Australia and related *Aname* species (Diplurinae : Dipluridae : Araneae). *Aust. J. Zool. Supp. Ser.* **96**: 1–51 [26 Jan. 1984] [15]. Type data: holotype, QM S780 M. adult, paratypes QM S781, S782, S785, S786, S787, S788, S789 8M. adult, from Bunya Mts., Qld.

Distribution: NE coastal, Qld.; Bunya Mts., D'Aguilar Range. Ecology: closed forest, terrestrial, burrowing, predator.

Namea calcaria Raven, 1984

Namea calcaria Raven, R.J. (1984). A new diplurid genus from eastern Australia and related *Aname* species (Diplurinae : Dipluridae : Araneae). *Aust. J. Zool. Supp. Ser.* **96**: 1–51 [26 Jan. 1984] [16]. Type data: holotype,

QM S790 M. adult, paratypes QM S791 F. adult, QM S792, S793, S794 3M. adult, from Bulburin State Forest, Qld.

Distribution: NE coastal, Qld.; Many Peaks and Jimna Ranges. Ecology: closed forest, terrestrial, predator.

Namea callemonda Raven, 1984

Namea callemonda Raven, R.J. (1984). A new diplurid genus from eastern Australia and related *Aname* species (Diplurinae : Dipluridae : Araneae). *Aust. J. Zool. Supp. Ser.* **96**: 1–51 [26 Jan. 1984] [19]. Type data: holotype, QM S795 M. adult, paratype QM S796 F. adult (described), from Bulburin State Forest, Qld.

Distribution: NE coastal, Qld.; Bulburin State Forest. Ecology: closed forest, terrestrial, burrowing, predator.

Namea capricornia Raven, 1984

Namea capricornia Raven, R.J. (1984). A new diplurid genus from eastern Australia and related *Aname* species (Diplurinae : Dipluridae : Araneae). *Aust. J. Zool. Supp. Ser.* **96**: 1–51 [26 Jan. 1984] [10]. Type data: holotype, QM S760 M. adult, paratypes QM S761 F. adult (described), QM S762 2M.,1F. adult, QM S763 4M.,1F. adult, QM S764 M. adult, from Crediton, Qld.

Distribution: NE coastal, Qld. Ecology: closed forest, terrestrial, burrowing, predator.

Namea cucurbita Raven, 1984

Namea cucurbita Raven, R.J. (1984). A new diplurid genus from eastern Australia and related *Aname* species (Diplurinae : Dipluridae : Araneae). *Aust. J. Zool. Supp. Ser.* **96**: 1–51 [26 Jan. 1984] [22]. Type data: holotype, QM S797 M. adult, paratypes QM S798 M. adult, QM S799 M. adult, from Wratten's Camp, Qld.

Distribution: NE coastal, Qld.; Wratten's Camp, near Widgee. Ecology: (terrestrial), (burrowing), (predator).

Namea dahmsi Raven, 1984

Namea dahmsi Raven, R.J. (1984). A new diplurid genus from eastern Australia and related *Aname* species (Diplurinae : Dipluridae : Araneae). *Aust. J. Zool. Supp. Ser.* **96**: 1–51 [26 Jan. 1984] [24]. Type data: holotype, QM S800 M. adult, paratypes QM S801 F. adult (described), QM S802, S803 2M. adult, QM S804 F. adult, QM S805 2F. adult, from Calamvale, Brisbane, Qld.

Distribution: NE coastal, Qld.; southeast Qld. Ecology: tall open forest, terrestrial, burrowing, predator.

Namea dicalcaria Raven, 1984

Namea dicalcaria Raven, R.J. (1984). A new diplurid genus from eastern Australia and related *Aname* species (Diplurinae : Dipluridae : Araneae). *Aust. J. Zool. Supp. Ser.* **96**: 1–51 [26 Jan. 1984] [26]. Type data: holotype, QM S806 M. adult, paratypes QM S807 F. adult (described), QM S808 2M. adult, QM S809, S810, S811, S812 4M. adult, from Mt. Nardi, N.S.W.

Distribution: SE coastal, N.S.W.; northeastern N.S.W., between Mt. Nardi and Lismore. Ecology: closed forest, terrestrial, burrowing, predator.

Namea excavans Raven, 1984

Namea excavans Raven, R.J. (1984). A new diplurid genus from eastern Australia and related *Aname* species (Diplurinae : Dipluridae : Araneae). *Aust. J. Zool. Supp. Ser.* **96**: 1–51 [26 Jan. 1984] [29]. Type data: holotype, QM S813 M. adult, paratypes QM S814 M. adult, QM S815 M. adult, QM S830 M. adult, from Mt. Glorious, Qld.

Distribution: NE coastal, Qld.; Mt. Glorious and Mt. Mee, D'Aguilar Range. Ecology: closed forest, terrestrial, burrowing, predator.

Namea flavomaculata (Rainbow and Pulleine, 1918)

Aname flavomaculata Rainbow, W.J. & Pulleine, R.H. (1918). Australian trap-door spiders. *Rec. Aust. Mus.* **12**: 81–169 pls 12–24 [24 Dec. 1918] [146] [Raven, R.J. (1984). A new diplurid genus from eastern Australia and a related *Aname* species (Diplurinae : Dipluridae : Araneae). *Aust. J. Zool. Suppl. Ser.* **96**: 1–51 (31) designated the specimen AM KS8189 as a lectotype; this specimen, however, was labelled by Rainbow as "TYPE"; as no other specimen, such as the "half-grown" one mentioned by Rainbow and Pulleine following their description can be traced and registration refers to a single specimen, AM KS8189 is regarded as the holotype]. Type data: holotype, AM K40942, KS8189 F. adult, from Tamborine Mt., Qld.

Distribution: NE coastal, SE coastal, Qld., N.S.W.; MacPherson and Tweed Ranges. Ecology: terrestrial, burrowing, predator, closed forest. Biological references: Main, B.Y. (1976). *Spiders.* Sydney : Collins 296 pp. (as *Ixamatus flavomaculatus*); Main, B.Y. (1981). Eco-evolutionary radiation of mygalomorph spiders. pp. 853–872 *in* Keast, A. (ed.) *Ecological Biogeography of Australia.* The Hague : Junk Vol. 2 [biology, species misidentified as *Ixamatus flavomaculatus* (Rainbow and Pulleine)]; Raven, R.J. (1984). A new diplurid genus from eastern Australia and a related *Aname* species (Diplurinae : Dipluridae : Araneae). *Aust. J. Zool. Suppl. Ser.* **96**: 1–51 (burrow, behaviour, generic placement, *nov. comb.*).

Namea jimna Raven, 1984

Namea jimna Raven, R.J. (1984). A new diplurid genus from eastern Australia and related *Aname* species (Diplurinae : Dipluridae : Araneae). *Aust. J. Zool. Supp. Ser.* **96**: 1–51 [26 Jan. 1984] [35]. Type data: holotype, QM S821 M. adult, paratypes QM S822 F. adult (described), QM S823 4M. adult, QM S824 M. adult,

QM S825, S826 2F. adult, QM S827 5M. adult, QM S828, S829 2F. adult, QM S831 5M. adult, QM S832 5F. adult, QM S833 M. adult, from Mt. Cabinet, Qld.

Distribution: NE coastal, Qld.; Jimna Range and Mt. Bauple, near Maryborough. Ecology: closed forest, terrestrial, burrowing, predator.

Namea nebulosa Raven, 1984

Namea nebulosa Raven, R.J. (1984). A new diplurid genus from eastern Australia and related *Aname* species (Diplurinae : Dipluridae : Araneae). *Aust. J. Zool. Supp. Ser.* **96**: 1–51 [26 Jan. 1984] [38]. Type data: holotype, QM S834 M. adult, paratypes QM S835 F. adult (described), QM S836 F. adult, QM S837 7M. adult, QM S838 5M. adult, QM S839 6M. adult, from Bellenden Ker Range, Qld.

Distribution: NE coastal, Qld.; Bellenden Ker, possibly Atherton Tableland and Kirrama Range. Ecology: closed forest, terrestrial, burrowing, predator; mist forest.

Namea olympus Raven, 1984

Namea olympus Raven, R.J. (1984). A new diplurid genus from eastern Australia and related *Aname* species (Diplurinae : Dipluridae : Araneae). *Aust. J. Zool. Supp. Ser.* **96**: 1–51 [26 Jan. 1984] [41]. Type data: holotype, QM S1163 M. adult, paratype QM S1164 M. adult, from Mt. Bartle Frere, Qld.

Distribution: NE coastal, Qld.; Mt. Bartle Frere. Ecology: closed forest, terrestrial, (burrowing), predator.

Namea salanitri Raven, 1984

Namea salanitri Raven, R.J. (1984). A new diplurid genus from eastern Australia and related *Aname* species (Diplurinae : Dipluridae : Araneae). *Aust. J. Zool. Supp. Ser.* **96**: 1–51 [26 Jan. 1984] [42]. Type data: holotype, QM S1166 M. adult, paratypes QM S1167 F. adult (described), QM S1168 M. adult, QM S1169 M.,F. adult, QM S1170, S1171 2M. adult, QM S1172 F. adult, QM S1173 M. adult, QM S1174 3M. adult, QM S1175 M.,F. adult, QM S1176 2M. adult, QM S1177, S1181 2F. adult, QM S1178, S1179, S1180, S1182, S1183 14M. adult, from Mt. Mee, Qld.

Distribution: NE coastal, SE coastal, Qld., N.S.W.; mountainous border area of Qld. and N.S.W. Ecology: closed forest, terrestrial, burrowing, predator.

Namea saundersi Raven, 1984

Namea saundersi Raven, R.J. (1984). A new diplurid genus from eastern Australia and related *Aname* species (Diplurinae : Dipluridae : Araneae). *Aust. J. Zool. Supp. Ser.* **96**: 1–51 [26 Jan. 1984] [45]. Type data: holotype, QM S1184 M. adult, paratype QM S1185 M. adult, from Mt. Spec Natl. Park, Qld.

Distribution: NE coastal, Qld.; Mt. Spec Natl. Park, N Qld. Ecology: closed forest, terrestrial, (burrowing), predator.

Namirea Raven, 1984

Namirea Raven, R.J. (1984). Systematics of the Australian Curtain-web spiders (Ischnothelinae : Dipluridae : Chelicerata). *Aust. J. Zool. Supp. Ser.* 93: 1–102 [5 Mar. 1984] [51]. Type species *Namirea planipes* Raven, 1984 by original designation.

Namirea eungella Raven, 1984

Namirea eungella Raven, R.J. (1984). Systematics of Australian Curtain-web Spiders (Ischnothelinae : Dipluridae : Chelicerata). *Aust. J. Zool. Supp. Ser.* **93**: 1–102 [5 Mar. 1984] [56]. Type data: holotype, QM S523 F. adult, paratypes QM S531 F. adult, QM S524 ?F. adult, from Homevale, Qld.

Distribution: NE coastal, Qld.; Eungella Natl. Park, Homevale and Finch Hatton. Ecology: closed forest, open forest, terrestrial, burrowing, predator; web over burrow (in rainforest), coastal vine thickets.

Namirea fallax Raven, 1984

Namirea fallax Raven, R.J. (1984). Systematics of Australian Curtain-web Spiders (Ischnothelinae : Dipluridae : Chelicerata). *Aust. J. Zool. Supp. Ser.* **93**: 1–102 [5 Mar. 1984] [57]. Type data: holotype, AM KS4449, from Bilpin, N.S.W.

Distribution: SE coastal, N.S.W.; Bilpin, Blue Mts. Ecology: open forest, terrestrial, (burrowing), predator; eucalypt and heath, probably web-weaving over cracks and crevices in ground and logs.

Namirea insularis Raven, 1984

Namirea insularis Raven, R.J. (1984). Systematics of Australian Curtain-web Spiders (Ischnothelinae : Dipluridae : Chelicerata). *Aust. J. Zool. Supp. Ser.* **93**: 1–102 [5 Mar. 1984] [59]. Type data: holotype, QM S519 M. adult, paratypes QM S520 F. adult (described), QM S521 M. adult, QM S522 F. adult, from Burleigh Heads, Qld.

Distribution: NE coastal, Qld.; Burleigh Heads. Ecology: closed forest, terrestrial, (burrowing), predator; probably sheet or curtain web over burrow.

Namirea montislewisi Raven, 1984

Namirea montislewisi Raven, R.J. (1984). Systematics of Australian Curtain-web Spiders (Ischnothelinae : Dipluridae : Chelicerata). *Aust. J. Zool. Supp. Ser.* **93**: 1–102 [5 Mar. 1984] [61]. Type data: holotype, QM S527 F. adult, paratypes QM S528 2F. adult, QM S529 2F. adult, QM S530 F. adult, from Mt. Lewis, Qld.

Distribution: NE coastal, Qld.; Cairns environs. Ecology: closed forest, terrestrial, (burrowing), predator; tube webs under logs, rocks and in the ground.

Namirea planipes Raven, 1984

Namirea planipes Raven, R.J. (1984). Systematics of Australian Curtain-web Spiders (Ischnothelinae : Dipluridae : Chelicerata). *Aust. J. Zool. Supp. Ser.* **93**: 1–102 [5 Mar. 1984] [52]. Type data: holotype, QM S498 M. adult, paratypes QM S515 (described), QM S499 M.,2F. adult, QM S500 F. adult, QM S501 F. adult, QM S502 M. adult, QM S503 M. adult, QM S504 M. adult, QM S505 F. adult and others in QM, from Rochedale State Forest, Brisbane, Qld.

Distribution: NE coastal, Qld.; southeast corner *e.g.* MacPherson Ranges and adjacent ranges. Ecology: open forest, terrestrial, burrowing, predator; curtain webs over cracks and crevices in the ground and rotting logs.

Stanwellia Rainbow and Pulleine, 1918

Stanwellia Rainbow, W.J. & Pulleine, R.H. (1918). Australian trap-door spiders. *Rec. Aust. Mus.* **12**: 81–169 pls 12–24 [24 Dec. 1918] [164]. Type species *Stanwellia decora* Rainbow and Pulleine, 1918 by monotypy.

Aparua Todd, V. (1945). Systematic and biological account of the New Zealand Mygalomorphae (Arachnida). *Trans. R. Soc. N.Z.* **74**: 375–407 pls 55–58 [390]. Type species *Aparua bipectinata* Todd, 1945 by original designation.

Taxonomic decision of Main, B.Y. (1983). Further studies on the systematics of Australian Dipluridae (Chelicerata : Mygalomorphae : Dipluridae): two new genera from south Western Australia. *J. Nat. Hist.* **17**: 923–949 [923].

This group is also found in New Zealand.

Stanwellia grisea (Hogg, 1901)

Aname grisea Hogg, H.R. (1901). On Australian and New Zealand spiders of the suborder Mygalomorphae. *Proc. Zool. Soc. Lond.* **1901(2)**: 218–279 [4 June 1901] [252]. Type data: lectotype, BMNH 30.2.10–15 juv., paralectotypes BMNH 30.2.10–15 2 juv., also catalogued under E21 and E17 & 18, from Macedon, Vic., designation by Main, B.Y. (1972). The mygalomorph spider genus *Stanwellia* Rainbow and Pulleine (Dipluridae) and its relationship to *Aname* Koch and certain other diplurine genera. *J. R. Soc. West. Aust.* **55**: 100–114 [105].

Aname arborea Hogg, H.R. (1901). On Australian and New Zealand spiders of the suborder Mygalomorphae. *Proc. Zool. Soc. Lond.* **1901(2)**: 218–279 [4 June 1901] [254]. Type data: lectotype, BMNH 03.2.10.17 F. adult, paralectotype BMNH 03.2.10.17 juv., from Macedon, Vic., designation by Main, B.Y. (1972). The Mygalomorph spider genus *Stanwellia* Rainbow and Pulleine (Dipluridae) and its relationship to *Aname* Koch and certain other diplurine genera. *J. R. Soc. West. Aust.* **55**: 100–114 [105].

Aname pellucida Hogg, H.R. (1901). On Australian and New Zealand spiders of the suborder Mygalomorphae. *Proc. Zool. Soc. Lond.* **1901(2)**: 218–279 [4 June 1901] [255]. Type data: lectotype, BMNH 03.2.10–16 F. adult, paralectotype BMNH 03.2.10–16 juv., from Bacchus Marsh, Vic., designation by Main, B.Y. (1972). The mygalomorph spider genus *Stanwellia* Rainbow and Pulleine (Dipluridae) and its relationship to *Aname* Koch and certain other diplurine genera. *J. R. Soc. West. Aust.* **55**: 100–114 [105].

Ixamatus gregori Hogg, H.R. (1901). On Australian and New Zealand spiders of the suborder Mygalomorphae. *Proc. Zool. Soc. Lond.* **1901(2)**: 218–279 [4 June 1901] [258]. Type data: holotype, BMNH 1903.2.10.14 M. adult, from Macedon, Vic.

Chenistonia major Hogg, H.R. (1901). On Australian and New Zealand spiders of the suborder Mygalomorphae. *Proc. Zool. Soc. Lond.* **1901(2)**: 218–279 [4 June 1901] [263]. Type data: lectotype, BMNH 03.2.10.8 F. adult, paralectotypes BMNH 6F. adult, from Macedon, Vic., designation by Main, B.Y. (1972). The mygalomorph spider genus *Stanwellia* Rainbow and Pulleine (Dipluridae) and its relationship to *Aname* Koch and certain other diplurine genera. *J. R. Soc. West. Aust.* **55**: 100–114 [105].

Aname butleri Rainbow, W.J. & Pulleine, R.H. (1918). Australian trap-door spiders. *Rec. Aust. Mus.* **12**: 81–169 pls 12–24 [24 Dec. 1918] [157]. Type data: lectotype, AM K41482, KS8212 M. immature, from Merri Creek, Melbourne, Vic., designation by Main, B.Y. (1972). The mygalomorph spider genus *Stanwellia* Rainbow and Pulleine (Dipluridae) and its relationship to *Aname* Koch and certain other diplurine genera. *J. R. Soc. West. Aust.* **55**: 100–114 [105]; the original type series includes 4 specimens of an undescribed ctenizid species of *Misgolas* Karsch mus. nos. AM K41614 KS8211, AM K41615 KS8211.

Taxonomic decision of Main, B.Y. (1972). The mygalomorph spider genus *Stanwellia* Rainbow and Pulleine (Dipluridae) and its relationship to *Aname* Koch and certain other diplurine genera. *J. R. Soc. West. Aust.* **55**: 100–114 [105].

Distribution: SE coastal, Vic.; Macedon, Melbourne environs, Grampian Mts. and Dandenong Range. Ecology: tall open forest, terrestrial, burrowing, predator; wet situations, burrow in forest litter.

Stanwellia hoggi (Rainbow, 1914)

Chenistonia hoggi Rainbow, W.J. (1914). Studies in Australian Araneidae. No. 6. The Terretelariae. *Rec. Aust. Mus.* **10**: 187–270 [15 Aug. 1914] [240]. Type data: holotype, AM K31010, KS8619 F. adult, from North Sydney, N.S.W.

Aname decora Rainbow, W.J. & Pulleine, R.H. (1918). Australian trap-door spiders. *Rec. Aust. Mus.* **12**: 81–169 pls 12–24 [24 Dec. 1918] [149]. Type data: holotype, AM K40923 F. adult, from Clifton Gardens, Sydney, N.S.W.

Stanwellia decora Rainbow, W.J. & Pulleine, R.H. (1918). Australian trap-door spiders. *Rec. Aust. Mus.* **12**: 81–169 pls 12–24 [24 Dec. 1918] [164]. [Main, B.Y. (1972). The mygalomorph spider genus *Stanwellia* Rainbow and Pulleine (Dipluridae) and its relationship to *Aname* Koch and certain other diplurine genera. *J. R. Soc. West. Aust.* **55**: 100–114 incorrectly designated AM KS8416 as lectotype and SAMA N198164 as paralectotype] Type data: holotype, AM K40955, KS8416 F. adult, paratype SAMA N198164 F. adult, from Stanwell Park, N.S.W.

Taxonomic decision of Main, B.Y. (1972). The mygalomorph spider genus *Stanwellia* Rainbow and Pulleine (Dipluridae) and its relationship to *Aname* Koch and certain other diplurine genera. *J. R. Soc. West. Aust.* **55**: 100–114 [103].

Distribution: SE coastal, N.S.W.; Sydney area only. Ecology: terrestrial, burrowing, predator.

Stanwellia inornata Main, 1972

Stanwellia inornata Main, B.Y. (1972). The mygalomorph spider genus *Stanwellia* Rainbow and Pulleine (Dipluridae) and its relationship to *Aname* Koch and certain other diplurine genera. *J. R. Soc. West. Aust.* **55**: 100–114 [Dec. 1972] [109]. Type data: holotype, AM K69299 KS6405 F. adult, paratypes AM K69300, KS6406 M. adult, NMV K-25 F. adult, from Rose's Gap, Grampian Mts., Vic.

Distribution: SE coastal, Vic.; Grampian Mts. only. Ecology: tall open forest, burrowing, terrestrial, predator; in areas of heath understorey with sandy soil.

Stanwellia nebulosa (Rainbow and Pulleine, 1918)

Aname nebulosa Rainbow, W.J. & Pulleine, R.H. (1918). Australian trap-door spiders. *Rec. Aust. Mus.* **12**: 81–169 pls 12–24 [24 Dec. 1918] [147]. Type data: lectotype, AM K40924, KS8209 M. adult, from Mallala, S.A., paralectotypes, AM K40926, KS8206 F. adult (Tea Gardens, Mt. Lofty, S.A.), AM K41460, KS8208 2F. adult (Mallala, S.A.), AM K40930, KS8207 F. adult (Meningal, S.A.), SAMA N198163 F. adult (Aldgate, S.A.), designation by Main, B.Y. (1972). The mygalomorph spider genus *Stanwellia* Rainbow and Pulleine (Dipluridae) and its relationship to *Aname* Koch and certain other diplurine genera. *J. R. Soc. West. Aust.* **55**: 100–114.

Aname confusa Rainbow, W.J. & Pulleine, R.H. (1918). Australian trap-door spiders. *Rec. Aust. Mus.* **12**: 81–169 pls 12–24 [24 Dec. 1918] [155]. Type data: holotype, AM K40947, KS8210 F. adult, from Mylor Road, S.A.

Taxonomic decision of Main, B.Y. (1972). The mygalomorph spider genus *Stanwellia* Rainbow and Pulleine (Dipluridae) and its relationship to *Aname* Koch and certain other diplurine genera. *J. R. Soc. West. Aust.* **55**: 100–114 [106]; Main, B.Y. (1964,1967). *Spiders of Australia,* Brisbane : Jacaranda 124 pp. (*nov. comb.*).

Distribution: S Gulfs, S.A.; E of Spencer Gulf only. Ecology: open forest, woodland, low woodland, burrowing, terrestrial, predator; burrower in loamy and limestone soils.

Stanwellia occidentalis Main, 1972

Stanwellia occidentalis Main, B.Y. (1972). The mygalomorph spider genus *Stanwellia* Rainbow and Pulleine (Dipluridae) and its relationship to *Aname* Koch and certain other diplurine genera. *J. R. Soc. West. Aust.* **55**: 100–114 [Dec. 1972] [108]. Type data: holotype, AM K69302, KS6407 F. adult, paratypes AM K6930, KS6408 F. adult (east of Cummins, S.A.), SAMA N19719 F. adult (12.8 kms E. Cummins, S.A.), from North Shields, 10 km N of Pt. Lincoln, S.A. (originally as "mouth of Todd River").

Distribution: S Gulfs, S.A.; Eyre Peninsula, southern area only, record from SW W.A. reported by Main, B.Y. (1972). The mygalomorph spider genus *Stanwellia* Rainbow and Pulleine (Dipluridae) and its relationship to *Aname* Koch and certain other diplurine genera. *J. R. Soc. West. Aust.* **55**: 100–114 [109] now recognized as an undescribed distinct species. Ecology: open scrub, supralittoral, burrowing, terrestrial, predator; wet sites only in shrubby understorey, may be close to shore, just above high tide level.

Stanwellia pexa (Hickman, 1930)

Aname pexa Hickman, V.V. (1930). Studies in Tasmanian spiders. Pt. 4. *Pap. Proc. R. Soc. Tasm.* **1929**: 87–122 pls 19–26 [19 Mar. 1930] [87]. Type data: lectotype, QVM 1957/13/18 *M. adult Type 30 from Derwent Park, Tas., paralectotype QVM 1957/13/19 *F. adult Type 31 from Prince of Wales Bay, designation by Main, B.Y. (1972). The mygalomorph spider genus *Stanwellia* Rainbow and Pulleine (Dipluridae) and its relationship to *Aname* Koch and certain other diplurine genera. *J. R. Soc. West. Aust.* **55**: 100–114 [108] (*nov. comb.*).

Distribution: Tas.; N Tas. and islands of Bass Strait *e.g.* Fisher Is. Ecology: supralittoral, burrowing, terrestrial, predator; near shore, above tide level. Biological references: Main, B.Y. (1972). The mygalomorph spider genus *Stanwellia* Rainbow and Pulleine (Dipluridae) and its relationship to *Aname* Koch and certain other diplurine genera. *J. R. Soc. West. Aust.* **55**: 100–114 (taxonomy).

Stenygrocercus Simon, 1892

Stenygrocercus Simon, E. (1892). *Histoire Naturelle des Araignées.* Paris : Roret Vol. 1 [185]. Type species *Macrothele silvicola* Simon, 1889 by original designation.

Stenygrocercus australiensis Raven, 1984

Stenygrocercus australiensis Raven, R.J. (1984). Systematics of Australian Curtain-web Spiders (Ischnothelinae : Dipluridae : Chelicerata). *Aust. J. Zool. Supp. Ser.* **93**: 1–102 [5 Mar. 1984] [63]. Type data: holotype, NMV K-057 M. adult, paratypes QM S591 F. adult (described), AM KS3669 F. adult, AM KS8374 M. adult, NMV K-058 F. adult, QM S592 4F. adult, from Cann River valley forest, Vic.

Distribution: SE coastal, Vic.; Gippsland. Ecology: tall open forest, terrestrial, predator; moist sclerophyll fern forest, flimsy sheet/tube web under logs and bark.

Teyl Main, 1975

Teyl Main, B.Y. (1975). The citrine spider: a new genus of trap-door spider (Mygalomorphae : Dipluridae). *West. Aust. Nat.* **13**: 73–78 [30 Nov. 1975] [74]. Type species *Teyl luculentus* Main, 1975 by monotypy.

Teyl luculentus Main, 1975

Teyl luculentus Main, B.Y. (1975). The citrine spider: a new genus of trapdoor spider (Mygalomorphae : Dipluridae). *West. Aust. Nat.* **13**: 73–78 [30 Nov. 1975] [74]. Type data: holotype, WAM 75–944 F. adult, paratypes WAM 75–951–2 2M. (75–951 described), WAM 75 945–950 6F. adult, AM KS3395 M. adult, from 14.5 km N of Bruce Rock, W.A.

Distribution: SW coastal, W.A.; inland on plateau. Ecology: low open woodland, open scrub, burrowing, terrestrial, predator; predominantly found in *Eucalyptus loxophleba* and *Acacia acuminata* associations around granite outcrops and *Melaleuca* summer-dry bogs.

Troglodiplura Main, 1969

Troglodiplura Main, B.Y. (1969). A blind mygalomorph spider from a Nullarbor Plain cave. *J. R. Soc. West. Aust.* **52**: 9–11 [9]. Type species *Troglodiplura lowryi* Main, 1969 by monotypy.

Troglodiplura lowryi Main, 1969

Troglodiplura lowryi Main, B.Y. (1969). A blind mygalomorph spider from a Nullarbor Plain cave. *J. R. Soc. West. Aust.* **52**: 9–11 [9]. Type data: holotype, WAM 68–342 ?adult, from Roaches' Rest Cave, 43 km NE Madura, W.A.

Distribution: W plateau, S.A., W.A.; Nullarbor Plain only. Ecology: caves, terrestrial, predator.

Xamiatus Raven, 1981

Xamiatus Raven, R.J. (1981). A review of the Australian genera of the mygalomorph spider subfamily Diplurinae (Dipluridae : Chelicerata). *Aust. J. Zool.* **29**: 321–363 [17 June 1981] [343]. Type species *Xamiatus rubrifrons* Raven, 1981 by original designation.

Xamiatus rubrifons Raven, 1981

Xamiatus rubrifrons Raven, R.J. (1981). A review of the Australian genera of the mygalomorph spider subfamily Diplurinae (Dipluridae : Chelicerata). *Aust. J. Zool.* **29**: 321–363 [344]. Type data: holotype, QM S704 M. adult, paratypes, QM S706 F. adult (described), QM S707 M. adult, QM S708 M. adult, QM S709 F. adult, QM S710 M. adult, QM S705 2M., juv., from Conondale Range, near Booloumba Creek, SE Qld.

Distribution: NE coastal, Qld.; Conondale and D'Aguilar Ranges, SE Qld. Ecology: closed forest, terrestrial, predator.

Xamiatus bulburin Raven, 1981

Xamiatus bulburin Raven, R.J. (1981). A review of the Australian genera of the mygalomorph spider subfamily Diplurinae (Dipluridae : Chelicerata). *Aust. J. Zool.* **29**: 321–363 [349]. Type data: holotype, QM S714 M. adult, from Bulburin State Forest, SE Qld.

Distribution: NE coastal, Qld.; known only from Many Peaks Range. Ecology: closed forest, terrestrial, predator.

Xamiatus kia Raven, 1981

Xamiatus kia Raven, R.J. (1981). A review of the Australian genera of the mygalomorph spider subfamily Diplurinae (Dipluridae : Chelicerata). *Aust. J. Zool.* **29**: 321–363 [358]. Type data: holotype, QM S719 M. adult, from Newee Creek Road via Macksville, N.S.W.

Distribution: SE coastal, N.S.W.; Macksville and Emerald Beach. Ecology: terrestrial, predator.

Xamiatus magnificus Raven, 1981

Xamiatus magnificus Raven, R.J. (1981). A review of the Australian genera of the mygalomorph spider subfamily Diplurinae (Dipluridae : Chelicerata). *Aust. J. Zool.* **29**: 321–363 [352]. Type data: holotype, QM S717 M. adult, paratype QM S718 F. adult, from Crater Natl. Park, NE Qld.

Distribution: NE coastal, Qld.; only from Evelyn Tableland. Ecology: terrestrial, predator.

Incertae sedis

Brachythele platypus Koch L. *in* Ausserer, A. (1875). Zweiter Beiträge zur Kenntniss der Arachniden-Familie der Territelariae Thorell. (Mygalidae Autor.). *Verh. Zool.-Bot. Ges. Wien* **25**: 125–206 pls 5–7 [159] [probable holotype in poor condition and possibly of the genus *Aname* or *Chenistonia*]. Type data: holotype, BMNH 1139 from "Koch Collection" F. adult (Note: this specimen not labelled as a type but deduced to be holotype, BYM), from Australia (as New Holland).

Idioctis palmarum Hogg, H.R. (1901). On Australian and New Zealand spiders of the suborder Mygalomorphae. *Proc. Zool. Soc. Lond.* **1901(2)**: 218–279 [4 June 1901] [242] [as the type specimen is lost, and the description inadequate, this species, which is probably a diplurid, cannot be placed satisfactorily]. Type data: holotype, F. adult missing, from Palm Creek, central Australia.

Aganippe bancrofti Rainbow, W.J. & Pulleine, R.H. (1918). Australian trap-door spiders. *Rec. Aust. Mus.* **12**: 81–169 pls 13–24 [24 Dec. 1918] [95] [this syntype series contains both diplurid and barychelid species; the original description is clearly of a barychelid but not attributable to a genus]. Type data: syntype, AM K40849 KS6152 juv., from Eidsvold, Qld.

HEXATHELIDAE

INTRODUCTION

With two or three pairs of spinnerets; terminal segment of posterior lateral pair digitiform. One or two rows of teeth on chelicerae. Labium squarish, anteriorly with dense cluster of short, stout cuspules. Maxillae with serrula and numerous cuspules. Females without scopula on legs. Paired tarsal claws with single row of teeth. Males with anterior tibiae (first and/or second pair) modified. Usually expansive sheet web around entrance of shallow burrows, arboreal or terrestrial. Distribution mainly southeast Asia, Australasia (although in Australia, southeast Australia and Tas. only), Chile and the Mediterranean. Three subfamilies occur in Australia; the Hexathelinae (*Bymainiella*, *Terania* and *Paraembolides*); the Macrothelinae (*Atrax* and *Hadronyche*); the Plesiothelinae (*Plesiothele*).

Atrax Cambridge, 1877

Atrax Cambridge, O.P. (1877). On some new genera and species of Araneidea. *Ann. Mag. Nat. Hist. (4)* **19**: 26–39 pls 6–7 [Jan. 1877] [26]. Type species *Atrax robustus* Cambridge, 1877 by monotypy.

Styphlopis Rainbow, W.J. (1913). Arachnida from the Solomon Islands. *Rec. Aust. Mus.* **10**: 1–16 [5]. Type species *Styphlopis insularis* Rainbow, 1913 by monotypy.

Euctimena Rainbow, W.J. (1914). Studies in Australian Araneidae. No. 6. The Terretelariae. *Rec. Aust. Mus.* **10**: 187–270 [15 Aug. 1914] [248]. Type species *Euctimena tibialis* Rainbow, 1914 by monotypy.

Pseudatrax Rainbow, W.J. (1914). Studies in Australian Araneidae. No. 6. The Terretelariae. *Rec. Aust. Mus.* **10**: 187–270 figs 67–70 [15 Aug. 1914] [260]. Type species *Pseudatrax moreaui* Rainbow, 1914 by monotypy.

Poikilomorpha Rainbow, W.J. (1914). Studies in Australian Araneidae. No. 6. The Terretelariae. *Rec. Aust. Mus.* **10**: 187–270 figs 67–70 [15 Aug. 1914] [264]. Type species *Poikilomorpha montana* Rainbow, 1914 by monotypy.

Anepsiada Rainbow, W.J. & Pulleine, R.H. (1918). Australian trap-door spiders. *Rec. Aust. Mus.* **12**: 81–169 pls 12–24 [24 Dec. 1918] [167]. Type species *Anepsiada ventricosa* Rainbow and Pulleine, 1918 by monotypy.

Taxonomic decision of Musgrave, A. (1927). Some poisonous Australian spiders. *Rec. Aust. Mus.* **16**: 33–46 pls 2–3 [33] (*Euctimena*); Gray, M.R. (1978). Venoms of Dipluridae. A. Systematics and distribtuion. pp. 120–126 *in* Bettini, S. (ed.) *Handbook of Experimental Pharmacology.* Vol. 48 Arthropod Venoms. Berlin : Springer Verl.

This group is also found in New Guinea? and Solomon Ils.?, see Rainbow, W.J. (1913). Arachnida from the Solomon Islands. *Rec. Aust. Mus.* **10**: 1–16 [5]; Rainbow, W.J. (1920). Trapdoor spiders of the *Chevert* Expedition. *Rec. Aust. Mus.* **13**: 77–85 [84] (probable error in locality).

Atrax bicolor (Rainbow, 1914)

Aname bicolor Rainbow, W.J. (1914). Studies in Australian Araneidae. No. 6. The Terretelariae. *Rec. Aust. Mus.* **10**: 187–270 [15 Aug. 1914] [233]. Type data: holotype, AM K19077, KS6331 juv., paratype AM K18738, KS6332 juv., from Antonio, near Rydal, N.S.W.

Distribution: SE coastal, N.S.W.; Ecology: terrestrial, predator. Biological references: Rainbow, W.J. & Pulleine, R.H. (1918). Australian trap-door spiders. *Rec. Aust. Mus.* **12**: 81–169 pls 12–24.

Atrax formidabilis Rainbow, 1914

Atrax formidabilis Rainbow, W.J. (1914). Studies in Australian Araneidae. No. 6. The Terretelariae. *Rec. Aust. Mus.* **10**: 187–270 [15 Aug. 1914] [255]. Type data: holotype, AM K35282 KS1038 M. adult, from Richmond River, N.S.W.

Distribution: SE coastal, NE coastal, Qld., N.S.W.; S Qld., N N.S.W. mainly MacPherson Ranges. Ecology: open forest, closed forest, arboreal, predator; eats coldblooded vetebrates as well as invertebrates. Biological references: Musgrave, A. (1927). Some poisonous Australian spiders. *Rec. Aust. Mus.* **12**: 33–46 pls 2–3 (medical); Gray, M.R. (1978) Venoms of Dipluridae. A. Systematics and distribution. pp. 120–126 *in* Bettini, S. (ed.) *Handbook of Experimental Pharmacology.* Vol. 48. Arthropod Venoms. Berlin : Springer Verl. (taxonomy, habitat, distribution).

Atrax infensus Hickman, 1964

Atrax infensus Hickman, V.V. (1964). On *Atrax infensus* sp.n. (Araneida : Dipluridae) its habits and a method of trapping the males. *Pap. Proc. R. Soc. Tasm.* **98**: 107–112 pls 1–2 [7 May 1964] [108]. Type data: holotype, AM K68385, KS0953 M. adult, paratype AM K68386, KS0954 F. adult, from Toowoomba, Qld.

Distribution: NE coastal, Qld.; Toowoomba. Ecology: open forest, burrowing, terrestrial, predator; with extended silk lining to burrow.

Atrax modestus Simon, 1891

Atrax modesta Simon, E. (1891). Études arachnologiques. 23 Mém. 38. Descriptiones d'espèces et de genres nouveaux de la famille des Aviculariidae. *Ann. Soc. Entomol. Fr.* **60**: 300–312 [302]. Type data: holotype, MNHP F. adult, from Melbourne, Vic.

Distribution: SE coastal, Vic.; Melbourne and W Gippsland area. Ecology: terrestrial, predator.

Atrax montanus (Rainbow, 1914)

Poikilomorpha montana Rainbow, W.J. (1914). Studies in Australian Araneidae. No. 6. The Terretelariae. *Rec. Aust. Mus.* **10**: 187–270 [15 Aug. 1914] [265]. Type data: holotype, AM K11956, KS1039 F. adult, from Wentworth Falls, Blue Mts., Jamieson Valley, N.S.W.

Distribution: SE coastal, N.S.W.; Blue Mts. Ecology: terrestrial, predator. Biological references: Gray, M.R. (1978). Venoms of Dipluridae. A. Systematics and distribution. pp. 120–126 *in* Bettini, S. (ed.) *Handbook of Experimental Pharmacology.* Vol. 48. Arthropod Venoms. Berlin : Springer Verl.

Atrax moreaui (Rainbow, 1914)

Pseudatrax moreaui Rainbow, W.J. (1914). Studies in Australian Araneidae. No. 6. The Terretelariae. *Rec. Aust. Mus.* **10**: 187–270 [15 Aug. 1914] [261]. Type data: holotype, AM K18737, KS1119 F. adult, paratype K34372, KS0994 F. adult (Narabeen), from Antonio, N.S.W.

Distribution: SE coastal, N.S.W.; Ecology: terrestrial, predator. Biological references: Gray, M.R. (1978). Venoms of Dipluridae. A. Systematics and distribution. pp. 120–126 *in* Bettini, S. (ed.) *Handbook of Experimental Pharmacology.* Vol. 48. Arthropod Venoms. Berlin : Springer Verl.

Atrax pulvinator Hickman, 1927

Atrax pulvinator Hickman, V.V. (1927). Studies in Tasmanian spiders. Pt. 1. *Pap. Proc. R. Soc. Tasm.* **1926**: 52–86 [24 Feb. 1927] [70 pl 7 figs 7–8]. Type data: holoytype, QVM 1957/13/20 *F. adult Type 17, from Cascades, Hobart, Tas.

Distribution: Tas.; Hobart region. Ecology: terrestrial, predator. Biological references: Hickman, V.V. (1927). Studies in Tasmanian spiders. Pt. 1. *Pap. Proc. R. Soc. Tasm.* **1926**: 52–86 pls 4–10 (biology, burrow, eggs).

Atrax robustus Cambridge, 1877

Atrax robustus Cambridge, O.P. (1877). On some new genera and species of Araneidea. *Ann. Mag. Nat. Hist. (4)* **19**: 26–39 pls 6–7 [Jan. 1877] [27]. Type data: holotype, BMNH 71 F. adult, from Australia (as New Holland).

Euctimena tibialis Rainbow, W.J. (1914). Studies in Australian Araneidae. No. 6. The Terretelariae. *Rec. Aust. Mus.* **10**: 187–270 [15 Aug. 1914] [248]. Type data: syntypes, AM K3363, KS4092 2M. adult, from Turramurra, N.S.W.

Taxonomic decision of Musgrave, A. (1927). Some poisonous Australian spiders. *Rec. Aust. Mus.* **16**: 33–46 pls 2–3 [34].

Distribution: SE coastal, N.S.W., Vic.; Newcastle-Hunter River region N.S.W. to E Gippsland, Vic. Ecology: closed forest, burrowing, terrestrial, predator; silk tubes in burrows. Biological references: Musgrave, A. (1927). Some poisonous Australian spiders. *Rec. Aust. Mus.* **16**: 33–46 pls 2–3 (distribution, medical); Kaire, G.H. (1963). Observations on some funnel-web spiders (*Atrax* species) and their venoms, with particular reference to *Atrax robustus. Med. J. Aust.* **1963**: 307–311 (venom, pharmacology); Levitt, V. (1961). The funnel-web spider in captivity. *R. Zool. Soc. N.S.W.* **1958–59**: 80–84 (biology); Sutherland, S.K. (1972). The Sydney funnel-web spider (*Atrax robustus*), 3: A review of some clinical records of human envenomation. *Med. J. Aust.* **2**: 643–647 (medical); Sutherland, S.K. (1976). Treatment of arachnid poisoning in Australia. *Aust. Fam. Physician* **5**: 305–312 (medical); Sutherland, S.K. (1978). Venoms of Dipluridae. B. Biology and venoms. pp. 126–148 *in* Bettini, S. (ed.) *Handbook of Experimental Pharmacology.* Vol. 48. Arthropod Venoms. Berlin : Springer Verl. (venom); Gray, M.R. (1978). Venoms of Dipluridae, A. Systematics and distribution. pp. 120–126 *in* Bettini, S. (ed.) *Handbook of Experimental Pharmacology.* Vol. 48. Arthropod Venoms. Berlin : Springer Verl. (taxonomy, habitat, distribution).

Atrax validus Rainbow and Pulleine, 1918

Atrax valida Rainbow, W.J. & Pulleine, R.H. (1918). Australian trap-door spiders. *Rec. Aust. Mus.* **12**: 81–169 pls 12–24 [24 Dec. 1918] [165]. Type data: holotype, AM K40961, KS955 F. adult, paratypes SAMA N198165 M.,F. adult (in same tube, no locality), AM K41452, KS956 3F. adult, AM K40963, KS958 F. adult, AM K40964, KS959 F. adult, AM K41457, KS957 2F. adult, from Tamborine Mt., Qld.

Anepsiada ventricosa Rainbow, W.J. & Pulleine, R.H. (1918). Australian trap-door spiders. *Rec. Aust. Mus.* **12**: 81–169 pls 12–24 [24 Dec. 1918] [167]. Type data: holotype, AM K40965 F. adult, from Cloncurry, Qld.; incorrect (published) record, emended to Tamborine Mt. in accordance with register record by M.R. Gray.

Taxonomic decision of Gray, M.R. (personal comm.) probably *Anepsiada ventricosa* Rainbow and Pulleine synonymous with *Atrax validus*; Cloncurry locality for *Anepsiada*, in error; Main, B.Y. (1982). Some geographic considerations of families of spiders occurring in New Guinea. pp. 583–602 *in* Gressitt, J.L. (ed.) *Biogeography and Ecology in New Guinea.* The Hague : Junk, also states record from Papua New Guinea probably in error.

Distribution: NE coastal, SE coastal, N.S.W., Qld.; MacPherson Ranges, N.S.W. Ecology: open forest, closed forest, burrowing, terrestrial, predator; burrower often under logs, silk expansion from burrow, predator of vertebrates as well as invertebrates. Biological references: Gray, M.R. (1978). Venoms of Dipluridae, A. Systematics and distribution. pp. 120–126 *in* Bettini, S. (ed). *Handbook of Experimental Pharmacology.* Vol. 48. Arthropod Venoms. Berlin : Springer Verl. (biology).

Atrax venenatus Hickman, 1927

Atrax venenatus Hickman, V.V. (1927). Studies in Tasmanian spiders. Pt. 1. *Pap. Proc. R. Soc. Tasm.* **1926**: 52–86 pl 6 figs 5–6 [24 Feb. 1927] [63]. Type data: syntypes, QVM 1957/13/16 *M. adult Type 28, QVM 1957/13/17 *F. adult Type 29, from Newtown Creek, Hobart, Tas.

Distribution: Tas.; NE and Hobart region. Ecology: open forest, terrestrial, burrowing, predator; also found in montane areas, makes silk-lined tubes.

Atrax versutus Rainbow, 1914

Atrax versuta Rainbow, W.J. (1914). Studies in Australian Araneidae. No. 6. The Terretelariae. *Rec. Aust. Mus.* **10**: 187–270 [15 Aug. 1914] [253]. Type data: syntypes, AM K12907, KS0969 F. adult, ?M. immature, from Jenolan, N.S.W.

Distribution: SE coastal, N.S.W.; Blue Mts. Ecology: predator; silk tubes.

Bymainiella Raven, 1978

Bymainiella Raven, R.J. (1978). Systematics of the spider subfamily Hexathelinae (Dipluridae : Mygalomorphae : Arachnida). *Aust. J. Zool. Suppl. Ser.* **65**: 1–75 [15 Sept. 1978] [12]. Type species *Hexathele terraereginae* Raven, 1976 by original designation.

Bymainiella lugubris Raven, 1978

Bymainiella lugubris Raven, R.J. (1978). Systematics of the spider subfamily Hexathelinae (Dipluridae : Mygalomorphae : Arachnida). *Aust. J. Zool. Supp. Ser.* **65**: 1–75 [15 Sept. 1978] [56]. Type data: holotype, QM W6016 *M. adult, paratypes AM KS181 F. adult (described), QM W6017–27, *11F. adult, AM KS180 *M. adult, AM KS182–193 *12F. adult, from New England Plateau, N.S.W.

Distribution: Murray-Darling basin, SE coastal, N.S.W.; New England Plateau. Ecology: open forest, alpine, terrestrial, predator; builds a ground web and lives in alpine, open sclerophyll and snow-gum forests.

Bymainiella monteithi Raven, 1978

Bymainiella monteithi Raven, R.J. (1978). Systematics of the spider subfamily Hexathelinae (Dipluridae : Mygalomorphae : Arachnida). *Aust. J. Zool. Supp. Ser.* **65**: 1–75 [15 Sept. 1978] [62]. Type data: holotype, QM W6029 M. adult, paratypes QM W6030 (described), W6034 F. adult, QM W6031–33 2M.,F. adult, from Cunningham's Gap, Qld.

Distribution: SE coastal, NE coastal, Qld., N.S.W.; found in SE Qld. and NE N.S.W. Ecology: closed forest, terrestrial, predator; found in montane rainforest, builds silk tubes under logs.

Bymainiella polesoni Raven, 1978

Bymainiella polesoni Raven, R.J. (1978). Systematics of the spider subfamily Hexathelinae (Dipluridae : Mygalomorphae : Arachnida). *Aust. J. Zool. Supp. Ser.* **65**: 1–75 [15 Sept. 1978] [66]. Type data: holotype, AM KS205 *M. adult, paratypes AM KS211 *F. adult (described), AM KS206–10 KS212 *6F. adult, KS213 *M. adult, from New England Natl. Park, N.S.W.

Distribution: SE coastal, N.S.W.; found at Mt. Boss, in the New England Natl. Park, Dorrigo. Ecology: closed forest, terrestrial, predator.

Bymainiella terraereginae (Raven, 1976)

Hexathele terraereginae Raven, R.J. (1976). A new spider of the genus *Hexathele* Ausserer (Dipluridae : Mygalomorphae) from Australia. *Proc. R. Soc. Qd.* **87**: 53–61 [30 June 1976] [53]. Type data: holotype, QM W4849 *M. adult, paratypes QM W4851, W4853 *2M. adult, W4863–65 *3M. juv., QM W4850 (described), W4852, W4854–62, W4866 *11F. adult, EUQ *2F. adult, NMV K26 *F. adult, AM KS64–65 *M.,F. adult, from Lamington Plateau, Qld.

Distribution: NE coastal, Qld.; Lamington Plateau. Ecology: closed forest, terrestrial, burrowing, predator; makes silk-lined excavations under rocks and logs. Biological references: Raven, R.J. (1978). Systematics of the spider subfamily Hexathelinae (Dipluridae : Mygalomorphae : Arachnida). *Aust. J. Zool. Supp. Ser.* **65**: 1–75 (taxonomy); Raven, R.J. (1980). The evolution and biogeography of the mygalomorph spider family Hexathelidae (Araneae, Chelicerata). *J. Arachnol.* **8**: 251–266 (evolution, biogeography).

Hadronyche L. Koch, 1873

Hadronyche Koch, L. (1873). *Die Arachniden Australiens.* Lf. 9 pp. 409–472 pls 33–36 Nürnberg :

Bauer & Raspe [463]. Type species *Hadronyche cerberea* L. Koch, 1873 by monotypy.

This group may occur in New Guinea, see Rainbow, W.J. (1920). Trapdoor spiders of the *Chevert* Expedition. *Rec. Aust. Mus.* **13**: 77–85 [84] (probable error in locality).

Hadronyche cerberea L. Koch, 1873

Hadronyche cerberea Koch, L. (1873). *Die Arachniden Australiens.* Lf. 9 pp. 409–472 pls 33–36 Nürnberg : Bauer & Raspe [463] [purported male of this species was described by Hogg in 1901, later named *H. meridiana* Hogg, 1902, see *Hadronyche meridiana*]. Type data: holotype, SMNS F. adult destroyed, from Sydney, N.S.W.

Distribution: SE coastal, N.S.W.; Sydney?. Ecology: terrestrial, predator.

Hadronyche hirsuta Rainbow, 1920

Hadronyche hirsuta Rainbow, W.J. (1920). Trapdoor spiders of the *Chevert* Expedition. *Rec. Aust. Mus.* **13**: 77–85 [4 Dec. 1920] [83]. Type data: holotype, AM K41705 KS992 *F. adult, from Papua.

Distribution: Papua New Guinea, doubtful record, see Main, B.Y. (1982). Some zoogeographic considerations of families of spiders occurring in New Guinea. pp. 583–602 *in* Gressitt, J.L. (ed.) *Biogeography and Ecology in New Guinea.* The Hague : Junk [588], may have been part of an Australian collection confused as collection from *Chevert* expedition, species status uncertain, may be *Atrax* species. Ecology: terrestrial, predator.

Hadronyche meridiana Hogg, 1902

Hadronyche meridiana Hogg, H.R. (1902). On some additions to the Australian spiders of the suborder Mygalomorphae. *Proc. Zool. Soc. Lond.* **1902**: 121–142 pl 13 [3 June 1902] [122] [holotype originally recorded as male of *Hadronyche cerberea* Koch by Hogg, H.R. (1901). On Australian and New Zealand spiders of the Suborder Mygalomorphae. *Proc. Zool. Soc. Lond.* **1901(2)**: 218–279 (274); later Hogg (1902) designated the new species *meridiana*]. Type data: holotype, BMNH 1903.2.10.6 M. adult, from Macedon, Vic.

Distribution: SE coastal, Vic.; Macedon. Ecology: terrestrial, predator.

Paraembolides Raven, 1980

Paraembolides Raven, R.J. (1980). The evolution and biogeography of the mygalomorph spider family Hexathelidae (Araneae, Chelicerata). *J. Arachnol.* **8**: 251–266 [Oct. 1980] [254]. Type species *Bymainiella boycei* Raven, 1978 by original designation.

Paraembolides boycei (Raven, 1978)

Bymainiella boycei Raven, R.J. (1978). Systematics of the spider subfamily Hexathelinae (Dipluridae : Mygalomorphae : Arachnida). *Aust. J. Zool. Supp. Ser.* **65**: 1–75 [15 Sept. 1978] [16]. Type data: holotype, QM W4869 *M. adult, paratypes QM W4870 (described), W6006, W6008 3F. adult, EUQ 2F. juv., from Boyce Reserve, Toowoomba, Qld.

Distribution: NE coastal, Qld.; Bunya Mts. and Toowoomba. Ecology: closed forest, arboreal, predator; forms silk tubes. Biological references: Raven, R.J. (1980). The evolution and biogeography of the mygalomorph spider family Hexathelidae (Araneae, Chelicerata). *J. Arachnol.* **8**: 251–266 (taxonomy).

Paraembolides boydi (Raven, 1978)

Bymainiella boydi Raven, R.J. (1978). Systematics of the spider subfamily Hexathelinae (Dipluridae : Mygalomorphae : Arachnida). *Aust. J. Zool. Supp. Ser.* **65**: 1–75 [15 Sept. 1978] [22]. Type data: holotype, AM KS150 M. adult, paratypes, AM KS151 F. adult (described), AM KS152, KS155, KS156, KS157 4F. adult, AM KS153, KS154 *M.,F. adult, AM KS158, KS159, K12914 *3M. adult, from Mt. Edwards, Boyd Plateau, N.S.W.

Distribution: SE coastal, Murray-Darling basin, N.S.W.; Boyd Plateau and Jenolan region. Ecology: open forest, terrestrial, predator; in logs. Biological references: Raven, R.J. (1980). The evolution and biogeography of the mygalomorph spider family Hexathelidae (Araneae, Chelicerata). *J. Arachnol.* **8**: 251–266.

Paraembolides brindabella (Raven, 1978)

Bymainiella brindabella Raven, R.J. (1978). Systematics of the spider subfamily Hexathelinae (Dipluridae : Mygalomorphae : Arachnida). *Aust. J. Zool. Supp. Ser.* **65**: 1–75 [15 Sept. 1978] [26]. Type data: holotype, AM KS164 M. adult, paratypes AM KS165 F. adult (described), AM KS167, M. adult, KS168 F. juv., AM KS166 M. adult, from W of Wyora to Canberra-Tumut Road, 148°42′E, 35°17′S, N.S.W.

Distribution: Murray-Darling basin, N.S.W., A.C.T.; Brindabella Ranges. Ecology: open forest, alpine, terrestrial, predator; in snow grass. Biological references: Raven, R.J. (1980). The evolution and biogeography of the mygalomorph spider family Hexathelidae (Araneae, Chelicerata). *J. Arachnol.* **8**: 251–266.

Paraembolides cannoni (Raven, 1978)

Bymainiella cannoni Raven, R.J. (1978). Systematics of the spider subfamily Hexathelinae (Dipluridae : Mygalomorphae : Arachnida). *Aust. J. Zool. Supp. Ser.* **65**: 1–75 [15 Sept. 1978] [34]. Type data: holotype, QM W6014 *M. adult, paratype QM W6015 *F. juv., from Lamington Natl. Park, Qld.

Distribution: NE coastal, Qld.; Lamington Plateau. Ecology: open forest, terrestrial, arboreal, predator. Biological references: Raven, R.J. (1980). The evolution and biogeography of the mygalomorph spider family Hexathelidae (Araneae, Chelicerata). *J. Arachnol.* **8**: 251–266.

Paraembolides grayi (Raven, 1978)

Bymainiella grayi Raven, R.J. (1978). Systematics of the spider subfamily Hexathelinae (Dipluridae : Mygalomorphae : Arachnida). *Aust. J. Zool. Supp. Ser.* **65**: 1–75 [15 Sept. 1978] [39]. Type data: holotype, AM KS176 *M. adult, paratype AM KS177 *M. adult, from Royal Natl. Park, N.S.W.

Distribution: SE coastal, N.S.W.; between Sydney and Wollongong. Ecology: open forest, terrestrial, predator; in dry sclerophyll forest. Biological references: Raven, R.J. (1980). The evolution and biogeography of the mygalomorph spider family Hexathelidae (Araneae, Chelicerata). *J. Arachnol.* **8**: 251–266.

Paraembolides montisbossi (Raven, 1978)

Bymainiella montisbossi Raven, R.J. (1978). Systematics of the spider subfamily Hexathelinae (Dipluridae : Mygalomorphae : Arachnida). *Aust. J. Zool. Supp. Ser.* **65**: 1–75 [15 Sept. 1978] [42]. Type data: holotype, QM W6036 M. adult, paratypes AM KS201 F. adult (described), QM W6037–38 2M. adult, QM W6039–43 5F. adult, AM KS202 M. adult, from Mt. Banda Banda Beech Reserve, near Mt. Boss, N.S.W.

Distribution: SE coastal, N.S.W.; Mt. Boss State Forest, Mt. Banda Banda Beech Reserve near Mt. Boss. Ecology: closed forest, terrestrial, predator; builds silk tubes under logs. Biological references: Raven, R.J. (1980). The evolution and biogeography of the mygalomorph spider family Hexathelidae (Araneae, Chelicerata). *J. Arachnol.* **8**: 251–266.

Paraembolides tubrabucca (Raven, 1978)

Bymainiella tubrabucca Raven, R.J. (1978). Systematics of the spider subfamily Hexathelinae (Dipluridae : Mygalomorphae : Arachnida). *Aust. J. Zool. Supp. Ser.* **65**: 1–75 [15 Sept. 1978] [52]. Type data: holotype, AM KS219 *M. adult, paratypes AM KS220 *F. adult (described), AM KS221, KS223 *2F. adult, AM KS222 *M. adult, EUQ *M. adult, from Tubrabucca, N.S.W.

Distribution: SE coastal, N.S.W.; Barrington Tops and Tubrabucca. Ecology: open forest, terrestrial, predator; wet sclerophyll forest. Biological references: Raven, R.J. (1980). The evolution and biogeography of the mygalomorph spider family Hexathelidae (Araneae, Chelicerata). *J. Arachnol.* **8**: 251–266.

Paraembolides variabilis (Raven, 1978)

Bymainiella variabilis Raven, R.J. (1978). Systematics of the spider subfamily Hexathelinae (Dipluridae : Mygalomorphae : Arachnida). *Aust. J. Zool. Supp. Ser.* **65**: 1–75 [15 Sept. 1978] [31]. Type data: holotype, AM KS169 M. adult, paratypes AM KS171 F. adult (described), AM KS170, KS172, KS173 3M. adult, KS174 F. adult, from Bilpin, N.S.W.

Distribution: SE coastal, N.S.W; Bilpin and Blue Gum forests. Ecology: open forest, terrestrial, predator; dry sclerophyll forest. Biological references: Raven, R.J. (1980). The evolution and biogeography of the mygalomorph spider family Hexathelidae (Araneae, Chelicerata). *J. Arachnol.* **8**: 251–266.

Plesiothele Raven, 1978

Plesiothele Raven, R.J. (1978). Systematics of the spider subfamily Hexathelinae (Dipluridae : Mygalomorphae : Arachnida). *Aust. J. Zool. Suppl. Ser.* **65**: 1–75 [15 Sept. 1978] [10]. Type species *Hexathele fentoni* Hickman, 1936 by original designation.

Plesiothele fentoni (Hickman, 1936)

Hexathele fentoni Hickman, V.V. (1936). A new mygalomorph spider belonging to the genus *Hexathele* Ausserer. *Pap. Proc. R. Soc. Tasm.* **1935**: 131–137 [17 Aug. 1936] [132]. Type data: holotype, AM K66981, KS6338 M. adult, paratype AM K66982, KS6339 F. adult, from Natl. Park, Lake Fenton, Tas.

Distribution: Tas; Lake Fenton, Lake Pedder. Ecology: closed forest, terrestrial, predator; in *Nothofagus* forest, forms silk-lined burrows. Biological references: Raven, R.J. (1978). Systematics of the spider subfamily Hexathelinae (Dipluridae : Mygalomorphae : Arachnida). *Aust. J. Zool. Supp. Ser.* **65**: 1–75 (taxonomy); Raven, R.J. (1980). The evolution and biogeography of the mygalomorph spider family Hexathelidae (Araneae, Chelicerata). *J. Arachnol.* **8**: 251–266 (evolution and biogeography).

Terania Raven, 1980

Terania Raven, R.J. (1980). The evolution and biogeography of the mygalomorph spider family Hexathelidae (Araneae, Chelicerata). *J. Arachnol.* **8**: 251–266 [Oct. 1980] [253]. Type species *Hexathele montanus* Hickman, 1927 by original designation.

Terania montana (Hickman, 1927)

Hexathele montanus Hickman, V.V. (1927). Studies in Tasmanian spiders. Pt 1. *Pap. Proc. R. Soc. Tasm.* **1926**: 52–86 pls 4–11 [24 Feb. 1927] [73]. Type data: syntypes, QVM 1957/13/5 *M. adult, QVM 1957/13/4 *F. adult, from Higgs Track, Western Tiers, Chudleigh, Tas.

Distribution: Tas., SE coastal, Vic. Ecology: closed forest, terrestrial, predator; forms silk-lined burrows in rotten logs and under stones. Biological references: Raven, R.J. (1980). The evolution and biogeography of the mygalomorph spider family Hexathelidae (Araneae, Chelicerata). *J. Arachnol.* **8**: 251–266 (evolution, biogeography, taxonomy).

Terania otwayensis (Raven, 1978)

Bymainiella otwayensis Raven, R.J. (1978). Systematics of the spider subfamily Hexathelinae (Dipluridae : Mygalomorphae : Arachnida). *Aust. J. Zool. Supp. Ser.*

65: 1–75 [15 Sept. 1978] [48]. Type data: holotype, AM KS204 M. adult*, paratypes NMV K39–41 3F. adult, K42 M. adult, from Otway Range, Vic.

Distribution: SE coastal, Vic.; southern Vic. only. Ecology: open forest, terrestrial, burrowing, predator; wet sclerophyll forest. Biological references: Raven, R.J. (1980). The evolution and biogeography of the mygalomorph spider family Hexathelidae (Araneae, Chelicerata). *J. Arachnol.* **8**: 251–266.

MIGIDAE

INTRODUCTION

Small spiders, not hairy. Eyes wide spaced (at least one third width of caput) but larger and not as widespread as in Actinopodidae. Chelicerae never with a rastellum. Labium in one piece with sternum *i.e.* without a marked suture; two pairs sternal sigilla. Legs never with scopulae; frequently heavily spinose. Spinnerets short, terminal segment of lateral pair short, rounded. Burrow with a trapdoor. Terrestrial or arboreal (*e.g.* in tree fern stems). Recorded from Tas., Qld. and southern N.S.W., also probably occurs in Vic. Elsewhere, southern continents only, including Madagascar, New Zealand and New Caledonia. Two subfamilies occur in Australia, the Miginae (*Migas*) and Calathotarsinae (*Heteromigas*).

Heteromigas Hogg, 1902

Heteromigas Hogg, H.R. (1902). On some additions to the Australian spiders of the suborder Mygalomorphae. *Proc. Zool. Soc. Lond.* **1902(2)**: 121–142 [3 June 1902] [123 pl 13]. Type species *Heteromigas dovei* Hogg, 1902 by original designation.

This group is also recorded from Chile, see Zapfe, H. (1961). La Familia Migidae en Chile. *Invest. Zool. Chil.* **7**: 151–157, identification considered doubtful, see Raven, R.J. (1984). Systematics and biogeography of the mygalomorph spider family Migidae (Araneae) in Australia. *Aust. J. Zool.* **32**: 379–390

Heteromigas dovei Hogg, 1902

Heteromigas dovei Hogg, H.R. (1902). On some additions to the Australian spiders of the suborder Mygalomorphae. *Proc. Zool. Soc. Lond.* **1902(2)**: 121–142 [3 June 1902] [123 pl 13]. Type data: syntypes, BMNH 1924.3.1.3–4 2F. adult, from Table Cape, Tas.

Distribution: Tas.; N areas. Ecology: terrestrial, predator. Biological references: Hickman, V.V. (1926). Notes on a Tasmanian Araneidae with a description of a new species. *Pap. Proc. R. Soc. Tasm.* **1925**: 171–186 pls 12–15 [5 Mar. 1926] (description of male).

Migas L. Koch, 1873

Migas Koch, L. (1873). *Die Arachniden Australiens.* Lf. 9 pp. 409–472 pls 33–36 Nürnberg : Bauer & Raspe [467]. Type species *Migas paradoxus* L. Koch, 1873 by monotypy.

This group is also found in New Zealand, New Caledonia and Chile, see Zapfe, H. (1961). La Familia Migidae en Chile. *Invest. Zool. Chil.* **7**: 151–157. Species now known not to occur in Australia: *Migas paradoxus* L. Koch, 1873, see Hogg, H.R. (1901). On Australian and New Zealand spiders of the suborder Mygalomorphae. *Proc. Zool. Soc. Lond.* **1901(2)**: 218–279 [228].

Migas nitens Hickman, 1927

Migas nitens Hickman, V.V. (1927). Studies in Tasmanian spiders. Pt. 1. *Pap. Proc. R. Soc. Tasm.* **1926**: 52–86 pls 4–10 [24 Feb. 1927] [52] [Hickman (1927) described *Migas nitens* but mentioned several specimens from two localities: Prince of Wales Bay, Derwent Park and Cornelian Bay, New Town, Tas.; the description, however, referred to a single specimen (gave measurements and counts) hence is designated as lectotype. In 1929 he described the adult male which had been collected as an immature (sex not realized) the same date as lectotype and the other specimens mentioned. It came from Prince of Wales Bay. QVM catalogue of types lists two specimens as "Holotype" but Hickman did not designate any sort of type in either of his papers. The former is considered the lectotype and the later described male a paralectotype because it was part of the original material cited]. Type data: lectotype, QVM 1957/13/22 *F. adult Type 33, from Cornelian Bay, New Town, Tas., paralectotype QVM 1957/13/23 *M. adult Type 34 (Prince of Wales Bay), designation by Main, B.Y., this work.

Distribution: Tas.; E region. Ecology: supralittoral, burrowing, terrestrial, predator; burrow on seashore, with door. Biological references: Hickman, V.V. (1929). Studies in Tasmanian spiders. Pt. 3. *Pap. Proc. R. Soc. Tasm.* **1928**: 96–118 pls 15–17 [96] (biology, laboratory care).

THERAPHOSIDAE

INTRODUCTION

Large, hairy spiders. Commonly called "Bird eating spiders" or "Tarantulas". Eyes always in a compact group of two rows, tuberculate. Labium with dense cluster of cuspules. Two tarsal claws and pronounced claw tufts; dense scopulae (usually on all legs as well as palps); legs not heavily spinose. Two pairs of spinnerets; spinnerets long; terminal segments of lateral pair long and tapering. Burrows usually deep and sinuous; never closed with a door. Generally terrestrial, some (unidentified forms) make silk nests under bark of tree butts (N Qld.). Tropical and arid inland Australia. An abundant and diverse group in the Americas (South America to southern states of U.S.A.), Africa, India to New Guinea - less diverse in Australia. The taxonomy of Australian forms is poorly known but all belong in the Selenocosmiinae.

Selenocosmia Ausserer, 1871

Selenocosmia Ausserer, A. (1871). Beiträge zur Kenntniss der Arachniden-Familie der Territelariae Thorell (Mygalidae Autor.). *Verh. Zool.-Bot. Ges. Wien* **21**: 117–224 [4 Jan. 1871] [204]. Type species *Mygale javanensis* Walkenaer, 1837 by monotypy.

Phrictus Koch, L. (1874). *Die Arachniden Australiens.* Lf. 10 pp. 473–528 pls 37–40 Nürnberg : Bauer & Raspe [488] [*non Phrictus* Spinola, 1839]. Type species *Phrictus crassipes* L. Koch, 1874 by monotypy.

Phlogius Simon, E. (1887). Quelques observations sur les Arachnides. *Bull. Soc. Entomol. Fr.* **1887**: 193–195 [195] [*nom. nov.* for *Phrictus* L. Koch, 1874].

Taxonomic decision of Simon, E. (1903). *Histoire Naturelle de Araignées.* Paris : Roret [955].

This group is also found from India to Australia, see Roewer, C.Fr. (1942). *Katalog der Araneae.* Bremen : Komissons - Verl. von "NATURA" Vol. 1 [265]; Bonnet, P. (1958). *Bibliographia Araneorum.* Toulouse : Douladoure Vol. 2 Pt. 4 [4011].

Selenocosmia crassipes (L. Koch, 1874)

Ischnocolus lucubrans Koch, L. (1874). *Die Arachniden Australiens.* Lf. 10 pp. 473–528 pls 37–40 Nürnberg : Bauer & Raspe [487]. Type data: holotype, ZMH (MGH 8103) F. immature, from Port McKay, Qld.

Phrictus crassipes Koch, L. (1874). *Die Arachniden Australiens.* Lf. 10 pp. 473–528 pls 37–40 Nürnberg : Bauer & Raspe [489]. Type data: holotype, ZMH (MGH 9929) F. adult, from Bowen, Qld.

Selenocosmia vulpina Hogg, H.R. (1901). On Australian and New Zealand spiders of the suborder Mygalomorphae. *Proc. Zool. Soc. Lond.* **1901(2)**: 218–279 [4 June 1901] [246]. Type data: holotype, BMNH 46/73 M. adult, from Cape Upstart, near Bowen, Qld.

Taxonomic decision of Simon, E. (1903). *Histoire Naturelle des Araignées.* Paris : Roret [955]; Main, B.Y. (1982). Some zoogeographic considerations of families of spiders occurring in New Guinea. pp. 583–602 *in* Gressitt, J.L. (ed.) *Biogeography and Ecology of New Guinea.* The Hague : Junk; Hogg, H.R. (1901). On Australian and New Zealand spiders of the suborder Mygalomorphae. *Proc. Zool. Soc. Lond.* **1901(2)**: 218–279; Main, B.Y., this work [*Ischnocolus lucubrans* Koch occurs on an earlier page of this publication than does *Phrictus crassipes* Koch (=*Selenocosmia crassipes*), but because of wide usage it is preferable to retain *crassipes* as the valid name].

Distribution: NE coastal, (N Gulf), N coastal, Qld., N.T.; possibly New Guinea. Ecology: tall shrubland, low open forest, burrowing, terrestrial, predator.

Selenocosmia stirlingi Hogg, 1901

Selenocosmia stirlingi Hogg, H.R. (1901). On Australian and New Zealand spiders of the suborder Mygalomorphae. *Proc. Zool. Soc. Lond.* **1901(2)**: 218–279 [4 June 1901] [245]. Type data: syntypes, BMNH 1898.6.19.2–3 M.,F. adult, from Crown Point Station, S.A.

Selenocosmia stalkeri Hirst, A.S. (1907). On two spiders of the genus *Selenocosmia. Ann. Mag. Nat. Hist. (7)* **19**: 522–524 [522]. Type data: holotype, BMNH 1907.2.2.2 F. adult, from Alexandria, N.T.

Taxonomic decision of Main B.Y., this work.

Distribution: NW coastal, W plateau, N coastal, N Gulf, Lake Eyre Basin, S Gulfs, Bulloo River basin, Murray-Darling basin, W.A., S.A., N.T., Qld., N.S.W., Vic.; throughout inland areas of foregoing regions. Ecology: desert, hummock grassland, open scrub, low shrubland, low open woodland, tall open shrubland, burrowing, terrestrial, predator; burrows in sandy and light loamy soils. Biological references: Spencer, B. (1896). *Report on the work of the Horn Scientific Expedition to Central Australia.* Pt. 2 Zoology. Appendix. London : Dulau & Co. [biology, morphology; species misidentified as *Phlogius crassipes* (Koch, 1874)]; Main, B.Y. & Main, A.R. (1956). Spider predator on a vertebrate. *West. Aust. Nat.* **5**: 139 [biology, feeding, prey; misidentified as *Selenocosmia crassipes* (L. Koch, 1874)].

Selenocosmia strenua (Thorell, 1881)

Phrictus strenua Thorell, T. (1881). Studi sui ragni Malesi e Papuani. Pt. 3. Ragni dell'Austro-Malesi e del Capo York, conservati del Museo Civico di Storia Naturale di Genova. *Ann. Mus. Civ. Stor. Nat. Genova* **17**: vii–xxvii 1–720 (Territelariae pp. 236–254) [6 May 1881] [253]. Type data: syntypes, MCG F. adult, juv., from Somerset, Cape York, Qld.

Distribution: NE Coastal, Qld.; Somerset, Cape York. Ecology: terrestrial, predator.

Selenocosmia subvulpina Strand, 1907

Selenocosmia subvulpina Strand, E. (1907). Aviculariidae und Atypidae des Kgl. Naturalien-kabinetts in Stuttgart. *Jahr. Ges. Naturkd. Würt.* **63**: 1–100 [83]. Type data: holotype, SMNS M. adult destroyed, from N Qld.

Distribution: (NE coastal), (N Gulf), Qld.; N Qld. Ecology: terrestrial, predator.

Selenotholus Hogg, 1902

Selenotholus Hogg, H.R. (1902). On some additions to the Australian spiders of the suborder Mygalomorphae. *Proc. Zool. Soc. Lond.* **1902(2)**: 121–142 pl 13 [3 June 1902] [134] Type species *Selenotholus foelschei* Hogg, 1902 by original designation.

Selenotholus foelschi Hogg, 1902

Selenotholus foelschi Hogg, H.R. (1902). On some additions to the Australian spiders of the suborder Mygalomorphae. *Proc. Zool. Soc. Lond.* **1902(2)**: 121–142 pl 13 [3 June 1902] [135]. Type data: holotype, BMNH 1924.III.1.22 M. adult, from Darwin (as Palmerston), N.T.

Distribution: N coastal, N.T. Ecology: terrestrial, predator.

Selenotypus Pocock, 1895

Selenotypus Pocock, R.I. (1895). On a new and natural grouping of some of the Oriental genera of Mygalomorphae, with descriptions of new genera and species. *Ann. Mag. Nat. Hist. (6)* **15**: 165–184 [Jan. 1895] [176 pl 10]. Type species *Selenotypus plumipes* Pocock, 1895 by original designation.

Selenotypus plumipes Pocock, 1895

Selenotypus plumipes Pocock, R.I. (1895). On a new and natural grouping of some of the Oriental genera of Mygalomorphae, with descriptions of new genera and species. *Ann. Mag. Nat. Hist. (6)* **15**: 165–184 [Jan. 1895] [176 pl 10]. Type data: holotype, BMNH 1892.5.15–1 F. adult, from Major's Creek, Townsville, Qld.

Distribution: NE coastal, Qld. Ecology: burrowing, terrestrial, predator. Biological references: *The Emu*, Apr. 1919, Jan. 1920, (feeding on chickens).

Incertae sedis

Mygale notasiana Walckenaer, C.A. (1805). *Tableau des Araneides* ou Caractères Essentiels des Tribus, Genres, Familles et Races que Renferme le Genre *Aranea* de Linné, avec la Designation des Espéces Comprises dans Chacune de ces Divisions. Paris : Dentu xii 88 pp. 9 pls [5] [*nom. dub.*]. Type data: holotype(?), whereabouts unkown, from Australia (as New Holland).

Mygale australiana Walckenaer, C.A. (1837). *Histoire Naturelle des Insectes.* Aptères. Paris Vol. 1 682 pp. [227] [*nom. dub.*]. Type data: holotype(?), whereabouts unknown, from Australia. Compiled from secondary source: Bonnet, P. (1957). *Bibliographia Araneorum.* Toulouse: Douladoure Vol. 2 Pt. 3 (2992); Ausserer, A. (1871). Beiträge zur Kenntniss der Arachniden-Familie der Territelariae Thorell (Mygalidae Autor.). *Verh. Zool.-Bot. Ges. Wien* **21**: 117–224 [217].

Mygale herculea Koch, C.L. (1839). *Die Arachniden.* Bd. 5 pp. 1–158 pls CXLV–CLXXX Nürnberg : Bauer & Raspe [21] [*nom. dub.*]. Type data: holotype(?), whereabouts unknown, from Australia (as New Holland). Compiled from secondary source: Bonnet, P. (1957). *Bibliographia Araneorum.* Toulouse : Douladoure Vol. 1 Pt. 3 (2993); Ausserer, A. (1871). Beiträge zur Kenntniss der Arachniden-Familie der Territelariae Thorell (Mygalidae Autor.). *Verh. Zool.-Bot. Ges. Wien* **21**: 117–224 [217].

Mygale whitei Bonnet, P. (1957). *Bibliographia Araneorum.* Toulouse : Douladoure Vol. 2 Pt. 3 [2994] Type data: an illustration in White, J. (1790). *Journal of a Voyage to New, South Wales.* London : Debrett from N.S.W. Compiled from secondary source: Chisholm, A. (ed.) (1962). J. White *Journal of a Voyage to New South Wales.* Sydney : Angus & Robertson [194–195 pl 32] [the illustration is of a sparassid (Sparassidae = Heteropodidae)].

Taxonomic decision of Main, B.Y., this work.

ARANEOMORPHAE (in part)

Valerie Todd Davies

(including the Lycosidae by R.J. McKay)

INTRODUCTION

The first catalogue of Australian spiders, *A Census of Australian Araneidae*, was compiled by W.J. Rainbow (1911). Since then three catalogues of spider species of the world have been published. Two of these, P. Bonnet's *Bibliographia Araneorum* (1945–61) and C.F. Roewer's *Katalog der Araneae* (1942,1954) were published concurrently. The third and supplementary catalogue, P.M. Brignoli's *A Catalogue of the Araneae described between 1940 and 1981* (1983) lists most species which have been described since publication of the previous two works. A short catalogue of the spiders of Tasmania, listing 86 araneomorph species, was compiled by A. Musgrave (1948). Arachnologists, especially taxonomists, are fortunate to have these catalogues as working tools. Despite their existence, however, it seems reasonable to compile a new catalogue of Australian spiders which will be readily accessible to students. As well as synonymies, this catalogue contains details of designation and location of type specimens, distribution of species within Australia and references to biological literature; information is also given on species described between 1981 and April 1984.

Thirty-seven of the 60 known Australian araneomorph spider families are listed. These are: Anapidae, Anyphaenidae, Archaeidae, Cryptothelidae, Cyatholipidae, Deinopidae, Dysderidae, Filistatidae, Gradungulidae, Hadrotarsidae, Hahniidae, Hersiliidae, Hickmaniidae, Holarchaeidae, Loxoscelidae, Lycosidae (data prepared by R.J. McKay), Micropholcommatidae, Mimetidae, Mysmenidae, Nicodamidae, Oecobiidae, Oonopidae, Orsolobidae, Pararchaeidae, Pholcidae, Pisauridae, Psechridae, Scytodidae, Segestriidae, Selenopidae, Symphytognathidae, Tetrablemmidae, Textricellidae, Theridiosomatidae, Uloboridae, Zodariidae and Zoridae. For the most part they are small and well defined families. In many instances researchers in other countries have carried out familial or generic revisions (sometimes including Australian spiders), and they have up-dated diagnoses of many of these families. Although those listed here represent more than half the families in Australia they contain only about 30% of the described species.

Eighteen families, representing approximately 70% of the described species, have not been included in the catalogue. They are: Amaurobiidae, Araneidae, Clubionidae, Ctenidae, Cycloctenidae, Dictynidae, Gnaphosidae, Heteropodidae, Linyphiidae, Metidae, Miturgidae, Oxyopidae, Philodromidae, Salticidae, Tetragnathidae, Theridiidae, Thomisidae and Toxopidae. There are several reasons, apart from the magnitude of the work, why they are not included here. Some (Araneidae, Salticidae, Theridiidae and Thomisidae) contain large numbers of species and are in great need of revision *e.g.* within the Araneidae about one hundred *Araneus* spp. are listed which, after revision, are likely to be allocated to several genera. As well, the major sub-divisions within these world-wide families still are to be resolved; some of these may lead to the erection of further families. Amaurobiids (*sensu lato*) are very well represented in Australia, especially in temperate rainforest. The few described genera are difficult to relate to one another or to place confidently in a sub-family or family, despite the recent revisions of Lehtinen (1967), Forster (1970) and Forster & Wilton (1973). The claw-tufted families, Clubionidae and Gnaphosidae, currently are receiving attention from researchers in the U.S.A. When their diagnoses of families, sub-families and genera become available a catalogue of Australian species in these families will be more meaningful. Spiders in the families Toxopidae and Cycloctenidae are difficult to separate at present, but it is hoped that clear diagnoses will emerge as more of these spiders are described.

There are a further five araneomorph families that are represented in Australia though no species have been described. These are: Stenochilidae, Nesticidae, Ochyroceratidae, Periegopidae and Agelenidae (*sensu* Forster). *Colopea* sp. (Stenochilidae) and *Nesticella* sp. have been collected from Iron Range, near the tip of Cape York Peninsula, Queensland. Ochyroceratids are found in the litter of rainforests in north Queensland. *Periegops* sp. has been collected in southeastern Queensland - previously the genus was known only from New Zealand. Agelenids have been found in Tasmania and are probably more widely distributed.

Many spider families still are diagnosed inadequately and there is no universal agreement as to the number of families in the world. In some measure this is due to the relatively small number of spiders, particularly in the southern continents, that are described. In Australia for example it is estimated that probably less than 20% of the existing species have been described, which means that at present we have only a skeletal outline of our spider fauna.

Taxonomy of Australia's spiders began when Fabricius described *Gasteracantha fornicata* in 1775. The spider was collected, probably by Banks or Solander, during the time that Captain Cook's vessel, *Endeavour* was beached for repairs at Cooktown in 1770. This is the only description of a spider to precede European settlement of Australia at Port Jackson, New South Wales, in 1778. In 1790 John White, the surgeon-general to the colony, produced his *Journal of a Voyage to New South Wales* in which he described and figured, but did not name, two spiders: one small and unidentifiable, the other a large sparassid or huntsman.

The second spider to be described and named was *Nephila edulis* by Labillardière (1799) a botanist who accompanied the French admiral, D'Entrecasteaux on his voyage to Australia. In 1802, the naturalists Péron and Lesueur, who accompanied Baudin to Australia, collected extensively. Unfortunately, little of this material has been described; *Delena cancerides, Missulena occatoria* (a trap-door spider) and *Storena cyanea*, described by Walckenaer (1805), were probably from this collection. Quoy and Gaimard (1825) described and figured *Dolophones notacantha*; they were surgeon-naturalists on board Freycinet's vessel *l'Uranie* during its exploratory journey to the Pacific in 1819–1920. W.S. MacLeay (1827) described 4 spiders: *Nephila cunninghamii*, later recognized by Dahl (1912) as a synonym of *N. plumipes* (Latreille 1804); *Uloborus canus*, recognized by Walckenaer (1837) as probably being a philodromid; *Linyphia deplanata*, regarded by Walckenaer (1841) as unidentifiable and *Thomisus morbillosus* which MacLeay later (1839) transferred to *Olios*, remarking at the time that "the type of this genus is *Aranea venatoria* Linnaeus" (= *Heteropoda venatoria*). The descriptions are of two or three lines each and provide no localities. As no spiders belonging to any of the last three species have been identified since their description and the types are lost, there appears to be a good case for the eventual suppression of these names. Between 1860–1875 more collections were made, mostly from areas near the ports of the eastern coast of Australia, and these form the basis of L. Koch's (1871–1883) *Die Arachniden Australiens* which was completed by Keyserling, who later added a Supplement (1884–1890). These volumes are illustrated beautifully and are still the standard taxonomic work on Australian spiders. Thorell (1881) described, but did not illustrate, 47 species from Cape York Peninsula. Simon (1908,1909) described the spiders collected by the Michaelsen and Hartmeyer Expedition to Southwest Australia and Strand (1913) described those from the Leonhardi expedition to Central Australia. The types of nearly all these spiders, where they still exist, are in European museums.

Between 1893 and 1920, W.J. Rainbow, an Englishman who was the Entomologist at the Australian Museum in Sydney, contributed greatly to the knowledge of Australian spiders. He produced over 50 papers, describing some 200 spiders, and in 1911 published the catalogue *A Census of Australian Araneidae* in which he listed 1102 araneomorph species. Included were 76 species described by H.R. Hogg, an Englishman who was engaged in commerce in Melbourne between 1873–1892. While in Australia he developed an interest in spiders and when he returned to England to live he took back with him a collection of spiders and worked on them (and other Australian collections) at the British Museum. Emeritus Professor V.V. Hickman, Professor of Zoology at the University of Tasmania until his retirement in 1959, has been a major contributor to Australian spider taxonomy since 1926 and is Australia's foremost araneologist. He is known best for his papers on minute litter spiders for which he established three families, Symphytognathidae, Textricellidae and Micropholcommatidae; he also established the Toxopidae. Professor V.V. Hickman died in Hobart, Tasmania on 20 November 1984.

There have been few revisions of Australian araneomorph families. Hogg's (1903) paper on the heteropodids (sparassids) is the only attempt made to revise this family. Dunn (1951) produced a key to *Storena* spp. and R.J. McKay, who has compiled data on Lycosidae in this catalogue, has written a series of papers on the family. The most important revision is that of the archaeids of the world by Forster and Platnick (1984) in which they establish two new Australasian families, Pararchaeidae and Holarchaeidae. Currently the same authors are revising the Orsolobidae and Anapidae. Heimer (pers. comm.) is redescribing and revising the Australian Mimetidae; Gray (pers. comm.) is redescribing and revising the Australian filistatids.

The families in this catalogue are arranged in alphabetical order and a short introduction to each family precedes the data for the species. It should be noted that these familial descriptions refer only to Australian spiders and may not hold true for spiders from other parts of the world. Arachnologists will be familiar with most of the terms used and only "haplogyne" and "entelegyne", as they refer to the structure of the female, are defined. "Haplogyne" refers to those spiders which have one duct for both insemination and fertilization. Their genital region usually is unsclerotized and sperm is introduced via the uterus externus and spermathecal ducts (sometimes just the neck of the spermathecae) to the spermathecae. "Entelegyne" refers to females which have separate insemination and fertilization ducts. The openings (gonopores) of the insemination ducts are on the sclerotized epigynum. These ducts lead to the spermathecae and from here fertilization ducts convey sperm to the uterus. It is generally believed that the entelegyne condition has been derived from the haplogyne several times in spiders.

Information on the arrangement of the tracheae may be included in the introduction as sometimes it is considered to be phylogenetically important and is not always easy to find in the literature. The usual abbreviations for the eyes are used: ALE, AME, PLE and PME. The use of "sedentary" in the ecological data indicates that the species spins a web for capture of prey; "nomadic" is used to indicate that it does not. An asterisk indicates that the types have not been examined by the compiler. In the size-classes used the approximate total lengths are as follows: "large" 8mm or more; "medium" 4–8mm; "small" 2–4mm; "tiny" 1–2mm; "minute" less than 1mm. This work does not pretend to be more than a catalogue of species, which by its nature and timing is an interim document. It is intended as a reference for Australian workers and should be used in conjunction with the world lists of Bonnet (1945–1961), Roewer (1942,1954) and Brignoli (1983).

ACKNOWLEDGEMENTS

Financial support for 2 months was provided by the Australian Biological Resources Study to employ Anne Windsor as cataloguer. My assistant, Robert Raven helped un-ravel many nomenclatorial tangles and I am indebted to the authors of recent revisions for their diagnoses of several families. I am especially grateful to the Director and the Trustees of the Queensland Museum who allowed me to do this compilation as part of my curatorial work.

V.T.D.

References

Bonnet, P. (1945–1961). *Bibliographia Araneorum.* Toulouse : Douladoure 3 vols

Brignoli, P.M. (1983). *A Catalogue of the Araneae described between 1940 and 1981.* Merret, P. (ed.) Manchester : Manchester Univ. Press 755 pp.

Dahl, F. (1912). Seidenspinne und Spinnenseide. *Mitt. Zool. Mus. Berl.* **6**: 1–90

Dunn, R.A. (1951). Spiders of the Russell Grimwade Expedition. *Mem. Natl. Mus. Vict.* **17**: 9–18

Fabricius, J.C. (1775). *Systema Entomologiae,* sistens Insectorum classes, ordines, genera, species, adiectis, synonymis, locis descriptionibus observationibus. Flensburgi et Lipsiae 832 pp. (Aranea 431–439)

Forster, R.R. (1970). The spiders of New Zealand. Part III. *Otago Mus. Bull. No. 3* 184 pp.

Forster, R.R. & Platnick, N.I. (1984). A review of the archaeid spiders and their relatives, with notes on the limits of the Superfamily Palpimanoidea (Arachnida, Araneae). *Bull. Am. Mus. Nat. Hist.* **178**: 1–106

Forster, R.R. & Wilton, C.L. (1973). The spiders of New Zealand. Part IV. *Otago Mus. Bull. No. 4* 309 pp.

Heimer, S. (in press). Remarks on the spider genus *Arcys* Walckenaer, 1837, with description of new species. *Entomol. Abh. Mus. Tierkd. Dresden*

Heimer, S. (in press). Notes on the spider family Mimetidae with description of a new genus from Australia. *Entomol. Abh. Mus. Tierkd. Dresden*

Hogg, H.R. (1903). On the Australasian spiders of the subfamily Sparassinae. *Proc. Zool. Soc. Lond.* **1902(2)**: 414–466

Keyserling, E. (1884–1890). *Die Arachniden Australiens.* Supplement. Nürnberg : Bauer & Raspe 279 pp. 24 pls

Koch, L. & Keyserling, E. (1871–1883). *Die Arachniden Australiens.* Nürnberg : Bauer & Raspe 2 vols 1489 pp. 123 pls

Labillardière, J. (1799). *Relation du Voyage à la Recherche de La Pérouse, fait par Ordre de l'Assemblée Constituante.* Paris 2 vols (*Aranea edulis* Vol. 2: 239–241)

Lehtinen, P.T. (1967). Classification of the cribellate spiders and some allied families, with notes on the evolution of the suborder Araneomorpha. *Ann. Zool. Fenn.* **4**: 199–468

MacLeay, W.S. (1827). Annulosa. Catalogue of insects, collected by Captain King. pp. 438–469 *in* King, P.P. *Narrative of a Survey in the Intertropical and Western Coasts of Australia.* London Vol. 2

MacLeay, W.S. (1839). On some new forms of Arachnida. *Ann. Mag. Nat. Hist. (1)* **2:** 1–14

Musgrave, A. (1948). A catalogue of the spiders of Tasmania. *Rec. Queen Vict. Mus.* **2**: 75–91

Quoy, [J.R.C.] & Gaimard, [J.P.] (1825). Zoologie. *in* Freycinet, L.C.D. de (ed.) *Voyage Autour du Monde,* entrepris par orde du Roi, sous le Ministère et conformément aux instructions de S. Exc. M. le Vicomte du Boucharge, secrétaire d'État au Département de la Marine, exécuté sur les corvettes de S.M. *l'Uranie* et *la Physicienne*, pendent les années 1817, 1818, 1819 et 1820; publié sous les auspices de S.E.M. le Comte Corbière, secrétaire d'État de l'Intérieur, pour la partie historique et les sciences naturelles, et de S.E.M. le Marquis de Clermont-Tonnerre, Secrétaire d'État de la Marine et des Colonies, pour la partie nautique; par M. Louis Freycinet, etc. Paris : Pillet Aîné Vol. 1+atlas iv 712 pp. 96 pls

Rainbow, W.J. (1911). A Census of Australian Araneidae. *Rec. Aust. Mus.* **9**: 107–391

Roewer, C.F. (1942, 1954). *Katalog der Araneae.* Bremen : Paul Budy Bd. 1 1–1040; Bruxelles : Inst. R. Sci. Nat. Belg. Bd. 2 (a,b) 1–1751

Simon, E. (1908). Araneae. 1re partie pp. 359–446 *in* Michaelsen, W. & Hartmeyer, R. (eds.) *Die Fauna Südwest-Australiens.* 1(12) Jena : Fischer

Simon, E. (1909). Araneae. 2[e] partie pp. 155–212 *in* Michaelsen, W. & Hartmeyer, R. (eds.) *Die Fauna Südwest-Australiens.* 2(13) Jena : Fischer

Strand, E. (1913). Über einige australische Spinnen des Senckenbergischen Museums. *Zool. Jb. (Syst.)* **35**: 599–624

Thorell, T. (1881). Studi sui ragni Malesi e Papuani. III. Ragni dell'Austro-Malesia e del Capo York, conservati nel Museo Civico di Storia Naturale di Genova. *Ann. Mus. Civ. Stor. Nat. Genova* **17**: 1–720

Walckenaer, C.A. (1805). *Tableau des Aranéides* ou Caractères essentiels des tribus, genres, familles et races que renferme le genre Aranea de Linné, avec la désignation des espèces comprises dans chacune de ces divisions. Paris 88 pp.

Walckenaer, C.A. (1837–1847). *Histoire Naturelle des Insectes Aptères.* Paris 4 vols+atlas

White, J. (1790). *Journey of a Voyage to New South Wales,* with sixty-five plates of non-descripted animals, birds, lizards, serpents, curious cones of trees and other natural productions. Appendix. London : Debrett 299 pp.

ANAPIDAE

INTRODUCTION

Minute to small, three-clawed spiders without book-lungs (exc. *Risdonius*). Dorsal abdominal scutum on the male only, epigastric scutum and ring around spinnerets in both sexes. Colulus present. Very high clypeus. Eight or six eyes, AME small or lost. Labral (rostral) spur behind chelicerae. Female palp reduced. Tarsi longer than metatarsi, no tarsal trichobothria, one or none on metatarsi. Anterior spiracles (tracheae to cephalothorax and abdomen) present. Posterior spiracle, if present, supplies tracheae to cephalothorax. Small horizontal orb-webs, with centre drawn up slightly, like a tent, with the spider on underside.

Brignoli (1981) transferred *Pseudanapis burra, Pseudanapis darlingtoni* and *Chasmocephalon crassifemoralis* to *Anapogonia*, the type species of which is *Anapogonia lyrata* from Java. He has done this by comparing the descriptions with *Anapogonia pilupilu* from New Caledonia. These transfers, however, cannot be upheld as the type *A. lyrata* is not an anapid (Platnick, N.I., pers. comm.). Brignoli also transferred *Pseudanapis octocula* to *Risdonius.* The male palp of *Risdonius parvus* has tibial and patellar apophyses, quite unlike *Pseudanapis octocula* and so it seems preferable to leave these species in *Pseudanapis* until the current revision of the Anapidae by Forster and Platnick is completed.

Acrobleps Hickman, 1979

Acrobleps Hickman, V.V. (1979). Some Tasmanian spiders of the families Oonopidae, Anapidae and Mysmenidae. *Pap. Proc. R. Soc. Tasm.* **113**: 53–79 [68]. Type species *Acrobleps hygrophilus* Hickman, 1979 by monotypy and original designation.

Acrobleps hygrophilus Hickman, 1979

Acrobleps hygrophilus Hickman, V.V. (1979). Some Tasmanian spiders of the families Oonopidae, Anapidae and Mysmenidae. *Pap. Proc. R. Soc. Tasm.* **113**: 53–79 [69]. Type data: holotype, AM KS2708 *M. adult, paratype AM KS9628 *F. adult, TMH J1438 *2M.,F. adult, from The Cascades, Tas.

Distribution: Tas. Ecology: sedentary, predator.

Chasmocephalon O.P. Cambridge, 1889

Chasmocephalon Cambridge, O.P. (1889). On some new species and a new genus of Araneidea. *Proc. Zool. Soc. Lond.* **1889**: 34–46 [45]. Type species *Chasmocephalon neglectum* O.P. Cambridge, 1889 by monotypy.

This group is reported from southern Africa, see Simon, E. (1892–1903). *Histoire Naturelle des Araignées.* 2nd edn. Paris : Roret Vol. 1 [928]; and California, U.S.A., see Gertsch, W.J. (1961). Descriptions of American spiders of the family Symphytognathidae. *Am. Mus. Novit.* 1981: 1–40. Brignoli, P.M. (1981). New or interesting Anapidae (Arachnida, Araneae). *Rev. Suisse Zool.* **88**: 109–134 doubts that the spiders from Africa and California are *Chasmocephalon* spp.

Chasmocephalon crassifemoralis Wunderlich, 1976

Chasmocephalon crassifemoralis Wunderlich, J. (1976). Spinnen aus Australien. 1. Uloboridae, Theridiosomatidae und Symphytognathidae. *Senckenberg. Biol.* **57**: 113–124 [123]. Type data: holotype, SMF 29061 *M. adult, from Ku-ring-gai Natl. Park, N.S.W.

Distribution: SE coastal, N.S.W. Ecology: sedentary, predator. Biological references: Brignoli, P.M. (1981). New or interesting Anapidae (Arachnida, Araneae). *Rev. Suisse Zool.* **88**: 109–134 (taxonomy; Brignoli's transfer of the species to *Anapogonia* is not upheld as *Anapogonia* is not an anapid, Platnick, N.I., pers. comm.).

Chasmocephalon minutum Hickman, 1944

Chasmocephalon minutum Hickman, V.V. (1944). On some new Australian Apneumonomorphae with notes on their respiratory system. *Pap. Proc. R. Soc. Tasm.* **1943**: 179–195 [180]. Type data: syntypes, AM K6425 *M.,F. adult, from Cascades, Tas.

Distribution: Tas. Ecology: sedentary, predator.

Chasmocephalon neglectum O.P. Cambridge, 1889

Chasmocephalon neglectum Cambridge, O.P. (1889). On some new species and a new genus of Araneidea. *Proc. Zool. Soc. Lond.* **1889**: 34–46 [45]. Type data: holotype, OUM *F. adult, from Swan River, W.A.; found on leg of *Idiommata blackwallii* O.P. Cambridge, 1870 many years after the description of this trap-door spider.

Distribution: SW coastal, W.A. Ecology: sedentary, predator.

Olgania Hickman, 1979

Olgania Hickman, V.V. (1979). Some Tasmanian spiders of the families Oonopidae, Anapidae and Mysmenidae. *Pap. Proc. R. Soc. Tasm.* **113**: 53–79 [71]. Type species *Olgania excavata* Hickman, 1979 by monotypy and original designation.

Olgania excavata Hickman, 1979

Olgania excavata Hickman, V.V. (1979). Some Tasmanian spiders of the families Oonopidae, Anapidae and Mysmenidae. *Pap. Proc. R. Soc. Tasm.* **113**: 53–79 [73]. Type data: holotype, AM KS2709 *M. adult, paratype AM KS2710 *F. adult, from Gordon River Valley, SW Tas.

Distribution: Tas. Ecology: predator; in moss.

Pseudanapis Simon, 1905

Pseudanapis Simon, E. (1905). Arachnides de Java, recueillis par le Prof. K. Kraepelin en 1904. *Mitt. Naturh. Mus. Hamb.* **22**: 51–73 [64]. Type species *Anapis paroculus* Simon, 1899 by monotypy.

This group is also found in New Caledonia, Java and Sumatra, see Brignoli, P.M. (1981). New or interesting Anapidae (Arachnida, Araneae). *Rev. Suisse Zool.* **88**: 109–134 has removed the Mediterranean species to *Zaugherella*. Platnick, N.I. & Shadab, M.U. (1979). A review of the spider genera *Anapisona* and *Pseudanapis* (Araneae, Anapidae). *Am. Mus. Novit.* 2672: 1–20 believe that the Australian *Pseudanapis* species are misplaced.

Pseudanapis burra Forster, 1959

Pseudanapis burra Forster, R.R. (1959). Spiders of the family Symphytognathidae. *Trans. R. Soc. N.Z.* **86**: 269–329 [309]. Type data: holotype, QM S97 M. adult, paratype QM S98 F. adult, from Binna Burra, Lamington Natl. Park Qld.

Distribution: NE coastal, Qld. Ecology: predator, litter, closed forest. Biological references: Brignoli, P.M. (1981). New or interesting Anapidae (Arachnida, Araneae). *Rev. Suisse Zool.* **88**: 109–134 (taxonomy; Brignoli's transfer of the species to *Anapogonia* is not upheld as *Anapogonia* is not an anapid, Platnick, N.I., pers. comm.).

Pseudanapis darlingtoni Forster, 1959

Pseudanapis darlingtoni Forster, R.R. (1959). Spiders of the family Symphytognathidae. *Trans. R. Soc. N.Z.* **86**: 269–329 [312]. Type data: holotype, MCZ *M. adult, paratype QM S99 F. adult, from Mt. Spurgeon, Qld.

Distribution: NE coastal, Qld. Ecology: predator, closed forest. Biological references: Brignoli, P.M. (1981). New or interesting Anapidae (Arachnida, Araneae). *Rev. Suisse Zool.* **88**: 109–134 (taxonomy; Brignoli's transfer of species to *Anapogonia* is not upheld as *Anapogonia* is not an anapid, Platnick, N.I., pers. comm.).

Pseudanapis octocula Forster, 1959

Pseudanapis octocula Forster, R.R. (1959). Spiders of the family Symphytognathidae. *Trans. R. Soc. N.Z.* **86**: 269–329 [310]. Type data: holotype, QM S101 M. adult, paratype QM S102 F. adult, from Binna Burra, Lamington Natl. Park, Qld.

Distribution: SE coastal, Qld. Ecology: litter, predator. Biological references: Brignoli, P.M. (1981). New or interesting Anapidae (Arachnida, Araneae). *Rev. Suisse Zool.* **88**: 109–134 (taxonomy; Brignoli's transfer of species to *Risdonius* is not upheld).

Risdonius Hickman, 1939

Risdonius Hickman, V.V. (1939). On a dipneumone spider (*Risdonius parvus*, gen. et sp. n.), the female of which has reduced palpi. *Proc. Zool. Soc. Lond.* **108**: 655–660 [655]. Type species *Risdonius parvus* Hickman, 1939 by monotypy and original designation.

Risdonius parvus Hickman, 1939

Risdonius parvus Hickman, V.V. (1939). On a dipneumone spider (*Risdonius parvus*, gen. et sp. n.), the female of which has reduced palpi. *Proc. Zool. Soc. Lond.* **108**: 655–660 [656]. Type data: syntypes, AM KS6674 *5M.,F. adult, AM KS6675 *M.,F. adult, from East Risdon, Tas.

Distribution: Tas. Ecology: sedentary, predator.

ANYPHAENIDAE

INTRODUCTION

Medium to large, two-clawed spiders with claw tufts composed of lamelliform setae. Tracheal spiracle a short distance forward of spinnerets (tracheal tubes extend into cephalothorax). Colulus present. Eyes in two rows. Trichobothria in a double row on tibiae, a single row on metatarsi, and a double row on tarsi. In Australia *Amaurobioides*, restricted to a small coastal stretch of Tasmania, lives in a nest of tough white silk in crevices and holes in rocks near high-tide mark. In *Amaurobioides* the trochanters are un-notched and the compact eye group is on a low projecting mound.

Amaurobioides O.P. Cambridge, 1883

Amaurobioides Cambridge, O.P. (1883). On some new genera and species of spiders. *Proc. Zool. Soc. Lond.* **1883**: 352–365 [356]. Type species *Amaurobioides maritimus* O.P. Cambridge, 1883 by monotypy.

This group is also found in South Africa, New Zealand, Campbell Is., Auckland Is. and Chile, see Forster, R.R. (1970). The spiders of New Zealand. Pt III. *Otago Mus. Zool. Bull. No. 3* 184 pp. Species now known not to occur in Australia: *Amaurobioides maritimus* O.P. Cambridge, 1883 see Forster, R.R. *loc. cit.* [167].

Amaurobioides litoralis Hickman, 1949

Amaurobioides litoralis Hickman, V.V. (1949). Tasmanian littoral spiders with notes on their respiratory systems, habits and taxonomy. *Pap. Proc. R. Soc. Tasm.* **1948**: 31–43 [31] [although later Hickman, V.V. (1951). The identity of spiders belonging to the genus *Amaurobioides* Cambridge. *Pap. Proc. R. Soc. Tasm.* **1951**: 1–2 considered *litoralis* a junior synonym of the New Zealand species *A. maritimus* O.P. Cambridge, 1883, Forster (1970). (*op. cit.*) established that *A. maritimus* is not in Australia and recognized two species from Tasmania, *A. litoralis* and a figured, unnamed species]. Type data: syntypes, AM KS6410 *M.,F. adult, AM KS6411 *2F. adult, from Eaglehawk Neck, Tas. Compiled from secondary source: Forster, R.R. (1970). The spiders of New Zealand. Pt III. *Otago Mus. Zool. Bull. No. 3* 184 pp. (figs 468,469,485–488).

Distribution: Tas. Ecology: sedentary, predator. Biological references: Hickman, V.V. (1951). The identity of spiders belonging to the genus *Amaurobioides* Cambridge. *Pap. Proc. R. Soc. Tasm.* **1951**: 1–2.

ARCHAEIDAE

INTRODUCTION

Small, three-clawed, haplogyne spiders with extremely long chelicerae, the bases of which are separated from the rest of the mouthparts by a neck-like extension of the carapace. Fang is less than one quarter length of chelicera. A row of long rod-like peg teeth on distal promargin of chelicera. Anterior book-lungs; two minute posterior spiracles, well in advance of spinnerets, each leading to a single abdominal tracheal tube. Colulus absent. Patella I conspicuously long. One trichobothrium on metatarsi, none on tarsi. Onychium present. Abdomen-petiole stridulating system. *Austrarchaea* is found in rainforest, often in moss on tree trunks where it enters webs and preys on other spiders.

Austrarchaea Forster and Platnick, 1984

Austrarchaea Forster, R.R. & Platnick, N.I. (1984). A review of the archaeid spiders and their relatives, with notes on the limits of the Superfamily Palpimanoidea (Arachnida, Araneae). *Bull. Am. Mus. Nat. Hist.* **178**: 1–106 [21]. Type species *Archaea nodosa* Forster, 1956 by original designation.

Austrarchaea daviesae Forster and Platnick, 1984

Austrarchaea daviesae Forster, R.R. & Platnick, N.I. (1984). A review of the archaeid spiders and their relatives, with notes on the limits of the Superfamily Palpimanoidea (Arachnida, Araneae). *Bull. Am. Mus. Nat. Hist.* **178**: 1–106 [22]. Type data: holotype, QM S1091 M. adult, paratype QM S1092 F. adult, from Major's Mt., Atherton Tableland, Qld.

Distribution: NE coastal, Qld. Ecology: nomadic, predator.

Austrarchaea hickmani (Butler, 1929)

Archaea hickmani Butler, L.S.G. (1929). Studies in Victorian spiders. No. 1 *Proc. R. Soc. Vict.* **42**: 41–52 [46]. Type data: holotype, NMV K097 *F. adult, from Vic.

Distribution: SE coastal, Vic. Ecology: nomadic, predator. Biological references: Forster, R.R. & Platnick, N.I. (1984). A review of the archaeid spiders and their relatives, with notes on the limits of the Superfamily Palpimanoidea (Arachnida, Araneae). *Bull. Am. Mus. Nat. Hist.* **178**: 1–106.

Austrarchaea nodosa (Forster, 1956)

Archaea nodosa Forster, R.R. (1956). A new spider of the genus *Archaea* from Australia. *Mem. Qd. Mus.* **13**: 151–154 [151]. Type data: holotype, QM W1955 F. adult, from Tallawallal, Lamington Natl. Park, Qld.

Distribution: NE coastal, Qld. Ecology: nomadic, predator; in moss. Biological references: Forster, R.R. & Platnick, N.I. (1984). A review of the archaeid spiders and their relatives, with notes on the limits of the Superfamily Palpimanoidea (Arachnida, Araneae). *Bull. Am. Mus. Nat. Hist.* **178**: 1–106.

CRYPTOTHELIDAE

INTRODUCTION

Slow moving, three-clawed, spiders from litter of tropical forests. Body and legs usually caked with soil. Eyes in three rows 2.2.4; posterior row slightly procurved, anterior row strongly procurved. High clypeus. Cheliceral teeth absent, fang weak. Serrula present. Tufts of thick hair in rows on body and legs; a ring of hair round spinnerets and anal tubercle. Legs robust, patellae and tibiae rotund. Female palpal tarsus paddle-shaped and smooth ventrally. A row of trichobothria on metatarsi and tarsi. This family is often merged with Zodariidae; it is kept separate here because *Cryptothele* has a serrula which is absent in Australian zodariids.

Cryptothele L. Koch, 1872

Cryptothele Koch, L. (1872). *in* Koch, L. & Keyserling, E. (1871–1883). *Die Arachniden Australiens.* Nürnberg : Bauer & Raspe Vol. 1 [239]. Type species *Cryptothele verrucosa* L. Koch, 1872 by monotypy.

This group is also found in Seychelle Ils., Sri Lanka, India, Mexico, Fiji, Samoa, Mariana Ils., New Caledonia, Singapore, Sumatra, Irian Jaya and Java, see Roewer, C.F. (1942, 1954). *Katalog der Araneae.* Bremen : Paul Budy Bd. 1 1–1040.

Cryptothele doreyana Simon, 1890

Cryptothele doreyana Simon, E. (1890). Études arachnologiques. 22e Mémoire. Arachnides recueillis aux îles Mariannes par M.A. Marche. *Ann. Soc. Entomol. Fr.* **10**: 131–136 [132]. Publication date established from Bonnet, P. (1945–1961). *Bibliographia Araneorum.* Toulouse : Douladoure Vol. 2. Type data: syntypes, M. adult, juv. whereabouts unknown, from Dore, Irian Jaya.

Distribution: NE coastal, Qld.; Fitzroy Is., Qld. Ecology: terrestrial, predator. Biological references: Main, B.Y. (1976). *Spiders.* Sydney : Collins 296 pp. (biology).

CYATHOLIPIDAE

INTRODUCTION

Small, three-clawed spiders with a wide tracheal spiracle (tracheal tubes to cephalothorax). Colulus present. One trichobothrium on metatarsi, none on tarsi. No tibial apophysis on male palp. Paracymbium is not a separate sclerite. Female palp without a claw. Cyatholipids spin fine horizontal sheet webs against tree trunks or on shrubs in rainforests and have also been found on tiny webs in litter; they run on the underside of the web.

Teemenaarus Davies, 1978

Teemenaarus Davies, V. Todd (1978). A new family of spiders (Araneae : Teemenaaridae). *Symp. Zool. Soc. Lond.* **42**: 293–302 [294]. Type species *Teemenaarus silvestris* Davies, 1978 by monotypy and original designation.

Teemenaarus silvestris Davies, 1978

Teemenaarus silvestris Davies, V. Todd (1978). A new family of spiders (Araneae : Teemenaaridae). *Symp. Zool. Soc. Lond.* **42**: 293–302 [294]. Type data: holotype, QM S1 F. adult, paratypes QM S2 F. adult, QM S3 F. adult, QM S6 M. adult, QM S7 M. adult, QM S8 M. adult, from Bulburin State Forest, Qld.

Distribution: NE coastal, Qld. Ecology: sedentary, predator.

Tekellatus Wunderlich, 1978

Tekellatus Wunderlich, J. (1978). Zur Kenntnis der Cyatholipinae Simon, 1894 (Arachnida : Araneida : ?Tetragnathidae). *Zool. Beitr.* **24**: 33–41 [38]. Type species *Tekellatus lamingtoniensis* Wunderlich, 1978 by monotypy and original designation.

Tekellatus lamingtoniensis Wunderlich, 1978

Tekellatus lamingtoniensis Wunderlich, J. (1978). Zur Kenntnis der Cyatholipinae Simon, 1894 (Arachnida : Araneida : ?Tetragnathidae). *Zool. Beitr.* **24**: 33–41 [39]. Type data: holotype, Wunderlich's Collection *M. adult, from Lamington Natl. Park, Qld.

Distribution: NE coastal, Qld. Ecology: sedentary, predator.

DEINOPIDAE

INTRODUCTION

The net-casting spiders, medium to large, three-clawed cribellate spiders with abdomen much longer than wide. Eyes in three rows 4.2.2. PME facing forwards, very large in *Deinopis.* Embolus long and coiled. Cribellum undivided. Ventral row of macrosetae on tarsi IV. One trichobothrium on metatarsi, none on tarsi. Recently it has been shown that in construction of frame, radii, non-sticky spiral and sticky spiral, the web is an orb-web with the same patterns of web construction as other orb-web spiders (Coddington at IX Congress of Arachnology, Panama, 1983).

Avella O.P. Cambridge, 1877

Avella Cambridge, O.P. (1877). On some new species of Araneidea, with characters of two new genera, and some remarks on the families Podophthalmides and Dinopides. *Proc. Zool. Soc. Lond.* **1877**: 557–578 [574]. Type species *Avella despiciens* O.P. Cambridge, 1877 by monotypy.

This group is also found in New Caledonia and New Guinea, see Lehtinen, P.T. (1967). Classification of the cribellate spiders and some allied families, with notes on the evolution of the suborder Araneomorpha. *Ann. Zool. Fenn.* **4**: 199–468.

Avella angulata L. Koch, 1878

Avella angulata Koch, L. (1878). *in* Koch, L. & Keyserling, E. (1871–1883). *Die Arachniden Australiens.* Nürnberg : Bauer & Raspe Vol. 2 [1037 pl 91 fig 1]. Type data: holotype, ZMH *F. adult (MGH 16502), from Rockhampton, Qld., see Rack, G. (1961). Die Entomologischen Sammlungen des Zoologischen Staatsinstituts und Zoologischen Museums Hamburg. II. *Mitt. Hamb. Zool. Mus. Inst.* **59**: 1–60.

Distribution: NE coastal, Qld. Ecology: sedentary, predator.

Avella despiciens O.P. Cambridge, 1877

Avella despiciens Cambridge, O.P. (1877). On some new species of Araneidea, with characters of two new genera and some remarks on the families Podophthalmides and Dinopides. *Proc. Zool. Soc. Lond.* **1877**: 557–578 [574]. Type data: holotype, OUM *F. adult, from Rockhampton, Qld.

Distribution: NE coastal, Qld. Ecology: sedentary, predator.

Avella insularis (Rainbow, 1920)

Dinopis insularis Rainbow, W.J. (1920). Arachnida from Lord Howe and Norfolk Islands. *Rec. S. Aust. Mus.* **1**: 229–272 [229]. Type data: syntypes, SAMA N1981232 *M. adult, AM KS6476 *F. adult, from Lord Howe Is.

Distribution: Lord Howe Is. Ecology: sedentary, predator. Biological references: Lehtinen, P.T. (1967). Classification of the cribellate spiders and some allied families, with notes on the evolution of the suborder Araneomorpha. *Ann. Zool. Fenn.* **4**: 199–468 (taxonomy).

Avella superciliosa Thorell, 1881

Avella superciliosa Thorell, T. (1881). Studi sui ragni Malesi e Papuani. III. Ragni dell'Austro-Malesia e del Capo York, conservati nel Museo Civico di Storia Naturale di Genova. *Ann. Mus. Civ. Stor. Nat. Genova* **17**: 1–720 [200]. Type data: holotype, MCG F. adult, from Somerset, Cape York Peninsula, Qld.

Distribution: NE coastal, Qld. Ecology: sedentary, predator.

Avella trinodosa (Rainbow, 1920)

Menneus trinodosus Rainbow, W.J. (1920). Arachnida from Lord Howe and Norfolk Islands. *Rec. S. Aust. Mus.* **1**: 229–272 [230]. Type data: holotype, SAMA N1981233 *F. adult, from Lord Howe Is.

Distribution: Lord Howe Is. Ecology: sedentary, predator. Biological references: Lehtinen, P.T. (1967). Classification of the cribellate spiders and some allied families, with notes on the evolution of the suborder Araneomorpha. *Ann. Zool. Fenn.* **4**: 199–468 (taxonomy).

Avella unifasciata L. Koch, 1878

Avella unifasciata Koch, L. (1878). *in* Koch, L. & Keyserling, E. (1871–1883). *Die Arachniden Australiens.*

Nürnberg : Bauer & Raspe Vol. 2 [1035 pl 90 fig 6]. Type data: holotype, ?ZFMK *M. adult, from Sydney, N.S.W.

Distribution: SE coastal, N.S.W. Ecology: sedentary, predator. Biological references: Mascord, R. (1970). *Australian Spiders in Colour.* Sydney : Reed 112 pp.; Mascord, R. (1980). *Spiders of Australia.* A field guide. Sydney : Reed 128 pp. (ecology); Austin, A.D. & Blest, A.D. (1979). The biology of two Australian species of dinopid spider. *J. Zool. Lond.* **189**: 145–156; Blest, A.D., Williams, D.S. & Kao, L. (1980). The posterior median eyes of the dinopid spider *Menneus. Cell Tissue Res.* **211**: 391–403.

Deinopis MacLeay, 1839

Deinopis MacLeay, W.S. (1839). On some new forms of Arachnida. *Ann. Mag. Nat. Hist.* **2**: 1–14 [8] [the emended name *Dinopis* employed by Agassiz (1848), Karsch (1878) and many other authors is regarded as an invalid emendation under Art. 32(a)(ii) of ICZN]. Type species *Deinopis lamia* MacLeay, 1839 by monotypy.

This group is found worldwide, see Bonnet, P. (1945–1961). *Bibliographia Araneorum.* Toulouse : Douladoure Vol. 2.

Deinopis fasciata L. Koch, 1879

Dinopis fasciatus Koch, L. (1879). *in* Koch, L. & Keyserling, E. (1871–1883). *Die Arachniden Australiens.* Nürnberg : Bauer & Raspe Vol. 2 [1045 pl 92 figs 1–2]. Type data: syntypes, ZMH (MGH 16504) *F. juv., from Gayndah, Qld., *M. juv. whereabouts unknown from Bowen, Qld., see Rack, G. (1961). Die Entomologischen Sammlungen des Zoologischen Staatsinstituts und Zoologischen Museums Hamburg. II. *Mitt. Hamb. Zool. Mus. Inst.* **59**: 1–60.

Distribution: NE coastal, Qld. Ecology: sedentary, predator.

Deinopis ravida L. Koch, 1878

Dinopis ravidus Koch, L. (1878). *in* Koch, L. & Keyserling, E. (1871–1883). *Die Arachniden Australiens.* Nürnberg : Bauer & Raspe Vol. 2 [1041 pl 91 fig 3]. Type data: holotype, ZMH (MGH 16503) *F. adult, from Gayndah, Qld., see Rack, G. (1961). Die Entomologischen Sammlungen des Zoologischen Staatsinstituts und Zoologischen Museums Hamburg. II. *Mitt. Hamb. Zool. Mus. Inst.* **59**: 1–60.

Distribution: NE coastal, Qld. Ecology: sedentary, predator. Biological references: Mascord, R. (1980). *Spiders of Australia.* A field guide. Sydney : Reed 128 pp. (biology).

Deinopis schomburgki Karsch, 1878

Dinopis schomburgki Karsch, F. (1878). Exotisch-araneologisches. *Z. Ges. Naturw.* **51**: 322–333 [333]. Type data: holotype, ?ZMB *F. adult, from Adelaide, S.A.

Distribution: S Gulfs, S.A. Ecology: sedentary, predator.

Deinopis subrufa L. Koch, 1878

Dinopis subrufus Koch, L. (1878). *in* Koch, L. & Keyserling, E. (1871–1883). *Die Arachniden Australiens.* Nürnberg : Bauer & Raspe Vol. 2 [1039 pl 91 fig 2]. Type data: syntypes, F. adult whereabouts unknown, from Bowen and Brisbane, Qld.

Dinopis bicornis Koch, L. (1879). *in* Koch, L. & Keyserling, E. (1871–1883). *Die Arachniden Australiens.* Nürnberg : Bauer & Raspe Vol. 2 [1049 pl 92 fig 4]. Type data: holotype, NHMW *M. adult, from Sydney, N.S.W.

Taxonomic decision of Davies, V.T., this work.

Distribution: NE coastal, SE coastal, Qld., N.S.W. Ecology: sedentary, predator. Biological references: Clyne, D. (1967). Notes on the construction of the net and sperm-web of a cribellate spider *Dinopis subrufus* (Koch) (Araneida : Dinopidae) *Aust. Zool.* **14**: 189–197; Mascord, R. (1980). *Spiders of Australia.* A field guide. Sydney : Reed 128 pp.; McKeown, K.C. (1936). *Spider Wonders of Australia.* Sydney : Angus & Robertson 270 pp.; McKeown, K.C. (1952). *Australian Spiders.* Their lives and habits. Sydney : Angus & Robertson 274 pp.; Baum, J. (1937). On the habits of the Australian spider *Dinopis subrufus* L. Koch. *Vest. Cesk. Spol. Zool.* **5**: 28–33; Austin, A.D. & Blest, A.D. (1979). The biology of two Australian species of dinopid spider. *J. Zool. Lond.* **189**: 145–156; Blest, A.D. & Land, M.F. (1977). The physiological optics of *Dinopis subrufus* L. Koch: a fish lens in a spider. *Proc. R. Soc. Lond.* B **196**: 198–222; Blest, A.D. (1978). The rapid synthesis and destruction of photoreceptor membrane by a dinopid spider: a daily cycle. *Proc. R. Soc. Lond.* B **200**: 463–483; Blest, A.D., Kao, L. & Powell, K. (1978). Photoreceptor membrane breakdown in the spider *Dinopis*: the fate of rhabdomere products. *Cell Tissue Res.* **195**: 425–444; Blest, A.D., Powell, K. & Kao, L. (1978). Photoreceptor membrane breakdown in the spider *Dinopis*: GERL differentiation in the intermediate segments. *Cell Tissue Res.* **195**: 227–297; Blest, A.D., Price, G.D. & Marples, J. (1979). Photoreceptor membrane breakdown in the spider *Dinopis*: localisation of acid phosphatases. *Cell Tissue Res.* **199**: 455–472; Blest, A.D. & Price, G.D. (1981). A new mechanism for transitory, local endocytosis in photoreceptors of a spider, *Dinopis. Cell Tissue Res.* **217**: 267–282.

Deinopis tabida L. Koch, 1879

Dinopis tabidus Koch, L. (1879). *in* Koch, L. & Keyserling, E. (1871–1883). *Die Arachniden Australiens.* Nürnberg : Bauer & Raspe Vol. 2 [1047 pl 92 fig 3]. Type data: holotype, F. adult whereabouts unknown, from Cape York, Qld.

Distribution: NE coastal, Qld. Ecology: sedentary, predator.

Deinopis unicolor L. Koch, 1878

Dinopis unicolor Koch, L. (1878). *in* Koch, L. & Keyserling, E. (1871–1883). *Die Arachniden Australiens.* Nürnberg : Bauer & Raspe Vol. 2 [1043 pl 91 fig 4]. Type data: holotype, F. adult whereabouts unknown, from King George Sound, W.A.

Distribution: SW coastal, W.A. Ecology: sedentary, predator.

DYSDERIDAE

INTRODUCTION

Medium-sized, two-clawed, haplogyne spiders with claw tufts. Six eyes or less. A pair of tracheal spiracles just behind book-lungs (tracheae extend to cephalothorax). Sternal extensions around coxae. Retromargin of chelicerae smooth. Colulus present. Tarsal claws pectinate in single row. One trichobothrium on metatarsi, none on tarsi. The family is represented in Australia by the introduced species, *Dysdera crocota*.

Dysdera Latreille, 1804

Dysdera Latreille, P.A. (1804). Tableau méthodique des Insectes. *N. Dic. Hist. Nat.* **24**: 129–200 [134]. Publication date established from Bonnet, P. (1945–1961). *Bibliographia Araneorum.* Toulouse : Douladoure Vol. 2. Type species *Aranea erythrina* Walckenaer, 1802 by monotypy.

This group is found worldwide, see Bonnet, P. (1945–1961). *Bibliographia Araneorum.* Toulouse : Douladoure Vol. 2.

Dysdera crocota C.L. Koch, 1839

Dysdera crocota Koch, C.L. (1839). *Die Arachniden.* Nürnberg Vol. 6 [81] [Emend. *crocata* employed by many authors is regarded as an invalid emend. under Art. 32(a)(ii) of ICZN]. Publication date established from Bonnet, P. (1945–1961). *Bibliographia Araneorum.* Toulouse : Douladoure Vol. 2. Type data: syntypes (probable), whereabouts unknown F.,M. adult, from Greece and Aegean Sea, see Bonnet, P. (1945–1961). *Bibliographia Araneorum.* Toulouse : Douladoure Vol. 2.

Dysdera australiensis Rainbow, W.J. (1900). Descriptions of some new Araneidae of New South Wales. No. 9 *Proc. Linn. Soc. N.S.W.* **25**: 483–494 [485]. Type data: holotype, AM KS8700 F. adult, from Sydney, N.S.W.

Taxonomic decision of Hickman, V.V. (1967). *Some Common Spiders of Tasmania.* Hobart : Tasmanian Museum and Art Gallery 112 pp. [39].

Distribution: SE coastal, S Gulfs, N.S.W., Vic., S.A., Tas. Ecology: nomadic, terrestrial, predator; introduced. Biological references: Mascord, R. (1970). *Australian Spiders in Colour.* Sydney : Reed 112 pp.; Mascord, R. (1980). *Spiders of Australia.* A field guide. Sydney : Reed 128 pp.; Main, B.Y. (1976). *Spiders.* Sydney : Collins 296 pp.; Hickman, V.V. (1967). *Some Common Spiders of Tasmania.* Hobart : Tasmanian Museum and Art Gallery 112 pp.

FILISTATIDAE

INTRODUCTION

Cribellate, three-clawed, spiders with simple haplogyne vulva. Calamistrum short and proximal. Lamellate chelicerae fused at base. Eyes in two rows, in cluster well back on carapace. Labium long and pointed, fused to sternum. Maxillae long, converging and meeting over labium. Wide spiracle well forward of sub-terminal spinnerets. Two or more trichobothria on metatarsi, none on tarsi. Autospasy common at tibial-patellar joint. *Pritha* is found in small webs under bark and has also been found in termite mounds. A revision of the Australian filistatids is in progress (Gray, M.R., pers. comm.).

Pritha Lehtinen, 1967

Pritha Lehtinen, P.T. (1967). Classification of the cribellate spiders and some allied families, with notes on the evolution of the suborder Araneomorpha. *Ann. Zool. Fenn.* **4**: 199–468 [300]. Type species *Filistata nana* Simon, 1868 by monotypy.

This group is also found in the Oriental, Ethiopian and Mediterranean regions, see Lehtinen, P.T. (1967). Classification of the cribellate spiders and some allied families, with notes on the evolution of the suborder Araneomorpha. *Ann. Zool. Fenn.* **4**: 199–468.

Pritha australiensis (L. Koch, 1873)

Filistata australiensis Koch, L. (1873). *in* Koch, L. & Keyserling, E. (1871–1883). *Die Arachniden Australiens.* Nürnberg : Bauer & Raspe Vol. 1 [451 pl 35 fig 4] [as *Filistata Australiensis*]. Type data: syntypes, ZMH (MGH 8098) *F. adult, MNHP *F. adult, NHMW *F. adult, from Rockhampton, Qld.

Taxonomic decision of Lehtinen, P.T. (1967). *op. cit.*

Distribution: NE coastal, Qld. Ecology: sedentary, predator; webs under bark.

GRADUNGULIDAE

INTRODUCTION

Medium to large three-clawed, haplogyne spiders with 2 pairs of book-lungs (like mygalomorphs). Cribellum most often absent. Superior claws of legs I and II markedly dissimilar; proclaw long and strongly developed, retroclaw normal. Eyes in two rows, AME smallest. One trichobothrium on metatarsi, none on tarsi. *Gradungula* is a vagrant found in forest litter. *Progradungula*, the cribellate genus, spins an extensive web consisting of an upper retreat tangle with connecting threads to scaffolding which supports the ladder-like catching platform which in turn is guyed to the ground. *Progradungula* is a large spider with very long legs; it resembles *Hickmania* (Hickmaniidae). An undescribed progradungulid has been found in high altitude forest in north Qld.

Gradungula Forster, 1955

Gradungula Forster, R.R. (1955). A new family of spiders of the sub-order Hypochilomorphae. *Pac. Sci.* **9**: 277–285 [277]. Type species *Gradungula sorenseni* Forster, 1955 by original designation.

This group is also found in New Zealand, see Forster, R.R. (1955). A new family of spiders of the sub-order Hypochilomorphae. *Pac. Sci.* **9**: 277–285.

Gradungula woodwardi Forster, 1955

Gradungula woodwardi Forster, R.R. (1955). A new family of spiders of the sub-order Hypochilomorphae. *Pac. Sci.* **9**: 277–285 [282]. Type data: holotype, CMNZ *F. juv., [QUE 2F.,M. adult no type status], from Mt. Hobwee, Lamington Natl. Park, Qld.

Distribution: NE coastal, Qld. Ecology: nomadic, predator. Biological references: Davies, V. Todd (1969). The mature female and male *Gradungula woodwardi* Forster (Araneae: Hypochilomorphae: Gradungulidae). *J. Aust. Entomol. Soc.* **8**: 95–97 (taxonomy).

Progradungula Forster and Gray, 1979

Progradungula Forster, R.R. & Gray, M.R. (1979). *Progradungula*, a new cribellate genus of the spider family Gradungulidae (Araneae). *Aust. J. Zool.* **27**: 1051–1071 [1053]. Type species *Progradungula carraiensis* Forster and Gray, 1979 by monotypy and original designation.

Progradungula carraiensis Forster and Gray, 1979

Progradungula carraiensis Forster, R.R. & Gray, M.R. (1979). *Progradungula*, a new cribellate genus of the spider family Gradungulidae (Araneae). *Aust. J. Zool.* **27**: 1051–1071 [1053]. Type data: holotype, AM KS1583 *F. adult, [AM KS6740 M. adult no type status], paratypes OMNZ *2F. adult, AMNH *1 juv., from Carrai Bat Cave, 60 km W of Kempsey, N.S.W.

Distribution: SE coastal, N.S.W.; Carrai Bat Cave. Ecology: sedentary, predator; in cave. Biological references: Gray, M.R. (1973). Survey of the spider fauna of Australian caves. *Helictite, J. Austral. Cave Res.* **11**: 47–75 (biology); Gray, M.R. (1983). The male of *Progradungula carraiensis* Forster and Gray (Araneae : Gradungulidae) with observations on the web and prey capture. *Proc. Linn. Soc. N.S.W.* **107**: 51–58.

HADROTARSIDAE

INTRODUCTION

Tiny, sclerotized litter spiders. Smooth or punctate dorsal abdominal scutum with anterior median cleft and ventral scuta. Tarsi longer than metatarsi. PME large, flat and reniform. Onychium with two claws, no tufts. Female palpal claw fan-shaped. Colulus absent. One trichobothrium on metatarsi, none on tarsi. Tracheae confined to abdomen.

Gmogala Keyserling, 1890

Gmogala Keyserling, E. (1890). *Die Arachniden Australiens.* Supplement. Nürnberg : Bauer & Raspe (1884–1890) [269]. Type species *Gmogala scarabaeus* Keyserling, 1890 by monotypy.

Gmogala scarabaeus Keyserling, 1890

Gmogala scarabaeus Keyserling, E. (1890). *Die Arachniden Australiens.* Supplement. Nürnberg : Bauer & Raspe (1884–1890) [270] [as *Imogala scarabaea; lapsus* Bonnet, P. (1945–1961). *Bibliographia Araneorum.* Toulouse : Douladoure Vol. 2 (2000)]. Type data: holotype, BMNH 1970.79 F. adult, from Sydney, N.S.W.

Distribution: SE coastal, N.S.W. Ecology: terrestrial, predator.

Hadrotarsus Thorell, 1881

Hadrotarsus Thorell, T. (1881). Studi sui ragni Malesi e Papuani. III. Ragni dell'Austro-Malesia e del Capo York, conservati nel Museo Civico di Storia Naturale di Genova. *Ann. Mus. Civ. Stor. Nat. Genova* **17**: 1–720 [190]. Type species *Hadrotarsus babirussa* Thorell, 1881 by monotypy.

This group is also found in New Guinea, see Levi, H.W. (1968). The spider family Hadrotarsidae and the genus *Hadrotarsus. Trans. Am. Microsc. Soc.* **87**: 141–145.

Hadrotarsus fulvus Hickman, 1943

Hadrotarsus fulvus Hickman, V.V. (1943). On some new Hadrotarsidae (Araneae) with notes on their internal anatomy. *Pap. Proc. R. Soc. Tasm.* **1942**: 147–159 [152]. Type data: holotype, AM KS6530 *M. adult, paratype AM KS6531 *F. adult, from Fingal, Tas.

Distribution: SE coastal, Vic., Tas. Ecology: terrestrial, predator.

Hadrotarsus ornatus Hickman, 1943

Hadrotarsus ornatus Hickman, V.V. (1943). On some new Hadrotarsidae (Araneae) with notes on their internal anatomy. *Pap. Proc. R. Soc. Tasm.* **1942**: 147–159 [148]. Type data: holotype, AM KS6532 *F. adult, paratype AM KS6533 *M. adult, from The Domain, Hobart, Tas.

Distribution: Tas. Ecology: terrestrial, predator.

Hadrotarsus setosus Hickman, 1943

Hadrotarsus setosus Hickman, V.V. (1943). On some new Hadrotarsidae (Araneae) with notes on their internal anatomy. *Pap. Proc. R. Soc. Tasm.* **1942**: 147–159 [155]. Type data: holotype, AM KS6584 *M. adult, from Vic.

Distribution: SE coastal, Vic. Ecology: terrestrial, predator.

HAHNIIDAE

INTRODUCTION

Small, three-clawed spiders with six spinnerets in a transverse line. Tracheal spiracle wide, in advance of spinnerets (tracheal branches to cephalothorax). Posterior eye row usually procurved. Colulus reduced or absent. A row of trichobothria (usually 2) on metatarsi and tarsi. Male palp with patellar as well as tibial apophysis. Epigynum not strongly sclerotized. Hahniids construct small sheet webs in litter and in foliage.

Alistra Thorell, 1894

Alistra Thorell, T. (1894). Förtecking öfver Arachnider fran Java och närgrändsande öar, insamlade af Carl Aurivillius; jemte beskrifningar a nagra sydasiatiska och sydamerikanska Spindlar. *Bih. Svenska Vet.-Akad. Handl.* **20**(4): 1–63 [4]. Type species *Alistra longicauda* Thorell, 1894 by monotypy. Compiled from secondary source: Lehtinen, P.T. (1967). Classification of the cribellate spiders and some allied families, with notes on the evolution of the suborder Araneomorpha. *Ann. Zool. Fenn.* **4**: 199–468.

Nannonymphaeus Rainbow, W.J. (1920). Arachnida from Lord Howe and Norfolk Islands. *Rec. S. Aust. Mus.* **1**: 229–272 [259]. Type species *Nannonymphaeus pusillus* Rainbow, 1920 by monotypy.

Taxonomic decision of Lehtinen, P.T. (1967). Classification of the cribellate spiders and some allied families, with notes on the evolution of the suborder Araneomorpha. *Ann. Zool. Fenn.* **4**: 199–468 [210].

This group is also found in Sumatra.

Alistra astrolomae (Hickman, 1948)

Hahnia astrolomae Hickman V.V. (1948). Tasmanian Araneae of the family Hahniidae with notes on their respiratory systems. *Pap. Proc. R. Soc. Tasm.* **1947**: 21–35 [22]. Type data: syntypes, AM KS6586 *M.,F. adult, from Hobart, Tas.

Distribution: Tas. Ecology: sedentary, terrestrial, predator; sheet web. Biological references: Lehtinen, P.T. (1967). Classification of the cribellate spiders and some allied families, with notes on the evolution of the suborder Araneomorpha. *Ann. Zool. Fenn.* **4**: 199–468 (taxonomy).

Alistra pusillus (Rainbow, 1920)

Nannonymphaeus pusillus Rainbow, W.J. (1920). Arachnida from Lord Howe and Norfolk Islands. *Rec. S. Aust. Mus.* **1**: 229–272 [260]. Type data: holotype, SAMA N1981337 *M. adult, from Lord Howe Is.

Distribution: Lord Howe Is. Ecology: sedentary, terrestrial, predator. Biological references: Lehtinen, P.T. (1967). Classification of the cribellate spiders and some allied families, with notes on the evolution of the suborder Araneomorpha. *Ann. Zool. Fenn.* **4**: 199–468 (taxonomy).

Neoaviola Butler, 1929

Neoaviola Butler, L.S.G. (1929). Studies in Victorian spiders. No. 1. *Proc. R. Soc. Vict. (ns)* **42**: 41–52 [45]. Type species *Neoaviola insolens* Butler, 1929 by monotypy.

Neoaviola insolens Butler, 1929

Neoaviola insolens Butler, L.S.G. (1929). Studies in Victorian spiders. No. 1. *Proc. R. Soc. Vict. (ns)* **42**: 41–52 [45]. Type data: holotype, NMV K107 F. adult, from Whittlesea, 25 mi from Melbourne, Vic.

Distribution: SE coastal, Vic. Ecology: arboreal, predator.

Neoaviola wellingtoni Hickman, 1948

Neoaviola wellingtoni Hickman V.V. (1948). Tasmanian Araneae of the family Hahniidae with notes on their respiratory systems. *Pap. Proc. R. Soc. Tasm.* **1947**: 21–35 [28]. Type data: syntypes, AM KS6626 *M.,F. adult, from Mt. Wellington, Tas.

Distribution: Tas. Ecology: predator; in moss.

Scotospilus Simon, 1886

Scotospilus Simon, E. (1886). Descriptions de quelques espèces nouvelles de la famille des Agelenidae. *Ann. Soc. Entomol. Belg.* **30**C.R.: 56–61 [61]. Type species *Scotospilus bicolor* Simon, 1886 by monotypy.

Scotospilus ampullaria (Hickman, 1948)

Hahnia ampullaria Hickman V.V. (1948). Tasmanian Araneae of the family Hahniidae with notes on their respiratory systems. *Pap. Proc. R. Soc. Tasm.* **1947**: 21–35 [25]. Type data: syntypes, AM KS6585 *M.,F. adult, from Hobart, Tas.

Distribution: Tas. Ecology: sedentary, predator; sheet web. Biological references: Lehtinen, P.T. (1967). Classification of the cribellate spiders and some allied families, with notes on the evolution of the suborder Araneomorpha. *Ann. Zool. Fenn.* **4**: 199–468 (taxonomy).

Scotospilus bicolor Simon, 1886

Scotospilus bicolor Simon, E. (1886). Descriptions de quelques espèces nouvelles de la famille Agelenidae. *Ann. Soc. Entomol. Belg.* **30**C.R.: 56–61 [61 pl]. Type data: holotype, MNHP *F. adult, from Launceston, Tas.

Distribution: Tas. Ecology: sedentary, predator; sheet web.

Scotospilus hickmani Lehtinen, 1967

Scotospilus hickmani Lehtinen, P.T. (1967). Classification of the cribellate spiders and some allied families, with notes on the evolution of the suborder Araneomorpha. *Ann. Zool. Fenn.* **4**: 199–468 [264] [description is that of *Scotopsilus* (*sic*) *bicolor* Hickman, 1948, see Hickman V.V. (1948). Tasmanian Araneae of the family Hahniidae with notes on their respiratory systems. *Pap. Proc. R. Soc. Tasm.* **1947**: 21–35]. Type data: holotype, AM KS6677 *M. adult, from Risdon, Tas., paratype *F. adult whereabouts unknown, from Hobart, Tas.

Distribution: Tas. Ecology: sedentary, predator; sheet web, under bark. Biological references: Hickman V.V. (1948). Tasmanian Araneae of the family Hahniidae with notes on their respiratory systems. *Pap. Proc. R. Soc. Tasm.* **1947**: 21–35.

HERSILIIDAE

INTRODUCTION

Small to medium, three-clawed spiders with long, thin, laterigrade legs. Extremely long posterior lateral spinnerets with spigots along the length of the inside edge; these spinnerets are used in the capture of prey. Eyes on a mound, posterior and anterior rows strongly recurved, AME larger than ALE. Claws on onychium. One trichobothrium on metatarsi, none on tarsi. Trochanters un-notched. Colulus present. No tibial apophysis on male palp. Hersiliids live on bark and hunt over it at night. *Hersilia* sp. has been collected from the Northern Territory; this genus differs from *Tama* in having two-segmented metatarsi. The family appears to be closely related to the Oecobiidae.

Tama Simon, 1882

Rhadine Simon, E. (1882). Étude sur les Arachnides du Yémen méridional *in* Viaggio ad Assab nel Mar Rosso dei signori C. Doria ed O. Beccari con il r. Aviso exploratero del 16 Nov. 1879 ad 26 Feb. 1881. *Ann. Mus. Civ. Stor. Nat. Genova* **18**: 207–260 [255] [*non Rhadine* Le Conte, 1846]. Type species *Chalinura novaehollandiae* L. Koch, 1876 by monotypy.

Tama Simon, E. (1882). Étude sur les Arachnides du Yémen méridional *in* Viaggio ad Assab nel Mar Rosso dei signori C. Doria ed O. Beccari con il r. Aviso exploratero del 16 Nov. 1879 ad 26 Feb. 1881. *Ann. Mus. Civ. Stor. Nat. Genova* **18**: 207–260 [256]. Type species *Hersilia edwardsii* Lucas, 1846 by monotypy.

Taxonomic decision of Simon, E. (1892–1903). *Histoire Naturelle des Araignées.* 2nd edn. Paris : Roret Vol. 1 [446].

This group is also found in South America, Central America, Southern Africa and Sri Lanka, see Bonnet, P. (1945–1961). *Bibliographia Araneorum.* Toulouse : Douladoure Vol. 2. Species now known not to occur in Australia: *Tama sumatrana* (Thorell) incorrectly attributed to Australia by Bonnet (1945–1961) *Loc. cit.*

Tama brachyura Simon, 1908

Tama brachyura Simon, E. (1908). Araneae. 1re partie pp. 359–446 *in* Michaelsen, W. & Hartmeyer, R. (eds.) *Die Fauna Südwest-Australiens.* 1(12) Jena : Fischer [406]. Type data: syntypes, WAM F. juv., whereabouts unknown of other F. juvs., from Geraldton, Karrakatta, East Fremantle, Brunswick and Boyanup, W.A.

Distribution: SW coastal, NW coastal, S Gulfs, W.A., S.A. Ecology: arboreal, predator; on bark.

Tama eucalypti Rainbow, 1900

Tama eucalypti Rainbow, W.J. (1900). Descriptions of some new Araneidae of New South Wales. No. 9. *Proc. Linn. Soc. N.S.W.* **25**: 483–494 [486]. Type data: syntypes, AM KS6699 *M. adult, AM KS6679 *F. adult, AM KS6698 *F. adult, AM KS6700 *M. adult, from Sydney, N.S.W.

Distribution: SE coastal, N.S.W. Ecology: arboreal, predator; on bark.

Tama fickerti (L. Koch, 1876)

Chalinura fickerti Koch, L. (1876). *in* Koch, L. & Keyserling, E. (1871–1883). *Die Arachniden Australiens.* Nürnberg : Bauer & Raspe Vol. 2 [830 pl 71 fig 2] (as *Chalinura Fickerti*). Type data: holotype, ZMH F. adult missing, from Sydney, N.S.W.

Distribution: SE coastal, N.S.W. Ecology: arboreal, predator; on bark. Biological references: Mascord, R. (1970). *Australian Spiders in Colour.* Sydney : Reed 112 pp.

Tama novaehollandiae (L. Koch, 1876)

Chalinura novaehollandiae Koch, L. (1876). *in* Koch, L. & Keyserling, E. (1871–1883). *Die Arachniden Australiens.* Nürnberg : Bauer & Raspe Vol. 2 [828 pl 71 fig 1] [as *Chalinura novae Hollandiae*]. Type data: holotype (probable), ZMH *F. adult, from Sydney, N.S.W., see Rack, G. (1961). Die Entomologischen Sammlungen des Zoologischen Staatsinstituts und Zoologischen Museums Hamburg. II. *Mitt. Hamb. Zool. Mus. Inst.* **59**: 1–60.

Distribution: SE coastal, NE coastal, N.S.W., Qld. Ecology: arboreal, predator; on bark. Biological references: Clyne, D. (1969). *Australian Spiders.* Sydney : Nelson 168 pp. (biology).

HICKMANIIDAE

INTRODUCTION

A large, three-clawed, haplogyne spider with two pairs of book-lungs (like mygalomorphs) and a cribellum. Long slender legs. Eyes in two rows, AME smallest. One trichobothrium on metatarsi, none on tarsi. The only species, *Hickmania troglodytes*, is found on large horizontal sheet-webs in caves and other cool, dark habitats in Tasmania.

Hickmania Gertsch, 1958

Hickmania Gertsch, W.J. (1958). The spider family Hypochilidae. *Am. Mus. Novit.* 1912: 1–28 [16]. Type species *Theridion troglodytes* Higgins and Petterd, 1884 by original designation.

Hickmania troglodytes (Higgins and Petterd, 1884)

Theridion troglodytes Higgins, E.T. & Petterd, W.F. (1884). Description of a new cave-inhabiting spider, together with notes on mammalian remains from a recently discovered cave in the Chudleigh district. *Pap. Proc. R. Soc. Tasm.* **1883**: 191–192 [191]. Type data: syntypes, TMH *F.,M. adult, from Mole Creek Caves, Chudleigh, Tas.

Ectatosticta australis Simon, E. (1902). Descriptions de quelques Arachnides nouveaux de la section des Cribellatés. *Bull. Soc. Entomol. Fr.* **15**: 240–243 [240]. Type data: holotype, MNHP *juv., from Tas.

Taxonomic decision of Gertsch, W.J. (1958). The spider family Hypochilidae. *Am. Mus. Novit.* 1912: 1–28 [16].

Distribution: Tas. Ecology: sedentary, predator. Biological references: Hickman, V.V. (1967). *Some Common Spiders of Tasmania.* Hobart : Tasmanian Museum and Art Gallery 112 pp.; Main, B.Y. (1976). *Spiders.* Sydney : Collins 296 pp.

HOLARCHAEIDAE

INTRODUCTION

Tiny, three-clawed, entelegyne (only slight thickening of epigastric region) spiders with long chelicerae, the bases of which are encircled by the carapace. Viewed laterally, carapace triangular with highest area behind eyes. Anterior portion of carapace between eyes and chelicerae strongly swollen to form ridge projecting over base of chelicerae. Fang almost as long as chelicera. No peg teeth on chelicera. Anterior spiracles with tracheae to abdomen; no posterior spiracle. Linguiform colulus present. One trichobothrium on metatarsi, none on tarsi. Female palp without claw. Male palpal cymbium without processes. Tarsi longer than metatarsi; tarsi I and II swollen with a clump of hairs distally. *Holarchaea* has so far been recorded only from moss in dense Tasmanian rainforest.

Holarchaea Forster, 1955

Holarchaea Forster, R.R. (1955). Spiders of the family Archaeidae from Australia and New Zealand. *Trans. R. Soc. N.Z.* **83**: 391–403 [392]. Type species *Archaea novaeseelandiae* Forster, 1949 by original designation.

This group is also found in New Zealand, see Forster, R.R. & Platnick, N.I. (1984). A review of the archaeid spiders and their relatives, with notes on the limits of the Superfamily Palpimanoidea (Arachnida, Araneae). *Bull. Am. Mus. Nat. Hist.* **178**: 1–106.

Holarchaea globosa (Hickman, 1981)

Zearchaea globosa Hickman, V.V. (1981). New Tasmanian spiders of the families Archaeidae, Cycloctenidae, Amaurobiidae and Micropholcommatidae. *Pap. Proc. R. Soc. Tasm.* **115**: 47–68 [47]. Type data: holotype, AM KS6987 F. adult, from Strathgordon, SW Tas.

Distribution: Tas. Ecology: predator; in moss. Biological references: Forster, R.R. & Platnick, N.I. (1984). A review of the archaeid spiders and their relatives, with notes on the limits of the Superfamily Palpimanoidea (Arachnida, Araneae). *Bull. Am. Mus. Nat. Hist.* **178**: 1–106.

LOXOSCELIDAE

INTRODUCTION

Medium-sized, haplogyne spiders known for the high toxicity of their venom, which may produce a cutaneous necrosis, known as "loxoscelism". Median posterior spiracle. Six eyes in three diads. Female palp without claw. Two tarsal claws on onychium, single row of teeth on claws; no claw tufts. One trichobothrium on metatarsi, none on tarsi. Chelicerae fused at base, membraneous median lamella ends distally in strong tooth. Colulus long and pointed. Two species have been introduced into Australia; there has been no record of envenomation from bites. *L. rufipes* was recorded from a single male in Sydney in 1974 and has not been reported since.

Loxosceles Heineken and Lowe, 1832

Loxosceles Heineken, C. & Lowe, R.T. (1832). *in* Lowe, R.T. (1832). Descriptions of two species of Araneidae, natives of Madeira. *Zool. J.* **5**: 320–323 [321]. Publication date established from Bonnet, P. (1945–1961). *Bibliographia Araneorum.* Toulouse : Douladoure Vol. 2. Type species *Loxosceles citigrada* Heineken and Lowe, 1832 (=*Scytodes rufescens* Dufour, 1820) by monotypy.

Omosites Walckenaer, C.A. (1833). Mémoire sur une nouvelle classification des Aranéides. *Ann. Soc. Entomol. Fr.* **2**: 414–446 [445]. Publication date established from Bonnet, P. (1945–1961). *Bibliographia Araneorum.* Toulouse : Douladoure Vol. 2. Type species *Scytodes rufescens* Dufour, 1820 by monotypy.

Taxonomic decision of Bonnet, P. (1945–1961). *Bibliographia Araneorum.* Toulouse : Douladoure Vol. 2 [2572].

This group is found worldwide.

Loxosceles rufescens (Dufour, 1820)

Scytodes rufescens Dufour, L. (1820). Description de cinq Arachnides nouvelles. *Ann. Gén. Sci. Phys.* **5**: 198–209 [203]. Type data: lost, from Sagunto, Valencia Province, Spain. Compiled from secondary source: Gertsch, W.J. (1967). The spider genus *Loxosceles* in South America (Araneae, Scytodidae). *Bull. Am. Mus. Nat. Hist.* **136**: 117–173; Gertsch, W.J. & Ennik, F. (1983). The spider genus *Loxosceles* in North America, Central America, and the West Indies (Araneae, Loxoscelidae). *Bull. Am. Mus. Nat. Hist.* **175**: 264–360.

Distribution: S Gulfs, S.A. Ecology: sedentary, predator; introduced. Biological references: Gray, M.R. (1974). Records of loxosceline spiders from Australia. *J. Entomol. Soc. Aust. (N.S.W.)* **8**: 46; Southcott, R.V. (1976). Arachnidism and allied syndromes in the Australian region. *Rec. Adel. Child. Hosp.* **1**: 97–186; Sutherland, S.K. (1983). *Australian Animal Toxins.* Melbourne : Oxford Univ. Press 527 pp.

Loxosceles rufipes (Lucas, 1834)

Scytodes rufipes Lucas, H. (1834). [Description de *Scytodes rufipes* Lucas]. *Mag. Zool. Guérin* **4**: Classe 8 [pl 6 & facing page]. Type data: holotype, F. adult presumed lost, from Guatemala, South America. Compiled from secondary source: Gertsch, W.J. (1967). The spider genus *Loxosceles* in South America (Araneae, Scytodidae). *Bull. Am. Mus. Nat. Hist.* **136**: 117–173.

Distribution: SE coastal, N.S.W.; Sydney. Ecology: sedentary, predator; introduced. Biological references: Gray, M. (1974). Records of loxosceline spiders from Australia. *J. Entomol. Soc. Aust. (N.S.W.)* **8**: 46.

LYCOSIDAE

by R.J. McKay

INTRODUCTION

The wolf spiders, small to large, three-clawed, ground living, hunting spiders. Eyes 4.2.2. Trochanters notched. Colulus absent. Male palp without tibial apophysis. Leg IV longest. Female lycosids carry the egg sac attached to the spinnerets and the young, after hatching, climb on back of the female. Lycosids are found in a very wide range of habitats - under logs and stones, in litter, in grasslands, sand-dunes, beaches, on salt-lakes and claypans. Some do not construct retreats, others have simple or complex burrows with turrets of stones and twigs, leaf palisades or well hinged trap-doors. *Venonia* builds small, horizontal sheet webs on the ground.

Anomalomma Simon, 1890

Anomalomma Simon, E. (1890). *in* van Hasselt, A.W.M. Araneae ex Archipelago Malayano. *in* Weber, M. *Zoologische Ergebnisse einer Reise in Niederländisch Ost-Indien.* **2**: 193–210 [199]. Type species *Anomalomma lycosinum* Simon, 1890 by monotypy. Compiled from secondary source: Bonnet, P. (1955). *Bibliographia Araneorum.* Toulouse : Douladoure Vol. 2.

This group is also found in southern Africa, India, Java, and Luzon, see Bonnet, P. (1955). *Bibliographia Araneorum.* Toulouse : Douladoure Vol. 2 [329].

Anomalomma cinctipes Simon, 1898

Anomalomma cinctipes Simon, E. (1898). Descriptions d'Arachnides nouveaux des familles des Agelenidae, Pisauridae, Lycosidae et Oxyopidae. *Ann. Soc. Entomol. Belg.* **42**: 5–34 [24]. Type data: holotype, MNHP *F. subadult, from Cooktown, Qld.

Distribution: NE coastal, Qld. Ecology: terrestrial, predator.

Anomalosa Roewer, 1960

Anomalosa Roewer, C.F. (1960). Araneae Lycosiformia. 2. (Lycosidae). *Explor. Parc. Natl. Upemba* Fasc. 55 pp. 519–1040 [977]. Type species *Anomalomma kochi* Simon, 1898 by original designation.

Anomalosa kochi (Simon, 1898)

Anomalomma kochi Simon, E. (1898). Descriptions d'Arachnides nouveaux des familles des Agelenidae, Pisauridae, Lycosidae et Oxyopidae. *Ann. Soc. Entomol. Belg.* **42**: 5–34 [23]. Type data: syntypes, M.,F. adult, whereabouts unknown, from Cooktown, Qld.

Distribution: NE coastal, Qld. Ecology: terrestrial, predator. Biological references: Roewer, C.F. (1960). Araneae Lycosiformia. 2. (Lycosidae). *Explor. Parc. Natl. Upemba* Fasc. 55 pp. 519–1040 (taxonomy, generic change).

Anoteropsis L. Koch, 1878

Anoteropsis Koch, L. (1878). *in* Koch, L. & Keyserling, E. (1871–1883). *Die Arachniden Australiens.* Nürnburg : Bauer & Raspe Vol. 2 [971]. Type species *Anoteropsis flavescens* L. Koch, 1878 by subsequent designation, see Simon, E. (1898). *Histoire Naturelle des Araignées.* 2nd edn. Paris : Roret [314] (as *Anoteropis*).

This group is also found in New Zealand and Papua New Guinea, see Simon, E. (1898). *Histoire Naturelle des Araignées.* 2nd edn. Paris : Roret [314]; Forster, R.R. (1979). The spiders of New Zealand. Part V. *Otago Mus. Bull. No. 5* 173 pp. [11].

Anoteropsis longipes L. Koch, 1878

Anoteropsis longipes Koch, L. (1878). *in* Koch, L. & Keyserling E. (1871–1883). *Die Arachniden Australiens.* Nürnburg : Bauer & Raspe Vol. 2 [973–974]. Type data: holotype, F. adult whereabouts unknown, from Australia.

Distribution: no known locality. Ecology: terrestrial, predator.

Artoria Thorell, 1877

Artoria Thorell, T. (1877). Studi sui ragni Malesi e Papuani. I Ragni di Selebes raccolti nel 1874 dal Dott. O. Beccari. *Ann. Mus. Civ. Stor. Nat. Genova* **10**: 341–634 [531]. Type species *Artoria parvula* Thorell, 1877 by monotypy and original designation. Compiled from secondary source: Bonnet, P. (1955). *Bibliographia Araneorum.* Toulouse : Douladoure Vol. 2.

This group is also found in Polynesia, Celebes and South Africa, see Bonnet, P. (1955). *Bibliographia Araneorum.* Toulouse : Douladoure Vol. 2 [750].

Artoria cingulipes Simon, 1909

Artoria cingulipes Simon, E. (1909). *in* Michaelsen, W. & Hartmeyer, R. (eds.) *Die Fauna Südwest-Australiens.* Jena : Fischer Vol. 2 [193]. Type data: holotype, ZMB 10547 F. adult, from Collie, W.A.

Distribution: SW coastal, W.A. Ecology: terrestrial, predator, woodland.

Artoria flavimana Simon, 1909

Artoria flavimanus Simon, E. (1909). *in* Michaelsen, W. & Hartmeyer, R. (eds.) *Die Fauna Südwest-Australiens.* Jena : Fischer Vol. 2 [192] [as *A. flavimanus*]. Type data: holotype, ZMB 10548 F. adult, from Mundaring Weir, W.A.

Distribution: SW coastal, W.A. Ecology: terrestrial, predator, woodland.

Artoria taeniifera Simon, 1909

Artoria taeniifera Simon, E. (1909). *in* Michaelsen, W. & Hartmeyer, R. (eds.) *Die Fauna Südwest-Australiens.* Jena : Fischer Vol. 2 [193]. Type data: holotype, ZMB 10549 F. adult, from Bunbury, W.A.

Distribution: SW coastal, W.A. Ecology: terrestrial, predator, woodland.

Lycosa Latreille, 1804

Lycosa Latreille, P.A. (1804). Tableau méthodique des Insectes. *N. Dic. Hist. Nat.* **24**: 129–200 [135]. Type species *Aranea tarentula* Linné, 1758 by original designation. Compiled from secondary source: Bonnet, P. (1957). *Bibliographia Araneorum.* Toulouse : Douladoure Vol. 2.

Tarantula Koch, C.L. (1837). *Übersicht des Arachnidensystems.* Nürnberg Vol. 1 pp. 1–39 [21] [*non Tarantula* Fabricius, 1793]. Type species *Tarantula apuliae* C.L. Koch, 1837 by monotypy. Compiled from secondary source: Bonnet, P. (1959). *Bibliographia Araneorum.* Toulouse : Douladoure Vol. 2.

Taxonomic decision of Bonnet, P. (1957). *Bibliographia Araneorum.* Toulouse : Douladoure Vol. 2 [2601].

This group is found worldwide, see Bonnet, P. (1957). *Bibliographia Araneorum.* Toulouse : Douladoure Vol. 2 [2584].

Lycosa abmingani Hickman, 1944

Lycosa abmingani Hickman, V.V. (1944). The Simpson Desert Expedition, 1939. Scientific Report 1. Biology — Scorpions and spiders. *Trans. R. Soc. S. Aust.* **68**: 18–48 [28] [probably a synonym of *Lycosa propitia* Simon, 1909]. Type data: holotype, AM K67697 F. adult, from Finke River, 25 mi from Abminga, S.A.

Distribution: Lake Eyre Basin, S.A. Ecology: salt marsh, terrestrial, predator.

Lycosa albopilata Urquhart, 1893

Lycosa albopilata Urquhart, A.T. (1893). On new species of Tasmanian Araneae. *Pap. Proc. R. Soc. Tasm.* **1892**: 94–130 [123]. Type data: syntypes, F.,M. adult lost, from Tas.

Distribution: Tas. Ecology: terrestrial, predator.

Lycosa albosparsa L. Koch, 1876

Lycosa albosparsa Koch, L. (1876). *in* Koch, L. & Keyserling, E. (1871–1883). *Die Arachniden Australiens.* Nürnberg : Bauer & Raspe Vol. 2 [886 pl 76 fig 4]. Type data: holotype, ZMH (MGH No. 14557) F. juv., from Rockhampton, Qld. Compiled from secondary source: Rack, G. (1961). Die Entomologischen Sammlungen des Zoologischen Staatsinstituts und Zoologischen Museums Hamburg. II. *Mitt. Hamb. Zool. Mus. Inst.* **59**: 1–60.

Distribution: NE coastal, Lake Eyre basin, Qld., N.T. Ecology: terrestrial, supralittoral, predator.

Lycosa alteripa McKay, 1976

Lycosa alteripa McKay, R.J. (1976). The Wolf Spiders of Australia (Araneae : Lycosidae): 8. Two new species inhabiting salt lakes of Western Australia. *Mem. Qd. Mus.* **17**: 417–423 [418]. Type data: holotype, WAM 71–40 M. adult, from Fitzgerald River, W.A.

Distribution: SW coastal, W.A.; Fitzgerald River and Israelite Bay. Ecology: salt marsh, terrestrial, burrowing, predator.

Lycosa arenaris Hogg, 1905

Lycosa arenaris Hogg, H.R. (1905). On some South Australian spiders of the family Lycosidae. *Proc. Zool. Soc. Lond.* **1905**: 569–590 [586]. Type data: lectotype, SAMA F. adult, paralectotype SAMA F. penultimate juv., subsequent designation by McKay, R.J. (1974). The Wolf Spiders of Australia (Araneae : Lycosidae): 2. The *arenaris* group. *Mem. Qd. Mus.* **17**: 1–19, from MacDonnell Ranges, N.T.

Lycosa celaenica Rainbow, W.J. (1917). Araneidae. Results of the South Australian Museum Expedition to Strzelecki and Cooper Creeks. *Trans. Proc. R. Soc. S. Aust.* **41**: 482–489 [488]. Type data: holotype, SAMA F. adult, from Farina, S.A.

Taxonomic decision of McKay, R.J. (1974). The Wolf Spiders of Australia (Araneae : Lycosidae): 2. The *arenaris* group. *Mem. Qd. Mus.* **17**: 1–19 [1].

Distribution: W plateau, Lake Eyre basin, S.A. Ecology: terrestrial, burrowing, predator; from wet sand on river banks in arid areas. Biological references: Main, B.Y. (1976). *Spiders.* Sydney : Collins 296 pp. (biology).

Lycosa ariadnae McKay, 1979

Lycosa ariadnae McKay, R.J. (1979). The Wolf Spiders of Australia (Araneae : Lycosidae): 12. Descriptions of some Western Australian species. *Mem. Qd. Mus.* **19**: 241–275 [241]. Type data: holotype, WAM 72–246 F. adult, from Darkan, W.A.

Distribution: SW coastal, W.A. Ecology: terrestrial, burrowing, predator; from sandy soil plains.

Lycosa aurea Hogg, 1896

Lycosa aurea Hogg, H.R. (1896). Araneidae. pp. 309–356 *Rep. Horn Expedition to Central Australia.* Pt. 2 Zoology [345]. Type data: holotype, F. adult whereabouts unknown, from Ellery Creek, N.T.

Distribution: Lake Eyre basin, N.T. Ecology: terrestrial, desert, predator.

Lycosa australicola (Strand, 1913)

Tarentula australicola Strand, E. (1913). Über einige australische Spinnen des Senckenbergischen Museums. *Zool. Jb. Syst.* **35**: 599–624 [619]. Type data: holotype, SMF 2255 F. adult, from central Australia.

Distribution: W plateau, Lake Eyre basin, W.A., N.T. Ecology: terrestrial, burrowing, predator; from clay soils and sandy soils. Biological references: Bonnet, P. (1957). *Bibliographia Araneorum.* Toulouse : Douladoure Vol. 2 (taxonomy, generic change); McKay, R.J. (1979). The Wolf Spiders of Australia (Araneae : Lycosidae): 12. Descriptions of some Western Australian species. *Mem. Qd. Mus.* **19**: 241–275 (ecology).

Lycosa berenice L. Koch, 1877

Lycosa berenice Koch, L. (1877). *in* Koch, L. & Keyserling, E. (1871–1883). *Die Arachniden Australiens.* Nürnberg : Bauer & Raspe Vol. 2 [937 pl 81 fig 3] [as *Lycosa Berenice*]. Type data: holotype, F. adult whereabouts unknown, from unknown locality.

Distribution: SE coastal, N.S.W., Vic. Ecology: terrestrial, predator.

Lycosa bicolor Hogg, 1905

Lycosa bicolor Hogg, H.R. (1905). On some South Australian spiders of the family Lycosidae. *Proc. Zool. Soc. Lond.* **1905**: 569–590 [580]. Type data: lectotype, SAMA F. adult, paralectotypes SAMA F. adult, M. juv., designation by McKay, R.J. (1973). The Wolf Spiders of Australia (Araneae : Lycosidae): 1. The *bicolor* group. *Mem. Qd. Mus.* **16**: 375–398, from unknown locality.

Distribution: W plateau, S.A., N.T., W.A. Ecology: desert, hummock grasslands, low open woodland, terrestrial, burrowing, predator; from arid red earth and desert loam soils. Biological references: McKay, R.J. (1973). The Wolf Spiders of Australia (Araneae : Lycosidae): 1. The *bicolor* group. *Mem. Qd. Mus.* **16**: 375–398 (ecology); Main, B.Y. (1976). *Spiders.* Sydney : Collins 296 pp. (biology, ecology).

Lycosa brisbanae L. Koch, 1878

Lycosa brisbanae Koch, L. (1878). *in* Koch, L. & Keyserling, E. (1871–1878). *Die Arachniden Australiens.* Nürnberg : Bauer & Raspe Vol. 2 [976 pl 85 fig 4]. Type data: holotype, F. adult whereabouts unknown, from unknown locality (?Brisbane).

Distribution: NE coastal, Qld. Ecology: terrestrial, predator.

Lycosa burti Hickman, 1944

Lycosa burti Hickman, V.V. (1944). The Simpson Desert Expedition, 1939. Scientific Report 1. Biology — Scorpions and spiders. *Trans. R. Soc. S. Aust.* **68**: 18–48 [29]. Type data: holotype, AM K67699 F. adult, from Burt's Waterhole, S.A. (55 mi S of Birdsville).

Distribution: Lake Eyre basin, S.A., Qld. Ecology: desert, low open woodland, terrestrial, predator; from sandy soils near creeks.

Lycosa castanea Hogg, 1905

Lycosa castanea Hogg, H.R. (1905). On some South Australian spiders of the family Lycosidae. *Proc. Zool. Soc. Lond.* **1905**: 569–590 [577]. Type data: holotype, SAMA F. adult, from unknown locality.

Distribution: S.A.; locality given only as "South Australia". Ecology: desert, terrestrial, predator.

Lycosa clara L. Koch, 1877

Lycosa clara Koch, L. (1877). *in* Koch, L. & Keyserling, E. (1871–1883). *Die Arachniden Australiens.* Nürnberg : Bauer & Raspe Vol. 2 [912 pl 79 fig 1]. Type data: holotype, ZMH (MGH No. 14559) F. adult, from Bowen, Qld. Compiled from secondary source: Rack, G. (1961). Die Entomologischen Sammlungen des Zoologischen Staatsinstituts und Zoologischen Museums Hamburg. II. *Mitt. Hamb. Zool. Mus. Inst.* **59**: 1–60.

Distribution: NE coastal, NW coastal, N coastal, Qld., N.T., W.A. Ecology: terrestrial, burrowing, predator; from sandy loam soils near creeks. Biological references: McKay, R.J. (1979). The Wolf Spiders of Australia (Araneae : Lycosidae): 12. Descriptions of some Western Australian species. *Mem. Qd. Mus.* **19**: 241–275 (ecology).

Lycosa corallina McKay, 1974

Lycosa corallina McKay, R.J. (1974). The Wolf Spiders of Australia (Araneae : Lycosidae): 3. A coral shingle inhabiting species from Western Australia. *Mem. Qd. Mus.* **17**: 21–26 [21]. Type data: holotype, WAM 71-1645 F. adult, from Wooded Is., Houtman Abrolhos, W.A.

Distribution: NW coastal, W.A.; Abrolhos and Rosemary Ils., Dampier Archipelago. Ecology: terrestrial, burrowing, predator; from coral shingle. Biological references: Main, B.Y. (1976). *Spiders.* Sydney : Collins 296 pp. (biology).

Lycosa cowlei Hogg, 1896

Lycosa cowlei Hogg, H.R. (1896). Araneidae. pp. 309–356 *Rep. Horn Expedition to Central Australia.* Pt. 2 Zoology [349]. Type data: holotype, F. adult whereabouts unknown, from Ayers Rock, N.T.

Distribution: W plateau, N.T. Ecology: terrestrial, predator.

Lycosa crispipes L. Koch, 1877

Lycosa crispipes Koch, L. (1877). *in* Koch, L. & Keyserling, E. (1871–1883). *Die Arachniden Australiens.* Nürnberg : Bauer & Raspe Vol. 2 [923 pl 79 fig 8, pl 80 fig 1]. Type data: syntypes, BMNH 1919.9.18.222 F. adult, ZMH (MGH No. 14572) F. adult, from Bowen, Qld. Compiled from secondary source: Rack, G. (1961). Die Entomologischen Sammlungen des Zoologischen Staatsinstituts und Zoologischen Museums Hamburg. II. *Mitt. Hamb. Zool. Mus. Inst.* **59**: 1–60.

Distribution: NW coastal, W plateau, Lake Eyre basin, Qld., N.T., W.A. Ecology: terrestrial, burrowing, predator; from dry sands of river beds. Biological references: McKay, R.J. (1979). The Wolf Spiders of Australia (Araneae : Lycosidae): 12. Descriptions of some Western Australian species. *Mem. Qd. Mus.* **19**: 241–275 (ecology).

Lycosa dimota Simon, 1909

Lycosa dimota Simon, E. (1909). *in* Michaelsen, W. & Hartmeyer, R. (eds.) *Die Fauna Südwest-Australiens.* Jena : Fischer Vol. 2 [184]. Type data: holotype, F. adult whereabouts unknown, from Day Dawn, W.A.

Tarentula laeta protruda Strand, E. (1913). Über einige australische Spinnen des Senckenbergischen Museums. *Zool. Jb. Syst.* **35**: 599–624 [620] [described as a variety]. Type data: holotype, SMF F. adult, from central Australia.

Taxonomic decision of McKay, R.J., this work.

Distribution: SW coastal, W.A. Ecology: terrestrial, predator, woodland, low open woodland.

Lycosa duracki McKay, 1975

Lycosa duracki McKay, R.J. (1975). The Wolf Spiders of Australia (Araneae : Lycosidae): 5. Two new species of the *bicolor* group. *Mem. Qd. Mus.* **17**: 313–318 [316]. Type data: holotype, WAM 74–494 F. adult, from Old Argyle Downs Station (now submerged), Ord River, W.A.

Distribution: N coastal, W.A.; Kimberley region. Ecology: terrestrial, burrowing, predator; from gravel slopes.

Lycosa egena L. Koch, 1877

Lycosa egena Koch, L. (1877). *in* Koch, L. & Keyserling, E. (1871–1883). *Die Arachniden Australiens.* Nürnberg : Bauer & Raspe Vol. 2 [935 pl 81 fig 2]. Type data: holotype, F. adult whereabouts unknown, from Cape York, N Qld.

Distribution: N Gulf, NE coastal, N coastal, Qld., W.A., (N.T.). Ecology: terrestrial, predator; from clay pans.

Lycosa errans Hogg, 1905

Lycosa errans Hogg, H.R. (1905). On some South Australian spiders of the family Lycosidae. *Proc. Zool. Soc. Lond.* **1905**: 569–590 [579]. Type data: holotype, SAMA F. adult, from unknown locality.

Distribution: Lake Eyre basin, W plateau, S.A. Ecology: terrestrial, hummock grasslands, open woodland, predator; in arid areas.

Lycosa excusor L. Koch, 1867

Lycosa excusor Koch, L. (1867). Beschreibungen neuer Arachniden und Myriapoden. *Verh. Zool.-Bot Ges. Wien* **17**: 173–250 [202] [as *L. excursor* erroneously listed in Bonnet, P. (1957). *Bibliographia Araneorum.* Toulouse : Douladoure Vol. 2 (2640)]. Publication date established from Roewer, C.F. (1954). *Katalog der Araneae.* Bremen : Schünemann Vol. 2 [271]. Type data: syntypes, ZMH? (MGH No. 2263) *M. adult, BMNH 1919.9.18.334 F. adult, BMNH 1919.9.18.335–336 M. adult, ZMH (MGH No. 2263) 2F. adult, from Brisbane, Qld. Compiled from secondary source: Rack, G. (1961). Die Entomologischen Sammlungen des Zoologischen Staatsinstituts und Zoologischen Museums Hamburg. II. *Mitt. Hamb. Zool. Mus. Inst.* **59**: 1–60.

Distribution: NE coastal, Qld. Ecology: terrestrial, predator.

Lycosa eyrei (Hickman, 1944)

Pardosa eyrei Hickman, V.V. (1944). The Simpson Desert Expedition, 1939. Scientific Report 1. Biology — Scorpions and spiders. *Trans. R. Soc. S. Aust.* **68**: 18–48 [24]. Type data: holotype, AM K67695 M. adult, from Lake Eyre, S.A.

Distribution: Lake Eyre basin, S.A. Ecology: terrestrial, predator, salt lake. Biological references: McKay, R.J., this work, *nov. comb.*

Lycosa fallax L. Koch, 1877

Lycosa fallax Koch, L. (1877). *in* Koch, L. & Keyserling, E. (1871–1883). *Die Arachniden Australiens.* Nürnberg : Bauer & Raspe Vol. 2 [934 pl 80 fig 8]. Type data: holotype, ZMH F. adult, from Bowen, Qld. Compiled from secondary source: Rack, G. (1961). Die Entomologischen Sammlungen des Zoologischen Staatsinstituts und Zoologischen Museums Hamburg. II. *Mitt. Hamb. Zool. Mus. Inst.* **59**: 1–60.

Distribution: NE coastal, Qld. Ecology: terrestrial, predator.

Lycosa finkei Hickman, 1944

Lycosa finkei Hickman, V.V. (1944). The Simpson Desert Expedition, 1939. Scientific Report 1. Biology — Scorpions and spiders. *Trans. R. Soc. S. Aust.* **68**: 18–48 [31]. Type data: holotype, AM F. adult, from Finke River, 25 mi from Abminga, S.A.

Distribution: Lake Eyre basin, S.A. Ecology: terrestrial, predator.

Lycosa flavisternis L. Koch, 1877

Lycosa flavisternis Koch, L. (1877). *in* Koch, L. & Keyserling, E. (1871–1877). *Die Arachniden Australiens.* Nürnberg : Bauer & Raspe Vol. 2 [950 pl 82 fig 4–5]. Type data: syntypes, BMNH 1919.9.18.361–362 M. adult, ZMH (MGH No. 14575) F. adult, ZMH M. adult, from Rockhampton, Port MacKay, Qld. and Sydney, N.S.W. Compiled from secondary source: Rack, G. (1961). Die Entomologischen Sammlungen des Zoologischen Staatsinstituts und Zoologischen Museums Hamburg. II. *Mitt. Hamb. Zool. Mus. Inst.* **59**: 1–60.

Distribution: NE coastal, SE coastal, Qld., N.S.W. Ecology: terrestrial, predator.

Lycosa forresti McKay, 1973

Lycosa forresti McKay, R.J. (1973). The Wolf Spiders of Australia (Araneae : Lycosidae): 1. The *bicolor* group. *Mem. Qd. Mus.* **16**: 375–398 [385]. Type data: holotype, WAM 70–44 F. adult, from 8 mi W of Moorine Rock, W.A.

Distribution: W plateau, W.A. Ecology: terrestrial, burrowing, predator; from red clay loam soils. Biological references: Main, B.Y. (1976). *Spiders.* Sydney : Collins 296 pp. (biology).

Lycosa funesta C.L. Koch, 1848

Lycosa (Tarantula) funesta Koch, C.L. (1848). *Die Arachniden.* Nürnberg Vol. 14 [148]. Publication date established from Bonnet, P. (1957). *Bibliographia Araneorum.* Toulouse : Douladoure Vol. 2 [2608]. Type data: holotype, whereabouts unknown, from unknown locality. Compiled from secondary source: Bonnet, P. (1957). *Bibliographia Araneorum.* Toulouse : Douladoure Vol. 2.

Distribution: Tas. Ecology: terrestrial, predator.

Lycosa furcillata L. Koch, 1867

Lycosa furcillata Koch, L. (1867). Beschreibungen neuer Arachniden und Myriapoden. *Verh. Zool.-Bot. Ges. Wien* **17**: 173–250 [201]. Publication date established from Koch, L. *in* Koch, L. & Keyserling, E. (1871–1883). *Die Arachniden Australiens.* Nürnberg : Bauer & Raspe Vol. 2 [903]. Type data: lectotype, BMNH 1919.9.18.363 F. adult, paralectotyes BMNH 1919.9.18.364 F. adult, BMNH 1919.9.18.365 M. adult, ZMH (MGH No. 3504), designation by McKay, R.J. (1974). The Wolf Spiders of Australia (Araneae : Lycosidae): 2. The *arenaris* group. *Mem. Qd. Mus.* **17**: 1–19 [16], from Sydney, N.S.W.

Distribution: NE coastal, SE coastal, Qld., N.S.W. Ecology: terrestrial, predator. Biological references: McKay, R.J. (1974). The Wolf Spiders of Australia (Araneae : Lycosidae): 2. The *arenaris* group. *Mem. Qd. Mus.* **17**: 1–19 (biology); Mascord, R. (1970). *Australian Spiders in Colour.* Sydney : Reed 112 pp. (biology).

Lycosa gibsoni McKay, 1979

Lycosa gibsoni McKay, R.J. (1979). The Wolf Spiders of Australia (Araneae : Lycosidae): 12. Descriptions of some Western Australian species. *Mem. Qd. Mus.* **19**: 241–275 [255]. Type data: holotype, WAM 70–1 M. adult, from Millstream Station, Fortescue River, W.A.

Distribution: NW coastal, W.A. Ecology: terrestrial, burrowing, predator; from stony areas.

Lycosa gilberta Hogg, 1905

Lycosa gilberta Hogg, H.R. (1905). On some South Australian spiders of the family Lycosidae. *Proc. Zool. Soc. Lond.* **1905**: 569–590 [582]. Type data: lectotype, SAMA F. adult, paralectotypes SAMA 4F. adult, designation by McKay, R.J. (1975). The Wolf Spiders of Australia (Araneae : Lycosidae): 6. The *leuckartii* group. *Mem. Qd. Mus.* **17**: 319–328 [325], from Gilbert River, Riverina, S.A.

Distribution: Murray-Darling basin, S.A., Vic., N.S.W. Ecology: woodland, terrestrial, predator. Biological references: McKay, R.J. (1975). The Wolf Spiders of Australia (Araneae : Lycosidae): 6. The *leuckartii* group. *Mem. Qd. Mus.* **17**: 319–328 (ecology).

Lycosa gloriosa Rainbow, 1920

Lycosa gloriosa Rainbow, W.J. (1920). Arachnida from Lord Howe and Norfolk Islands. *Rec. S. Aust. Mus.* **1**: 229–272 [261]. Type data: holotype, SAMA I11538 F. adult, from Lord Howe Is.

Distribution: Lord Howe Is. Ecology: supralittoral, closed forest, terrestrial, predator; lowland palm groves, lowland mixed rainforest to highland rainforest. Biological references: McKay, R.J. (1979). The Wolf Spiders of Australia (Araneae : Lycosidae): 11. A new species from Lord Howe Island. *Mem. Qd. Mus.* **19**: 237–240 (ecology).

Lycosa godeffroyi L. Koch, 1865

Lycosa godeffroyi Koch, L. (1865). Beschreibungen neuer Arachniden und Myriapoden. *Verh. Zool.-Bot. Ges. Wien* **15**: 857–892 [867]. Publication date established from Koch, L. & Keyserling, E. (1871–1883). *Die Arachniden Australiens.* Nürnberg : Bauer & Raspe Vol. 2 [957]. Type data: syntypes, BMNH 1890.7.1.2371 F. adult, BMNH 1919.9.18.396 F. adult, BMNH 1919.9.18.397 F. adult, ZMH (MGH No. 300), from Wollongong, Sydney, N.S.W., Peak Downs, Qld. Compiled from secondary source: Rack, G. (1961). Die Entomologischen Sammlungen des Zoologischen Staatsinstituts und Zoologischen Museums Hamburg. II. *Mitt. Hamb. Zool. Mus. Inst.* **59**: 1–60.

Lycosa bellatrix Koch, L. (1865). Beschreibungen neuer Arachniden und Myriapoden. *Verh. Zool.-Bot. Ges. Wien* **15**: 857–892 [866]. Publication date established from Koch, L. (1865). *in* Koch, L. & Keyserling, E. (1871–1883). *Die Arachniden Australiens.* Nürnberg : Bauer & Raspe Vol. 2 [957]. Type data: holotype (possible), BMNH 1919.9.18.1187 F. adult (labelled *Tarentula vecatrix*, -possibly could = *bellatrix*), from Sydney, N.S.W.

Taxonomic decision of Koch, L. *in* Koch, L. & Keyserling, E. (1871–1883). *Die Arachniden Australiens.* Nürnberg : Bauer & Raspe Vol. 2 [957].

Distribution: SW coastal, W plateau, S Gulfs, Murray-Darling basin, SE coastal, NE coastal, S.A., Vic., A.C.T., N.S.W., Qld., W.A.; widespread, inland to coastal. Ecology: terrestrial, predator. Biological references: Humphreys, W.F. (1974). Behavioural thermoregulation in a wolf spider. *Nature* **251**: 502–503; Humphreys, W.F. (1975). The respiration of *Geolycosa godeffroyi* (Araneae : Lycosidae) under conditions of constant and cyclic temperature. *Physiol. Zool.* **48**: 269–281; Humphreys, W.F. (1975). The food consumption of a wolf spider *Geolycosa godeffroyi* (Araneae : Lycosidae) in the Australian Capital Territory. *Oecologia* **18**: 343–358; Humphreys, W.F. (1975). The influence of burrowing and thermoregulatory behaviour on the water relations of *Geolycosa godeffroyi* (Araneae : Lycosidae), an Australian wolf spider. *Oecologia* **21**: 291–311; Humphreys, W.F. (1976). The population dynamics of an Australian wolf spider, *Geolycosa godeffroyi* (L. Koch, 1865)(Araneae : Lycosidae). *J. Anim. Ecol.* **45**: 59–80; Main, B.Y. (1976). *Spiders.* Sydney : Collins 296 pp. (biology).

Lycosa goyderi Hickman, 1944

Lycosa goyderi Hickman, V.V. (1944). The Simpson Desert Expedition, 1939. Scientific Report 1. Biology — Scorpions and spiders. *Trans. R. Soc. S. Aust.* **68**: 18–48 [33]. Type data: holotype, F. adult whereabouts unknown, from Goyders Lagoon Bore, S.A.

Distribution: Lake Eyre basin, S.A. Ecology: swamp, terrestrial, predator; swamps and damp watercourses in arid areas.

Lycosa habilis (Hogg, 1905)

Dolomedes habilis Hogg, H.R. (1905). On some South Australian spiders of the family Lycosidae. *Proc. Zool. Soc. Lond.* **1905**: 569–590 [588]. Type data: syntypes, 3F. adult whereabouts unknown, from Gilbert River, Riverina, N.S.W.

Distribution: Murray-Darling basin, N.S.W. Ecology: terrestrial, predator. Biological references: Rainbow, W.J. (1911). A Census of Australian Araneidae. *Rec. Aust. Mus.* **9**: 107–391 (taxonomy, generic change).

Lycosa halei Hickman, 1944

Lycosa halei Hickman, V.V. (1944). The Simpson Desert Expedition, 1939. Scientific Report 1. Biology — Scorpions and spiders. *Trans. R. Soc. S. Aust.* **68**: 18–48 [34]. Type data: holotype, F. adult (possibly penultimate) whereabouts unknown, from 11 mi E of Hale River, Simpson Desert, N.T.

Distribution: W plateau, N.T. Ecology: terrestrial, predator; arid areas.

Lycosa hasseltii L. Koch, 1877

Lycosa hasseltii Koch, L. (1877). *in* Koch, L. & Keyserling, E. (1871–1883). *Die Arachniden Australiens.* Nürnberg : Bauer & Raspe Vol. 2 [960 pl 83 fig 5] [as *Lycosa Hasseltii*]. Type data: syntypes, BMNH 1919.9.18.399 F. adult, ZMH (MGH No. 14556) F. adult, from Peak Downs, Qld., see Rack, G. (1961). Die Entomologischen Sammlungen des Zoologischen Staatsinstituts und Zoologischen Museums Hamburg. II. *Mitt. Hamb. Zool. Mus. Inst.* **59**: 1–60.

Distribution: NE coastal, Qld. Ecology: terrestrial, predator, woodland.

Lycosa hickmani (Roewer, 1954)

Lycosa fletcheri Hickman, V.V. (1944). The Simpson Desert Expedition, 1939. Scientific Report 1. Biology — Scorpions and spiders. *Trans. R. Soc. S. Aust.* **68**: 18–48 [32] [*non Lycosa fletcheri* Gravely, 1924]. Type data: holotype, AM F. adult, from Charlotte Waters, N.T.

Varacosa hickmani Roewer, C.F. (1954). *Katalog der Araneae.* Bruxelles : Inst. R. Sci. Nat. Belg. Bd. 2a [305] [*nom. nov.* for *Lycosa fletcheri* Hickman, 1944].

Distribution: Lake Eyre basin, N.T. Ecology: swamp, terrestrial, predator.

Lycosa howensis McKay, 1979

Lycosa howensis McKay, R.J. (1979). The Wolf Spiders of Australia (Araneae : Lycosidae): 11. A new species from Lord Howe Island. *Mem. Qd. Mus.* **19**: 237–240 [237]. Type data: holotype, AM KS60 F. adult, from Lord Howe Is.

Distribution: Lord Howe Is. Ecology: closed forest, terrestrial, predator.

Lycosa immansueta Simon, 1909

Lycosa immansueta Simon, E. (1909). *in* Michaelsen, W. & Hartmeyer, R. (eds.) *Die Fauna Südwest-Australiens.* Jena : Fischer Vol. 2 [183]. Type data: syntypes, M.,F. adult whereabouts unknown, from Wooroloo, Cannington, W.A.

Distribution: SW coastal, S Gulfs, SE coastal, NE coastal, W.A., Vic., N.S.W., S.A., Qld. Ecology: terrestrial, woodland, predator.

Lycosa infensa L. Koch, 1877

Lycosa infensa Koch, L. (1877). *in* Koch, L. & Keyserling, E. (1871–1883). *Die Arachniden Australiens.* Nürnberg : Bauer & Raspe Vol. 2 [932 pl 80 fig 7]. Type data: syntypes, BMNH 1919.9.18.412 M. juv., F. juv. whereabouts unknown, from Sydney, N.S.W.; Rockhampton, Qld.

Distribution: NE coastal, SE coastal, Qld., N.S.W. Ecology: terrestrial, predator.

Lycosa kosciuskoensis McKay, 1974

Lycosa kosciuskoensis McKay, R.J. (1974). The Wolf Spiders of Australia (Araneae : Lycosidae): 4. Three new species from Mt. Kosciusko, N.S.W. *Mem. Qd. Mus.* **17**: 27–36 [31]. Type data: holotype, AM K61675 F. adult, paratypes AM K61672 F. adult, AM K61673 3F. adult, AM K61674 3F. adult, from Lake Albina, Mt. Kosciusko, N.S.W.

Distribution: SE coastal, N.S.W.; Mt. Kosciusko. Ecology: alpine, terrestrial, burrowing, predator.

Lycosa koyuga McKay, 1979

Lycosa koyuga McKay, R.J. (1979). The Wolf Spiders of Australia (Araneae : Lycosidae): 12. Descriptions of some Western Australian species. *Mem. Qd. Mus.* **19**: 241–275 [258]. Type data: holotype, WAM 71–1968 F. adult, paratypes WAM 71–48–51 4F. adult, WAM 70–183–5 3F. adult, WAM 71–743–4 2F. adult, WAM 71–189 F. adult, WAM 71–1957–1967 8F.,1M. adult, 4 juv., WAM 71–1748–1750 3F. adult, WAM 69–809 F. adult, WAM 69–806 F. adult, WAM 71–1751 F. adult, from Guilderton, Moore River, W.A.

Distribution: W plateau, SW coastal, W.A. Ecology: terrestrial, predator; from sandy soils.

Lycosa lacertosa L. Koch, 1877

Lycosa lacertosa Koch, L. (1877). *in* Koch, L. & Keyserling, E. (1871–1883). *Die Arachniden Australiens.* Nürnberg : Bauer & Raspe Vol. 2 [952 pl 82 fig 6]. Type data: holotype, ZMH (MGH No. 14566) F. adult, from Adelaide, S.A. Compiled from secondary source: Rack, G. (1961). Die Entomologischen Sammlungen des Zoologischen Staatsinstituts und Zoologischen Museums Hamburg. II. *Mitt. Hamburg Zool. Mus. Inst.* **59**: 1–60.

Distribution: S Gulfs, S.A. Ecology: terrestrial, predator.

Lycosa laeta L. Koch, 1877

Lycosa laeta Koch, L. (1877). *in* Koch, L. & Keyserling, E. *Die Archniden Australiens.* Nürnberg : Bauer & Raspe [944–946]. Type data: 2 syntypes, BMNH 1919.9.18.475 F. adult (Peak Downs), ZMH F. adult, from Rockhampton, Bowen & Peak Downs, Qld.

Tarentula laeta curticeps Strand, E. (1913). Über einige australische Spinnen des Senckenbergischen Museums. *Zool. Jb. (Syst.)* **35**: 599–624 [621] [described as a variety]. Type data: holotype, SMF F. adult, from central Australia.

Taxonomic decision of McKay, R.J., this work.

Distribution: NE coastal, N Gulf, SE coastal, Qld., N.T., N.S.W. Ecology: open woodland, terrestrial, predator.

Lycosa lapidosa McKay, 1974

Lycosa lapidosa McKay, R.J. (1974). The Wolf Spiders of Australia (Araneae : Lycosidae): 2. The *arenaris* group. *Mem. Qd. Mus.* **17**: 1–19 [13]. Type data: holotype, QM W3865 F. adult, paratypes QM W3864 6F.,5M. adult, 4 juv., QM W3867 F. adult, QM W3868 6F.,2M. adult, 3 juv., QM W3869 F. adult, from Black Duck Creek, near Junction View, SE Qld.

Distribution: NE coastal, SE coastal, Qld., N.S.W. Ecology: terrestrial, predator; from under rocks of dry creek beds.

Lycosa leonhardii (Strand, 1913)

Tarentula leonhardii Strand, E. (1913). Über einige australische Spinnen des Senckenbergischen Museums. *Zool. Jb. (Syst.)* **35**: 599–624 [618]. Type data: syntypes, SMF 2237 2F. adult, from central Australia.

Distribution: Lake Eyre basin, N.S.W., N.T., S.A., Qld. Ecology: terrestrial, desert, predator; on sand dunes. Biological references: Bonnet, P. (1957). *Bibliographia Araneorum.* Toulouse : Douladoure Vol. 2 (taxonomy, generic change).

Lycosa leuckartii (Thorell, 1870)

Tarentula leuckartii Thorell, T. (1870), Araneae nonnullae Novae Hollandiae, descriptae. *Öfvers. K. Vetensk Akad. Förh.* **27**: 367–389 [388]. Type data: holotype, F. adult whereabouts unknown, from unknown locality.

Lycosa christopheri Simon, E. (1909). *in* Michaelsen, W. & Hartmeyer, R. (eds.) *Die Fauna Südwest-Australiens.* Jena : Fischer Vol. 2 [188]. Type data: holotype, ZMH F. adult, from Fremantle, W.A. Compiled from secondary source: Rack, G. (1961). Die Entomologischen Sammlungen des Zoologischen Staatsinstituts und Zoologischen Museums Hamburg. II. *Mitt. Hamb. Zool. Mus. Inst.* **59**: 1–60.

Taxonomic decision of McKay, R.J. (1975). The Wolf Spiders of Australia (Araneae : Lycosidae): 6. The *leuckartii* group. *Mem. Qd. Mus.* **17**: 319–328 [320].

Distribution: SW coastal, S Gulfs, SE coastal, Murray-Darling basin, NE coastal, W.A., S.A., Vic., N.S.W., Qld. Ecology: terrestrial, burrowing, predator; from clay soils near streams. Biological references: Main, B.Y. (1976). *Spiders.* Sydney : Collins 296 pp. (biology).

Lycosa leucophaeoides (Roewer, 1951)

Lycosa leucophaea Koch, L. (1877). *in* Koch, L. & Keyserling, E. (1871–1883). *Die Arachniden Australiens.* Nürnberg : Bauer & Raspe Vol. 2 [948 pl 82 fig 3] [*non Lycosa leucophaea* Blackwall, 1837]. Type data: holotype, ZMH (MGH No. 14565) F. adult, from Rockhampton, Qld. Compiled from secondary source: Rack, G. (1961). Die Entomologischen Sammlungen des Zoologischen Staatsinstituts und Zoologischen Museums Hamburg. II. *Mitt. Hamb. Zool. Mus. Inst.* **59**: 1–60.

Hogna leucophaeoides Roewer, C.F. (1951) Neue Namen einiger Araneen-Arten. *Abh. Naturw. Ver. Bremen* **32**: 437–456 [440] [*nom. nov.* for *Lycosa leucophaea* L. Koch, 1877].

Distribution: NE coastal, Qld. Ecology: terrestrial, predator; from bare soil.

Lycosa madigani Hickman, 1944

Lycosa madigani Hickman, V.V. (1944). The Simpson Desert Expedition, 1939. Scientific Report 1. Biology — Scorpions and spiders. *Trans. R. Soc. S. Aust.* **68**: 18–48 [36]. Type data: syntypes, AM M. adult, immature, 3F. adult whereabouts unknown, from 6 mi N of junction of Todd and Hale Rivers; centre of Simpson Desert, Indinda Well near Andado Station, N.T.

Distribution: Lake Eyre basin, N.T. Ecology: terrestrial, desert, predator; on sand dunes.

Lycosa maini McKay, 1979

Lycosa maini McKay, R.J. (1979). The Wolf Spiders of Australia (Araneae : Lycosidae): 12. Descriptions of some Western Australian species. *Mem. Qd. Mus.* **19**: 241–275 [260]. Type data: holotype, WAM 69–115 F. adult, from 88 km N of Murchison River, W.A.

Distribution: W plateau, W.A. Ecology: terrestrial, burrowing, predator; from red loam soils.

Lycosa marcentior Simon, 1909

Lycosa marcentior Simon, E. (1909). *in* Michaelsen, W. & Hartmeyer, R. (eds.) *Die Fauna Südwest-Australiens.* Jena : Fischer Vol. 2 [185]. Type data: syntypes, ZMH 466 juv., F. adult whereabouts unknown, from Dongara, Boyanup, W.A. Compiled from secondary source: Rack, G. (1961). Die Entomologischen Sammlungen des Zoologischen Staatsinstituts und Zoologischen Museums Hamburg. II. *Mitt. Hamb. Zool. Mus. Inst.* **59**: 1–60.

Distribution: NW coastal, W.A. Ecology: litter, terrestrial, predator. Biological references: McKay, R.J. (1979). The Wolf Spiders of Australia (Araneae : Lycosidae): 12. Descriptions of some Western Australian species. *Mem. Qd. Mus.* **19**: 241–275 (ecology).

Lycosa mayama McKay, 1976

Lycosa mayama McKay, R.J. (1976). The Wolf Spiders of Australia (Araneae : Lycosidae): 7. Two new species from Victoria. *Mem. Qd. Mus.* **17**: 413–416 [413]. Type data: holotype, NMV F. adult, from Werribee, Vic.

Distribution: SE coastal, Vic.; Melbourne. Ecology: terrestrial, predator; from under stones.

Lycosa meracula Simon, 1909

Lycosa meracula Simon, E. (1909). *in* Michaelsen, W. & Hartmeyer, R. (eds.) *Die Fauna Südwest-Australiens.* Jena : Fischer Vol 2. [190]. Type data: holotype, (possibly MNHP) *M. adult, paratypes WAM 1 juv., ZMH 467 F. juv., from Denham, W.A.

Distribution: NW coastal, W.A.; Barrow Is., Bernier Is., Dorre Is., Rosemary Is., (Dampier Archipelago). Ecology: terrestrial, burrowing, predator; from coastal sandy soils near beaches. Biological references: McKay, R.J. (1979). The Wolf Spiders of Australia (Araneae : Lycosidae): 12. Descriptions of some Western Australian species. *Mem. Qd. Mus.* **19**: 241–275 (ecology).

Lycosa molyneuxi Hogg, 1905

Lycosa molyneuxi Hogg, H.R. (1905). On some South Australian spiders of the family Lycosidae. *Proc. Zool. Soc. Lond.* **1905**: 569–590 [575]. Type data: holotype, SAMA F. adult, from Gilbert River, Riverina, N.S.W.

Distribution: Murray-Darling basin, N.S.W. Ecology: terrestrial, predator.

Lycosa festina L. Koch, 1877

Lycosa festina Koch, L. (1877). *in* Koch, L. & Keyserling, E. (1871–1883). *Die Arachniden Australiens.* Nürnberg : Bauer & Raspe Vol. 2 [927 pl 80 figs 3–4]. Type data: syntypes, BMNH 1919.9.18.359–360 M. adult, ZMH (MGH No. 6530) *juv., from Peak Downs, Rockhampton, Qld. Compiled from secondary source: Rack, G. (1961). Die Entomologischen Sammlungen des Zoologischen Staatsinstituts und Zoologischen Museums Hamburg. II. *Mitt. Hamb. Zool. Mus. Inst.* **59**: 1–60.

Distribution: NE coastal, Qld. Ecology: terrestrial, predator.

Lycosa musgravei McKay, 1974

Lycosa musgravei McKay, R.J. (1974). The Wolf Spiders of Australia (Araneae : Lycosidae): 4. Three new species from Mt. Kosciusko, N.S.W. *Mem. Qd. Mus.* **17**: 27–36 [34]. Type data: holotype, AM KS23 F. adult, from Mt. Kosciusko, N.S.W.

Distribution: SE coastal, N.S.W.; Mt. Kosciusko. Ecology: alpine, terrestrial, predator.

Lycosa naeviella (Roewer, 1951)

Lycosa naevia Koch, L. (1878). *in* Koch, L. & Keyserling, E. (1871–1883). *Die Arachniden Australiens.* Nürnberg : Bauer & Raspe Vol. 2 [978 pl 85 fig 5] [*non Lycosa naevia* L. Koch, 1875]. Type data: holotype, ZMH *F. adult (MGH No. 14560), from Sydney, N.S.W. Compiled from secondary source: Rack, G. (1961). Die Entomologischen Sammlungen des Zoologischen Staatsinstituts und Zoologischen Museums Hamburg. II. *Mitt. Hamb. Zool. Mus. Inst.* **59**: 1–60.

Tarentula naeviella Roewer, C.F. (1951). Neue Namen einiger Araneen-Arten. *Abh. Naturw. Ver. Bremen* **32**: 437–456 [442] [*nom. nov.* for *Lycosa naevia* L. Koch, 1878].

Distribution: SE coastal, N.S.W. Ecology: terrestrial, predator.

Lycosa neboissi McKay, 1976

Lycosa neboissi McKay, R.J. (1976). The Wolf Spiders of Australia (Araneae : Lycosidae): 7. Two new species from Victoria. *Mem. Qd. Mus.* **17**: 413–416 [415]. Type data: holotype, NMV F. adult, paratype NMV M. adult, from Rosanna, Vic.

Distribution: SE coastal, Vic. Ecology: terrestrial, predator.

Lycosa neptunus (Rainbow, 1896)

Dolomedes neptunus Rainbow, W.J. (1896). Descriptions of some new Araneidae of New South Wales. No. 6 *Proc. Linn. Soc. N.S.W.* **21**: 320–344 [326]. Type data: holotype, F. adult whereabouts unknown, from shores of Port Jackson, N.S.W.

Distribution: SE coastal, N.S.W. Ecology: supralittoral, terrestrial, predator. Biological references: Rainbow, W.J. (1911). A Census of Australian Araneidae. *Rec. Aust. Mus.* **9**: 107–391 (taxonomy, generic change).

Lycosa nigropunctata Rainbow, 1915

Lycosa nigropunctata Rainbow, R.J. (1915). Arachnida. *Trans. R. Soc. S. Aust.* **39**: 772–793 [787]. Publication date established from Bonnet, P. (1957). *Bibliographia Araneorum.* Toulouse : Douladoure Vol. 2 [2654]. Type data: holotype, SAMA F. adult, from Flat Rock Hole, Musgrave Ranges, S.A.

Distribution: Lake Eyre basin, S.A. Ecology: terrestrial, burrowing, predator.

Lycosa obscurides (Strand, 1906)

Lycosa obscura Koch, L. (1877). *in* Koch, L. & Keyserling, E. (1871–1883). *Die Arachniden Australiens.* Nürnberg : Bauer & Raspe Vol. 2 [954 pl 83 fig 1–2] [*non Lycosa obscura* Blackwall, 1841]. Type data: syntypes, ZMH 3F. adult, from Sydney, N.S.W., Peak Downs, Gayndah, Qld. Compiled from secondary source: Rack, G. (1961). Die Entomologischen Sammlungen des Zoologischen Staatsinstituts und Zoologischen Museums Hamburg. II. *Mitt. Hamb. Zool. Mus. Inst.* **59**: 1–60.

Tarentula obscurides Strand, E. (1906). Die arktischen Araneae, Opiliones und Chernetes. pp. 431–478 *in, Fauna Artica.* Jena : G. Fischer Bd. 4 Lfg. 3 [467] [*nom. nov.* for *Lycosa obscura* L. Koch, 1877].

Distribution: NE coastal, SE coastal, Qld., N.S.W. Ecology: terrestrial, predator. Biological references: Bonnet, P. (1957). *Bibliographia Araneorum.* Toulouse : Douladoure Vol. 2 (taxonomy, generic change).

Lycosa ornatula L. Koch, 1877

Lycosa ornatula Koch, L. (1877). *in* Koch, L. & Keyserling, E. (1871–1883). *Die Arachniden Australiens.* Nürnberg : Bauer & Raspe Vol. 2 [902 pl 77k fig 6]. Type data: syntypes, BMNH 1919.9.18.712 F. adult, ZMH (MGH No. 14555) F. adult, from Rockhampton, Bowen, Qld. Compiled from secondary source: Rack, G. (1961). Die Entomologischen Sammlungen des Zoologischen Staatsinstituts und Zoologischen Museums Hamburg. II. *Mitt. Hamb. Zool. Mus. Inst.* **59**: 1–60.

Distribution: NE coastal, Qld. Ecology: terrestrial, predator.

Lycosa palabunda L. Koch, 1877

Lycosa palabunda Koch, L. (1877). *in* Koch, L. & Keyserling, E. (1871–1883). *Die Arachniden Australiens.* Nürnberg : Bauer & Raspe Vol. 2 [906 pl 78 fig 3–4]. Type data: syntypes, BMNH 1919.9.18.714 M. adult, BMNH 1919.9.18.715–717 F. adult, ZMH (MGH No. 14564) *2 specimens, from Sydney, N.S.W., Gayndah and Rockhampton, Qld. Compiled from secondary source: Rack, G. (1961). Die Entomologischen Sammlungen des Zoologischen Staatsinstituts und Zoologischen Museums Hamburg. II. *Mitt. Hamb. Zool. Mus. Inst.* **59**: 1–60.

Distribution: NE coastal, SE coastal, N coastal, Qld., N.S.W., N.T., W.A. Ecology: terrestrial, predator. Biological references: Mascord, R. (1970). *Australian Spiders in Colour.* Sydney : Reed 112 pp. (biology).

Lycosa percauta Simon, 1909

Lycosa percauta Simon, E. (1909). *in* Michaelsen, W. & Hartmeyer, R. (eds.) *Die Fauna Südwest-Australiens.* Jena : Fischer Vol. 2 [187]. Type data: holotype, F. adult whereabouts unknown, from Rottnest Is., W.A.

Distribution: SW coastal, W.A.; Rottnest Is. Ecology: terrestrial, predator.

Lycosa perinflata Pulleine, 1922

Lycosa perinflata Pulleine, R.H. (1922). Two new species of *Lycosa* from South Australia. *Trans R. Soc. S. Aust.* **46**: 83–84 [84]. Type data: holotype, SAMA F. adult, from Whyte-Yarcowie, S.A.

Distribution: Lake Eyre basin, W plateau, S.A. Ecology: terrestrial, burrowing, predator. Biological references: McKay, R.J. (1973). The Wolf Spiders of Australia (Araneae : Lycosidae): 1. The *bicolor* group. *Mem. Qd. Mus.* **16**: 375–398 (ecology).

Lycosa phyllis Hogg, 1905

Lycosa phyllis Hogg, H.R. (1905). On some South Australian spiders of the family Lycosidae. *Proc. Zool. Soc. Lond.* **1905**: 569–590 [573]. Type data: syntypes, SAMA 197414–15 2F. adult, from Gilbert River, Riverina, N.S.W.

Distribution: Murray-Darling basin, S Gulfs, N.S.W., S.A.; Riverina, Kangaroo Is. Ecology: terrestrial, predator.

Lycosa pictiventris L. Koch, 1877

Lycosa pictiventris Koch, L. (1877). *in* Koch, L. & Keyserling, E. (1871–1883). *Die Arachniden Australiens.* Nürnberg : Bauer & Raspe Vol. 2 [899 pl 77 figs 4–5]. Type data: syntypes, BMNH 1919.9.18.734 F. adult, ZMH (MGH No. 14573) 3 specimens, from Sydney, N.S.W. and Rockhampton, Qld. Compiled from secondary source: Rack, G. (1961). Die Entomologischen Sammlungen des Zoologischen Staatsinstituts und Zoologischen Museums Hamburg. II. *Mitt. Hamb. Zool. Mus. Inst.* **59**: 1–60.

Distribution: SE coastal, NE coastal, N.S.W., Qld. Ecology: terrestrial, predator. Biological references: Mascord, R. (1970). *Australian Spiders in Colour.* Sydney : Reed 112 pp. (biology).

Lycosa propitia Simon, 1909

Lycosa propitia Simon, E. (1909). *in* Michaelsen, W. & Hartmeyer, R. (eds.) *Die Fauna Südwest-Australiens.* Jena : Fischer Vol. 2 [186]. Type data: syntypes, ZMH F. adult, WAM F. adult poor condition, from Cannington, W.A. Compiled from secondary source: Rack, G. (1961). Die Entomologischen Sammlungen des Zoologischen Staatsinstituts und Zoologischen Museums Hamburg. II. *Mitt. Hamb. Zool. Mus. Inst.* **59**: 1–60.

Distribution: SW coastal, Lake Eyre basin, NE coastal, Great Barrier Reef, W.A., S.A., Qld. Ecology: terrestrial, predator; from mudflats, claypans, swamps and offshore islands of the Great Barrier Reef.

Lycosa pruinosa L. Koch, 1877

Lycosa pruinosa Koch, L. (1877). *in* Koch, L. & Keyserling, E. (1871–1883). *Die Arachniden Australiens.* Nürnberg : Bauer & Raspe Vol. 2 [925 pl 80 fig 2]. Type data: holotype, M. adult whereabouts unknown, from Sydney, N.S.W.

Distribution: SE coastal, N.S.W. Ecology: terrestrial, predator.

Lycosa pullastra Simon, 1909

Lycosa pullastra Simon, E. (1909). *in* Michaelsen, W. & Hartmeyer, R. (eds.) *Die Fauna Südwest-Australiens.* Jena : Fischer Vol. 2 [184]. Type data: syntypes, F. adult whereabouts unknown, from Mundaring Weir and Albany, W.A.

Lycosa segregis Simon, E. (1909). *in* Michaelsen, W. & Hartmeyer, R. (eds.) *Die Fauna Südwest-Australiens.* Jena : Fischer Vol. 2 [186]. Type data: holotype, F. adult whereabouts unknown, from Fremantle, W.A.

Taxonomic decision of McKay, R.J. (1974). The Wolf Spiders of Australia (Araneae : Lycosidae): 2. The *arenaris* group. *Mem. Qd. Mus.* **17**: 1–19 [6].

Distribution: SW coastal, W.A. Ecology: terrestrial, burrowing, predator; from grassy areas near streams. Biological references: Main, B.Y. (1976). *Spiders.* Sydney : Collins 296 pp. (biology).

Lycosa pulveresparsa L. Koch, 1877

Lycosa pulveresparsa Koch, L. (1877). *in* Koch, L. & Keyserling, E. (1871–1883). *Die Arachniden Australiens.* Nürnberg : Bauer & Raspe Vol. 2 [941 pl 81 fig 6]. Type data: syntypes, ZMH (MGH No. 14554) F. adult, whereabouts of others unknown, from Rockhampton, Qld. Compiled from secondary source: Rack, G. (1961). Die Entomologischen Sammlungen des Zoologischen Staatsinstituts und Zoologischen Museums Hamburg. II. *Mitt. Hamb. Zool. Mus. Inst.* **59**: 1–60.

Distribution: NE coastal, Qld. Ecology: terrestrial, predator.

Lycosa rainbowi (Roewer, 1951)

Lycosa strenua Rainbow, W.J. (1920). Arachnida from Lord Howe and Norfolk Islands. *Rec. S. Aust. Mus.* **1**: 229–272 [260] [*non Lycosa strenua* Nicolet, 1849]. Type data: holotype, SAMA I.11537 F. adult, from Norfolk Is.

Trochosa rainbowi Roewer, C.F. (1951) Neue Namen einiger Araneen-Arten. *Abh. Naturw. Ver. Bremen* **32**: 437–456 [442] [*nom. nov.* for *Lycosa strenua* Rainbow, 1920].

Taxonomic decision of Roewer, C.F. (1954). *Katalog der Araneae.* Bremen : Schünemann Vol. 2 [272].

Distribution: Norfolk Is. Ecology: terrestrial, predator.

Lycosa ramosa L. Koch, 1877

Lycosa ramosa Koch, L. (1877). *in* Koch, L. & Keyserling, E. (1871–1883). *Die Arachniden Australiens.* Nürnberg : Bauer & Raspe Vol. 2 [910 pl 78 fig 6]. Type data: holotype, F. adult whereabouts unknown, from New Holland - no details of occurrence.

Distribution: Murray-Darling basin, Qld., N.S.W., Vic. Ecology: terrestrial, predator.

Lycosa salifodina McKay, 1976

Lycosa salifodina McKay, R.J. (1976). The Wolf Spiders of Australia (Araneae : Lycosidae): 8. Two new species inhabiting salt lakes of Western Australia. *Mem. Qd. Mus.* **17**: 417–423 [421]. Type data: holotype, WAM 69–369 F. adult, paratypes WAM 69–370–394 3F.,2M. adult, 22 juvs., from Lake Lefroy, Widgiemooltha, W.A.

Distribution: W plateau, W.A. Ecology: salt lake, terrestrial, predator.

Lycosa semicincta L. Koch, 1877

Lycosa semicincta Koch, L. (1877). *in* Koch, L. & Keyserling, E. (1871–1883). *Die Arachniden Australiens.* Nürnberg : Bauer & Raspe Vol. 2 [908 pl 78 fig 5]. Type data: syntypes, BMNH 1919.9.18.950 juv. (Gayndah), ZMH ?F. juv., from Gayndah and Rockhampton, Qld.; F. adult probably Rockhampton now lost, only one juv. in ZMH from Sydney but this locality is doubtful and specimen may not be this species. Compiled from secondary source: Rack, G. (1961). Die Entomologischen Sammlungen des Zoologischen Staatsinstituts und Zoologischen Museums Hamburg. II. *Mitt. Hamb. Zool. Mus. Inst.* **59**: 1–60.

Distribution: NE coastal, Qld. Ecology: terrestrial, tussock grassland, open woodland, predator.

Lycosa senilis L. Koch, 1877

Lycosa senilis Koch, L. (1877). *in* Koch, L. & Keyserling, E. (1871–1883). *Die Arachniden Australiens.* Nürnberg : Bauer & Raspe Vol. 2 [946 pl 82 fig 2]. Type data: syntypes, BMNH 1919.9.18.951 F. adult, ZMH (MGH No. 14570) *2 specimens, from Sydney, N.S.W. and Rockhampton, Qld. Compiled from secondary source: Rack, G. (1961). Die Entomologischen

Sammlungen des Zoologischen Staatsinstituts und Zoologischen Museums Hamburg. II. *Mitt. Hamb. Zool. Mus. Inst.* **59**: 1–60.

Distribution: NE coastal, SE coastal, Qld., N.S.W. Ecology: terrestrial, open woodland, predator.

Lycosa simsoni Simon, 1898

Lycosa simsoni Simon, E. (1898). Descriptions d'Arachnides nouveaux des familles des Agelenidae, Pisauridae, Lycosidae et Oxyopidae. *Ann. Soc. Entomol. Belg.* **42**: 5–34 [29]. Type data: holotype, MNHP 4640 F. adult, from Launceston, Tas.

Distribution: Tas. Ecology: terrestrial, predator. Biological references: Mascord, R. (1970). *Australian Spiders in Colour.* Sydney : Reed 112 pp.; Main, B.Y. (1976). *Spiders.* Sydney : Collins 296 pp. (biology).

Lycosa skeeti Pulleine, 1922

Lycosa skeeti Pulleine, R.H. (1922). Two new species of *Lycosa* from South Australia. *Trans. R. Soc. S. Aust.* **46**: 83–84 [83]. Type data: holotype, SAMA F. adult, from Flinders Ranges, S.A.

Distribution: Lake Eyre basin, W plateau, S.A. Ecology: terrestrial, predator, open woodland, hummock grassland; arid woodland.

Lycosa snelli McKay, 1975

Lycosa snelli McKay, R.J. (1975). The Wolf Spiders of Australia (Araneae : Lycosidae): 5. Two new species of the *bicolor* group. *Mem. Qd. Mus.* **17**: 313–318 [313]. Type data: holotype, WAM 69–797 F. adult, paratypes WAM 74–498–9 2M. adult, from Towera Station, N of Lyndon River, W.A.

Distribution: NW coastal, W.A. Ecology: terrestrial, burrowing, predator; from clay loams. Biological references: Main, B.Y. (1976). *Spiders.* Sydney : Collins 296 pp. (biology).

Lycosa speciosa L. Koch, 1877

Lycosa speciosa Koch, L. (1877). *in* Koch, L. & Keyserling, E. (1871–1883). *Die Arachniden Australiens.* Nürnberg : Bauer & Raspe Vol. 2 [890 pl 77 fig 1]. Type data: holotype, M. adult whereabouts unknown, from Caigan, N.S.W.

Distribution: SE coastal, Murray-Darling basin, NE coastal, N.S.W., Vic., S.A., Qld. Ecology: terrestrial, predator.

Lycosa spinipes (Rainbow, 1896)

Dolomedes spinipes Rainbow, W.J. (1896). Descriptions of some new Araneidae of New South Wales. No. 6 *Proc. Linn. Soc. N.S.W.* **21**: 320–344 [327]. Type data: holotype, F. adult whereabouts unknown, from Port Jackson, N.S.W.

Distribution: SE coastal, N.S.W. Ecology: terrestrial, predator. Biological references: Rainbow, W.J. (1911). A Census of Australian Araneidae. *Rec. Aust. Mus.* **9**: 107–391 (taxonomy, generic change).

Lycosa stirlingae Hogg, 1905

Lycosa stirlingae Hogg, H.R. (1905). On some South Australian spiders of the family Lycosidae. *Proc. Zool. Soc. Lond.* **1905**: 569–590 [584]. Type data: syntypes, M.,F. adult whereabouts unknown, from Gilbert River, Riverina, N.S.W.

Distribution: Lake Eyre basin, W plateau, S Gulfs, N.S.W., S.A., W.A. Ecology: terrestrial, predator.

Lycosa storri McKay, 1973

Lycosa storri McKay, R.J. (1973). The Wolf Spiders of Australia (Araneae : Lycosidae): 1. The *bicolor* group. *Mem. Qd. Mus.* **16**: 375–398 [389]. Type data: holotype, WAM 70–240 F. adult, paratype WAM 39–2564 M. adult, from 38 mi S of Yellowdine, W.A.

Distribution: W plateau, W.A.; arid and semi-arid regions. Ecology: terrestrial, burrowing, predator; from clay or desert loam soils. Biological references: Main, B.Y. (1976). *Spiders.* Sydney : Collins 296 pp. (biology).

Lycosa subligata (L. Koch, 1877)

Pirata subligata Koch, L. (1877). *in* Koch, L. & Keyserling, E. (1871–1883). *Die Arachniden Australiens.* Nürnberg : Bauer & Raspe Vol. 2 [963 pl 84 fig 2]. Type data: holotype, ZMH *F. adult, from Bowen, Qld. Compiled from secondary source: Rack, G. (1961). Die Entomologischen Sammlungen des Zoologischen Staatsinstituts und Zoologischen Museums Hamburg. II. *Mitt. Hamb. Zool. Mus. Inst.* **59**: 1–60.

Distribution: NE coastal, Qld. Ecology: terrestrial, predator. Biological references: Rainbow, W.J. (1911). A Census of Australian Araneidae. *Rec. Aust. Mus.* **9**: 107–391 (taxonomy, generic change); Mascord, R. (1970). *Australian Spiders in Colour.* Sydney : Reed 112 pp. (biology).

Lycosa subrufa Karsch, 1878

Lycosa subrufa Karsch, F. (1878). Exotischaraneologisches. *Z. Ges. Naturw.* **51**: 771–826 [813]. Publication date established from Bonnet, P. (1957). *Bibliographia Araneorum.* Toulouse : Douladoure Vol. 2 [2666]. Type data: holotype, ZMB *F. adult, from Tas.

Distribution: Tas. Ecology: terrestrial, predator.

Lycosa summa McKay, 1974

Lycosa summa McKay, R.J. (1974). The Wolf Spiders of Australia (Araneae : Lycosidae): 4. Three new species from Mt. Kosciusko, N.S.W. *Mem. Qd. Mus.* **17**: 27–36 [28]. Type data: holotype, AM K61669 F. adult, paratypes AM K61699 F. adult, juv., from Mt. Kosciusko, N.S.W.

Distribution: SE coastal, N.S.W.; Mt. Kosciusko. Ecology: alpine, terrestrial, burrowing, predator.

Lycosa tasmanica Hogg, 1905

Lycosa tasmanica Hogg, H.R. (1905). On some South Australian spiders of the family Lycosidae. *Proc. Zool. Soc. Lond.* **1905**: 569–590 [571]. Type data: holotype, SAMA F. adult, from Table Cape, Tas.

Distribution: Tas. Ecology: terrestrial, burrowing, predator. Biological references: Main, B.Y. (1976). *Spiders.* Sydney : Collins 296 pp. (biology).

Lycosa tasmanicola Roewer, 1960

Lycosa festiva Urquhart, A.T. (1893). On new species of Tasmanian Araneae. *Pap. Proc. R. Soc. Tasm.* **1892**: 94–130 [126] [*non Lycosa festiva* Parvesi, 1880]. Type data: holotype, CMNZ lost, from Tas.

Tarentula tasmanica Roewer, C.F. (1951) Neue Namen einiger Araneen-Arten. *Abh. Naturw. Ver. Bremen* **32**: 437–456 [441] [*nom. nov.* for *Lycosa festiva* Urquhart, 1893; *non Lycosa tasmanica* Hogg, 1905].

Lycosa tasmanicola Roewer, C.F. (1960). Araneae Lycosiformia 2. (Lycosidae). *Explor. Parc. Natl. Upemba* Fasc. 55 pp. 519–1040 [540] [*nom. nov.* for *Tarentula tasmanica* Roewer, 1951].

Distribution: Tas. Ecology: terrestrial, predator.

Lycosa topaziopsis Hogg, 1896

Lycosa topaziopsis Hogg, H.R. (1896). Araneidae. pp. 309–356 *Rep. Horn Expedition to Central Australia.* Pt. 2 Zoology [347]. Type data: holotype, F. adult whereabouts unknown, from Stevenson River, S.A.

Distribution: Lake Eyre basin, S.A. Ecology: terrestrial, predator.

Lycosa tula (Strand, 1913)

Tarentula tula Strand, E. (1913). Über einige australische Spinnen des Senckenbergischen Museums. *Zool. Jb. Syst.* **35**: 599–624 [622]. Type data: lectotype, SMF 2769 M. adult, designation by McKay, R.J. (1979). The Wolf Spiders of Australia (Araneae : Lycosidae): 12. Descriptions of some Western Australian species. *Mem. Qd. Mus.* **19**: 241–275, from central Australia.

Distribution: Lake Eyre basin, W plateau, S.A., W.A. Ecology: terrestrial, predator; from red soils with sparse vegetation. Biological references: McKay, R.J. (1979). The Wolf Spiders of Australia (Araneae : Lycosidae): 12. Descriptions of some Western Australian species. *Mem. Qd. Mus.* **19**: 241–275 (ecology, taxonomy, generic change).

Lycosa waitei Rainbow, 1917

Lycosa waitei Rainbow, W.J. (1917). Araneidae. Results of the South Australian Museum Expedition to Strzelecki and Cooper Creeks. *Trans. R. Soc. S. Aust.* **41**: 482–489 [487]. Type data: holotype, SAMA F. adult, from Coopers Creek, S.A.

Distribution: S Gulfs, S.A. Ecology: terrestrial, predator.

Lycosa woodwardi Simon, 1909

Lycosa woodwardi Simon, E. (1909). *in* Michaelsen, W. & Hartmeyer, R. (eds.) *Die Fauna Südwest-Australiens.* Jena : Fischer Vol. 2 [182]. Type data: syntypes, WAM F. adult labelled Hamb. S.W. Austral Exp. 1905. 83 Dongarra, ZMH? *F. adult, from Northampton, Dongara, Beverley, W.A. Compiled from secondary source: Rack, G. (1961). Die Entomologischen Sammlungen des Zoologischen Staatsinstituts und Zoologischen Museums Hamburg. II. *Mitt. Hamb. Zool. Mus. Inst.* **59**: 1–60.

Distribution: NW coastal, W.A. Ecology: terrestrial, predator.

Lycosa woonda McKay, 1979

Lycosa woonda McKay, R.J. (1979). The Wolf Spiders of Australia (Araneae : Lycosidae): 12. Descriptions of some Western Australian species. *Mem. Qd. Mus.* **19**: 241–275 [269]. Type data: holotype, QM S34 M. adult, paratypes QM S35 2F. adult, from Albion Downs via Wiluna, W.A.

Distribution: W plateau, W.A. Ecology: terrestrial, predator; from desert loam soils.

Lycosa yalkara McKay, 1979

Lycosa yalkara McKay, R.J. (1979). The Wolf Spiders of Australia (Araneae : Lycosidae): 12. Descriptions of some Western Australian species. *Mem. Qd. Mus.* **19**: 241–275 [271]. Type data: holotype, QM S40 M. adult, paratype QM S44 F. adult, from 40 km S of Mt. Magnet, W.A.

Distribution: W plateau, W.A. Ecology: terrestrial, predator; from desert loam soils.

Lycosa zualella (Strand, 1907)

Tarentula zualella Strand, E. (1907). Einige Spinnen aus Kamerun, Java und Australien. *Jb. Nassau. Ver. Naturk.* **60**: 177–219 [219]. Publication date established from Bonnet, P. (1957). *Bibliographia Araneorum.* Toulouse : Douladoure Vol. 2 [2631]. Type data: holotype, ZMB M. adult, from Sydney, N.S.W. Compiled from secondary source: Bonnet, P. (1945–1961). *Bibliographia Araneorum.* Toulouse : Douladoure 3 vols.

Distribution: SE coastal, N.S.W. Ecology: terrestrial, predator.

Pardosa C.L. Koch, 1848

Pardosa Koch, C.L. (1848). *Die Arachniden.* Nürnberg Vol. 14 [100]. Type species *Lycosa hortensis* Thorell, 1872 by subsequent designation, see Bonnet, P. (1951). Difficultés de nomenclature chez les aranéides. *Bull. Soc. Hist. Nat. Toulouse* **86**: 295–307.

This group is found worldwide, see Bonnet, P. (1958). *Bibliographia Araneorum.* Toulouse : Douladoure Vol. 2 [3341].

Pardosa pexa Hickman, 1944

Pardosa pexa Hickman, V.V. (1944). The Simpson Desert Expedition, 1939. Scientific Report 1. Biology —

Scorpions and spiders. *Trans. R. Soc. S. Aust.* **68**: 18–48 [25]. Type data: holotype, AM K67696 M. adult, from Burt's Waterhole, 55 mi S of Birdsville, S.A.

Distribution: Lake Eyre basin, S.A. Ecology: terrestrial, predator.

Pardosa serrata (L. Koch, 1877)

Lycosa serrata Koch, L. (1877). *in* Koch, L. & Keyserling, E. (1871–1883). *Die Arachniden Australiens.* Nürnberg : Bauer & Raspe Vol. 2 [930 pl 80 figs 5–6]. Type data: syntypes, F.,M. adult whereabouts unknown, from Sydney, N.S.W.

Pardosa praevelox Simon, E. (1909). *in* Michaelsen, W. & Hartmeyer, R. (eds.) *Die Fauna Südwest-Australiens.* Jena : Fischer Vol. 2 [191]. Type data: holotype, F. adult whereabouts unknown, from Buckland Hill, near Fremantle, W.A.

Taxonomic decision of McKay, R.J. (1979). The Wolf Spiders of Australia (Araneae : Lycosidae): 9. *Pardosa serrata* (L. Koch, 1877). *Mem. Qd. Mus.* **19**: 225–229 [225].

Distribution: SE coastal, SW coastal, S Gulfs, N.S.W., S.A., W.A. Ecology: terrestrial, burrowing, predator; from clay, gravel and sandy soils on coastal heathlands. Biological references: Mascord, R. (1970). *Australian Spiders in Colour.* Sydney : Reed 112 pp. (biology).

Pardosa versicolor L. Koch, 1877

Pardosa versicolor Koch, L. (1877). *in* Koch, L. & Keyserling, E. (1871–1883). *Die Arachniden Australiens.* Nürnberg : Bauer & Raspe Vol. 2 [966A pl 84 figs 4–5]. Type data: syntypes, F.,M. adult whereabouts unknown, from Sydney, N.S.W.

Distribution: SE coastal, N.S.W. Ecology: terrestrial, predator.

Trabaeosa Roewer, 1960

Trabaeosa Roewer, (1960). Araneae Lycosaeformia. 2. (Lycosidae). *Explor. Parc. Natnl. Upemba* Fasc. 55 pp. 519–1040 [582]. Type species *Trabea lineata* L. Koch, 1877 by original designation.

Trabaeosa australiensis (L. Koch, 1877)

Trabaea australiensis Koch, L. (1877). *in* Koch, L. & Keyserling, E. (1871–1883). *Die Arachniden Australiens.* Nürnberg : Bauer & Raspe Vol. 2 [968 pl 84 fig 6] [as *Trabea Australiensis*]. Type data: holotype, juv. whereabouts unknown, from Shelley's Flats, N.S.W.

Distribution: SE coastal, N.S.W. Ecology: terrestrial, predator. Biological references: Russell-Smith, A. 1982. A revision of the genus *Trabaea* Simon (Araneae : Lycosidae). *Zool. J. Linn. Soc.* **74**: 69–91.

Trabaeosa lineata (L. Koch, 1877)

Trabaea lineata Koch, L. (1877). *in* Koch, L. & Keyserling, E. (1871–1883). *Die Arachniden Australiens.* Nürnberg : Bauer & Raspe Vol. 2 [970 pl 84 fig 7] [as *Trabea lineata* L. Koch, 1877]. Type data: holotype, whereabouts unknown, from Sydney, N.S.W.

Distribution: SE coastal, N.S.W. Ecology: terrestrial, predator. Biological references: Russell-Smith, A. 1982. A revision of the genus *Trabaea* Simon (Araneae : Lycosidae). *Zool. J. Linn. Soc.* **74**: 69–91.

Trochosa C.L. Koch, 1848

Trochosa Koch, C.L. (1848). *Die Arachniden.* Nürnberg Vol. 14 [95]. Publication date established from Bonnet, P. (1959). *Bibliographia Araneorum.* Toulouse : Douladoure Vol. 2 [4698]. Type species *Aranea ruricola* de Geer, 1778 by original designation.

Diahogna Roewer, C.F. (1954). *Katalog der Araneae.* Bruxelles : Inst. R. Sci. Nat. Belg. Bd. 2a 1–923 [239]. Type species *Lycosa martensii* Karsch, 1878 by monotypy and original designation. Compiled from secondary source: Roewer, C.F. (1954). *Katalog der Araneae.* Bremen : Schünemann Vol. 2.

Taxonomic decision of McKay, R.J. (1979). The Wolf Spiders of Australia (Araneae : Lycosidae): 13. The genus *Trochosa. Mem. Qd. Mus.* **19**: 277–298 [277].

This group is found worldwide, see Bonnet, P. (1959). *Bibliographia Araneorum.* Toulouse : Douladoure Vol. 2 [4698].

Trochosa alboguttulata (L. Koch, 1878)

Lycosa alboguttulata Koch, L. (1878). *in* Koch, L. & Keyserling, E. (1871–1883). *Die Arachniden Australiens.* Nürnberg : Bauer & Raspe Vol. 2 [975 pl 85 fig 3] [listed as *Lycosa alboguttata* by Rainbow, W.J. (1911). A Census of Australian Araneidae. *Rec. Aust. Mus.* **9**: 107–391 (265)]. Type data: holotype, ZMH (MGH No. 14558) *F. adult, from Bowen, Qld. Compiled from secondary source: Rack, G. (1961). Die Entomologischen Sammlungen des Zoologischen Staatsinstituts und Zoologischen Museums Hamburg. II. *Mitt. Hamb. Zool. Mus. Inst.* **59**: 1–60.

Distribution: NE coastal, Qld. Ecology: terrestrial, predator; from coastal sandy soils. Biological references: McKay, R.J. (1979). The Wolf Spiders of Australia (Araneae : Lycosidae): 13. The genus *Trochosa. Mem. Qd. Mus.* **19**: 277–298 (ecology, taxonomy, generic change).

Trochosa candicans (L. Koch, 1877)

Lycosa candicans Koch, L. (1877).*in* Koch, L. & Keyserling, E. (1871–1883). *Die Arachniden Australiens.* Nürnberg : Bauer & Raspe Vol. 2 [888 pl 76 figs 5–6]. Type data: syntypes, F.,M. adult whereabouts unknown, from Sydney N.S.W.

Distribution: SE coastal, S Gulfs, N.S.W., S.A. Ecology: terrestrial, predator. Biological references: McKay, R.J. (1979). The Wolf Spiders of Australia (Araneae : Lycosidae): 13. The genus *Trochosa. Mem. Qd. Mus.* **19**: 277–298 (taxonomy, generic change).

Trochosa exculta (L. Koch, 1876)

Lycosa exculta Koch, L. (1876). *in* Koch, L. & Keyserling, E. (1871–1883). *Die Arachniden Australiens.* Nürnberg : Bauer & Raspe Vol. 2 [881 pl 76 fig 1]. Type data: lectotype, BMNH 1919.9.18.333 F. adult, possible paralectotypes ZMH *F.,M. adult, designation by McKay, R.J. (1979). The Wolf Spiders of Australia (Araneae : Lycosidae): 13. The genus *Trochosa. Mem. Qd. Mus.* **19**: 277–298, from Sydney, N.S.W.

Distribution: SE coastal, NE coastal, N.S.W., Qld. Ecology: terrestrial, predator. Biological references: Koch, L. *in* Koch, L. & Keyserling, E. (1871–1883). *Die Arachniden Australiens.* Nürnberg : Bauer & Raspe Vol. 2 (ecology); Mascord, R. (1970). *Australian Spiders in Colour.* Sydney : Reed 112 pp. (biology); McKay, R.J. (1979). The Wolf Spiders of Australia (Araneae : Lycosidae): 13. The genus *Trochosa. Mem. Qd. Mus.* **19**: 277–298 (taxonomy, generic change).

Trochosa expolita (L. Koch, 1877)

Trochosa expolita expolita (L. Koch, 1877)

Lycosa expolita Koch, L. (1877).*in* Koch, L. & Keyserling, E. (1871–1883). *Die Arachniden Australiens.* Nürnberg : Bauer & Raspe Vol. 2 [917 pl 79 figs 4–5]. Type data: syntypes, BMNH 1919.9.18.337 M. adult, whereabouts of others unknown, from Brisbane, Port Denison (Bowen), Qld.

Distribution: NE coastal, SE coastal, Qld., Vic. Ecology: tussock grassland, terrestrial, burrowing, predator. Biological references: McKay, R.J. (1979). The Wolf Spiders of Australia (Araneae : Lycosidae): 13. The genus *Trochosa. Mem. Qd. Mus.* **19**: 277–298 (ecology, taxonomy, generic change).

Trochosa expolita impedita (Simon, 1909)

Lycosa impedita Simon, E. (1909). *in* Michaelsen, W. & Hartmeyer, R. (eds.) *Die Fauna Südwest-Australiens.* Jena : Fischer Vol. 2 [187]. Type data: holotype, F. adult whereabouts unknown, from Gooseberry Hill, W.A.

Distribution: SW coastal, S Gulfs, W.A., S.A. Ecology: terrestrial, predator; grassland. Biological references: McKay, R.J. (1979). The Wolf Spiders of Australia (Araneae : Lycosidae): 13. The genus *Trochosa. Mem. Qd. Mus.* **19**: 277–298 (ecology, taxonomy, generic change).

Trochosa glarea McKay, 1979

Trochosa glarea McKay, R.J. (1979). The Wolf Spiders of Australia (Araneae : Lycosidae): 13. The genus *Trochosa. Mem. Qd. Mus.* **19**: 277–298 [296]. Type data: holotype, QM S25 F. adult, from Boonooroo, Qld.

Distribution: NE coastal, Qld. Ecology: terrestrial, predator.

Trochosa martensii (Karsch, 1878)

Lycosa martensii Karsch, F. (1878). Exotischaraneologisches. *Z. Ges. Naturw.* **51**: 771–826 [812]. Publication date established from Bonnet, P. (1957). *Bibliographia Araneorum.* Toulouse : Douladoure Vol. 2 [2652]. Type data: holotype, ZMB 756 *juv., from Vic.

Distribution: SE coastal, Vic. Ecology: terrestrial, predator. Biological references: McKay, R.J. (1979). The Wolf Spiders of Australia (Araneae : Lycosidae): 13. The genus *Trochosa. Mem. Qd. Mus.* **19**: 277–298 (ecology, taxonomy, generic change).

Trochosa oraria (L. Koch, 1876)

Lycosa oraria Koch, L. (1876). *in* Koch, L. & Keyserling, E. (1871–1883). *Die Arachniden Australiens.* Nürnberg : Bauer & Raspe Vol. 2 [883 pl 76 figs 2–3]. Type data: syntypes, F.,M. adult whereabouts unknown, from King George Sound, W.A.

Lycosa sibyllina Simon, E. (1909). *in* Michaelsen, W. & Hartmeyer, R. (eds.) *Die Fauna Südwest-Australiens.* Jena : Fischer Vol. 2 [188]. Type data: holotype, F. adult whereabouts unknown, from Albany, W.A.

Taxonomic decision of McKay, R.J. (1979). The Wolf Spiders of Australia (Araneae : Lycosidae): 13. The genus *Trochosa. Mem. Qd. Mus.* **19**: 277–298 [279].

Distribution: SW coastal, W.A. Ecology: supralittoral, terrestrial, burrowing, predator.

Trochosa properipes (Simon, 1909)

Lycosa properipes Simon, E. (1909). *in* Michaelsen, W. & Hartmeyer, R. (eds.) *Die Fauna Südwest-Australiens.* Jena : Fischer Vol. 2 [189]. Type data: holotype, M. adult whereabouts unknown, from "Guildford and Subiaco", W.A.

Distribution: SW coastal, W.A. Ecology: terrestrial, predator. Biological references: McKay, R.J. (1979). The Wolf Spiders of Australia (Araneae : Lycosidae): 13. The genus *Trochosa. Mem. Qd. Mus.* **19**: 277–298 (taxonomy, generic change).

Trochosa tristicula (L. Koch, 1877)

Trochosa tristicula tristicula (L. Koch, 1877)

Lycosa tristicula Koch, L. (1877). *in* Koch, L. & Keyserling, E. (1871–1883). *Die Arachniden Australiens.* Nürnberg : Bauer & Raspe Vol. 2 [892 pl 77 fig 2]. Type data: holotype, F. adult whereabouts unknown, from Sydney, N.S.W.

Distribution: NE coastal, SE coastal, Qld., N.S.W. Ecology: closed forest, terrestrial, predator. Biological references: McKay, R.J. (1979). The Wolf Spiders of Australia (Araneae : Lycosidae): 13. The genus *Trochosa. Mem. Qd. Mus.* **19**: 277–298 (ecology, taxonomy, generic change).

Trochosa tristicula phegea (Simon, 1909)

Lycosa phegea Simon, E. (1909). *in* Michaelsen, W. & Hartmeyer, R. (eds.) *Die Fauna Südwest-Australiens.* Jena : Fischer Vol. 2 [189]. Type data: syntypes, WAM M. adult in spirit, ZMH 473 F. adult, from Cannington, W.A.

Distribution: SW coastal, W.A. Ecology: swamp, terrestrial, burrowing, predator. Biological references: McKay, R.J. (1979). The Wolf Spiders of Australia (Araneae : Lycosidae): 13. The genus *Trochosa. Mem. Qd. Mus.* **19**: 277–298 (ecology, taxonomy, generic change).

Trochosa wundurra McKay, 1979

Trochosa wundurra McKay, R.J. (1979). The Wolf Spiders of Australia (Araneae : Lycosidae): 13. The genus *Trochosa. Mem. Qd. Mus.* **19**: 277–298 [296]. Type data: holotype, QM S96 F. adult, from Hyden Lake, Hyden, W.A.

Distribution: W plateau, W.A. Ecology: terrestrial, predator; from sandy ground.

Venator Hogg, 1900

Venator Hogg, H.R. (1900). A contribution to our knowledge of the spiders of Victoria; including some new species and genera. *Proc. R. Soc. Vict.* **13**: 68–123 [115]. Type species *Venator spenceri* Hogg, 1900 by original designation.

Venatrix Roewer, C.F. (1954). *Katalog der Araneae.* Bruxelles : Inst. R. Sci. Nat. Belg. Bd. 2a 1–923 [307]. Type species *Venator fusca* Hogg, 1900 by original designation.

Taxonomic decision of McKay, R.J., this work.

Venator fuscus Hogg, 1900

Venator fuscus Hogg, H.R. (1900). A contribution to our knowledge of the spiders of Victoria; including some new species and genera. *Proc. R. Soc. Vict.* **13**: 68–123 [118]. Type data: holotype, BMNH 1907.2.24.5 F. adult, from Macedon, Vic.

Distribution: SE coastal, Vic. Ecology: terrestrial, predator.

Venator marginatus Hogg, 1900

Venator marginatus Hogg, H.R. (1900). A contribution to our knowledge of the spiders of Victoria; including some new species and genera. *Proc. R. Soc. Vict.* **13**: 68–123 [120]. Type data: holotype and paratypes, BMNH 1907.2.24.7–9 F. adult, from Macedon, Vic.

Distribution: SE coastal, Vic. Ecology: terrestrial, predator.

Venator spenceri Hogg, 1900

Venator spenceri Hogg, H.R. (1900). A contribution to our knowledge of the spiders of Victoria; including some new species and genera. *Proc. R. Soc. Vict.* **13**: 68–123 [116]. Type data: holotype, BMNH 1907.2.24.6 F. adult, from Macedon, Vic.

Distribution: SE coastal, Vic. Ecology: terrestrial, predator; sandy open woodlands.

Venonia Thorell, 1894

Venonia Thorell, T. (1894). Decas Aranearum in ins. Singapore a Cel. Th. Workman inventarum. *Boll. Soc. Entomol. Ital.* **26**: 321–355 [332]. Type species *Venonia coruscans* Thorell, 1894 by monotypy. Compiled from secondary source: Bonnet, P. (1959). *Bibliographia Araneorum.* Toulouse : Douladoure Vol. 2.

This group is also found in Singapore and India, see Bonnet, P. (1959). *Bibliographia Araneorum.* Toulouse : Douladoure Vol. 2 [4788].

Venonia micarioides (L. Koch, 1877)

Aulonia micarioides Koch, L. (1877). *in* Koch, L. & Keyserling, E. (1871–1883). *Die Arachniden Australiens.* Nürnberg : Bauer & Raspe Vol. 2 [961 pl 83 fig 6, pl 84 fig 1]. Type data: lectotype, ZMH *F. adult, paralectotype ZMH (MGH No. 14624) M. adult, designation by Lehtinen, P.T. & Hippa, H. (1979). Spiders of the Oriental-Australian region. 1. Lycosidae : Venoniinae and Zoicinae. *Ann. Zool. Fenn.* **16**: 1–22, from Port MacKay, Qld.

Distribution: NE coastal, Qld. Ecology: terrestrial, predator. Biological references: Simon, E. (1892–1903). *Histoire Naturelle des Araignées.* 2nd edn. Paris : Roret 2 vols (taxonomy, generic change).

Zoica Simon, 1898

Zobia Thorell, T. (1895). *Descriptive Catalogue of the Spiders of Burma.* London : British Museum 406 pp. [54] [*non Zobia* Saalmueller, 1891]. Publication date established from Bonnet, P. (1945–1961). *Bibliographia Araneorum.* Toulouse : Douladoure Vol. 2 [4981]. Type species *Zobia parvula* Thorell, 1895 by monotypy.

Zoica Simon, E. (1898). *Histoire Naturelle des Araignées.* 2nd edn. Paris : Roret Vol. 2 [248]. Type species *Zobia parvula* Thorell, 1895 by original designation.

Flanona Simon, E. (1898). *Histoire Naturelle des Araignées.* 2nd edn. Paris : Roret Vol. 2 [349]. Type species *Flanona puellula* Simon, 1898 by original designation.

Taxonomic decision of Lehtinen, P.T. & Hippa, H. (1979). Spiders of the Oriental-Australian region. 1. Lycosidae : Venoniinae and Zoicinae. *Ann. Zool. Fenn.* **16**: 1–22 [15].

This group is also found in Sri Lanka, Thailand, Burma, Malaysia, Papua New Guinea and Indonesia, see Lehtinen, P.T. & Hippa, H. (1979). Spiders of the Oriental-Australian region. 1. Lycosidae : Venoniinae and Zoicinae. *Ann. Zool. Fenn.* **16**: 1–22.

Zoica minuta (McKay, 1979)

Flanona minuta McKay, R.J. (1979). Thc Wolf Spiders of Australia (Araneae : Lycosidae): 10. A new species of the genus *Flanona* Simon. *Mem. Qd. Mus.* **19**: 231–235 [231]. Type data: holotype, WAM 76–98 F. adult, paratypes QM W5810 17F. adult, 16 juv., from Mound Spring, 5 km SW of Old Lissadel Station Homestead, Ord River area, W.A.

Distribution: NW coastal, W.A. Ecology: litter, terrestrial, predator.

MICROPHOLCOMMATIDAE

INTRODUCTION

Minute, three-clawed spiders without book-lungs. Dorsal abdominal scutum; epigastric scutum extending round pedicel and sclerotized ring round spinnerets in both male and female. Colulus present. Eight eyes. Female palp vestigial or with less than six segments. Male palp with patellar apophysis. Tarsi longer than metatarsi, no tarsal or metatarsal trichobothria. Anterior spiracles (tracheae to cephalothorax and abdomen); posterior spiracle (with unbranched tracheae to abdomen). Builds small irregular webs.

Micropholcomma Crosby and Bishop, 1927

Micropholcomma Crosby, C.R. & Bishop, S.C. (1927). New species of Erigoneae and Theridiidae. *J. N.Y. Entomol. Soc.* **35**: 147–153 [152]. Type species *Micropholcomma caeligenum* Crosby and Bishop, 1927 by monotypy and original designation.

Microlinypheus Butler, L.S.G. (1932). Studies in Australian spiders. No. 2. *Proc. R. Soc. Vict. (ns)* **44**: 103–117 [103]. Type species *Microlinypheus bryophilus* Butler, 1932 by monotypy and original designation.

Plectochetos Butler, L.S.G. (1932). Studies in Australian spiders. No. 2. *Proc. R. Soc. Vict. (ns)* **44**: 103–117 [106]. Type species *Plectochetos longissimus* Butler, 1932 by monotypy and original designation.

Taxonomic decision of Forster, R.R. (1959). Spiders of the family Symphytognathidae. *Trans. R. Soc. N.Z.* **86**: 269–329 [297].

Micropholcomma bryophilum (Butler, 1932)

Microlinypheus bryophilus Butler, L.S.G. (1932). Studies in Australian spiders. No. 2. *Proc. R. Soc. Vict. (ns)* **44**: 103–117 [104]. Type data: syntypes, NMV K101 *F. adult, NMV K100 *M. adult, from Lorne, Vic.

Distribution: SE coastal, NE coastal, Vic., N.S.W., Qld., Tas. Ecology: predator, closed forest. Biological references: Hickman, V.V. (1944). On some new Australian Apneumonomorphae with notes on their respiratory system. *Pap. Proc. R. Soc. Tasm.* **1943**: 179–195 (anatomy, taxonomy).

Micropholcomma caeligenum Crosby and Bishop, 1927

Micropholcomma caeligenum Crosby, C.R. & Bishop, S.C. (1927). New species of Erigoneae and Theridiidae. *J. N.Y. Entomol. Soc.* **35**: 147–153 [152]. Type data: holotype, CU *M. adult, from Emerald, Vic.

Distribution: SE coastal, NE coastal, Vic., N.S.W., Qld., Tas. Ecology: predator; in moss. Biological references: Hickman, V.V. (1944). On some new Australian Apneumonomorphae with notes on their respiratory system. *Pap. Proc. R. Soc. Tasm.* **1943**: 179–195 (anatomy).

Micropholcomma longissimum (Butler, 1932)

Plectochetos longissimus Butler, L.S.G. (1932). Studies in Australian spiders. No. 2. *Proc. R. Soc. Vict. (ns)* **44**: 103–117 [107]. Type data: holotype, NMV K103 *M. adult, from Mt. Donna Buang, Vic.

Distribution: SE coastal, NE coastal, N.S.W., Vic., Qld. Ecology: predator.

Micropholcomma mirum Hickman, 1944

Micropholcomma mira Hickman, V.V. (1944). On some new Australian Apneumonomorphae with notes on their respiratory system. *Pap. Proc. R. Soc. Tasm.* **1943**: 179–195 [191]. Type data: syntypes, AM KS6620 *2M.,F. adult, from New Town, Hobart, Tas.

Distribution: Tas. Ecology: predator; in moss.

Micropholcomma parmatum Hickman, 1944

Micropholcomma parmata Hickman, V.V. (1944). On some new Australian Apneumonomorphae with notes on their respiratory system. *Pap. Proc. R. Soc. Tasm.* **1943**: 179–195 [188]. Type data: syntypes, AM KS6621 *2M.,F. adult, from Mt. Wellington, Tas.

Distribution: Tas. Ecology: sedentary, predator; in moss.

Micropholcomma turbans Hickman, 1981

Micropholcomma turbans Hickman, V.V. (1981). New Tasmanian spiders of the families Archaeidae, Cycloctenidae, Amaurobiidae and Micropholcommatidae. *Pap. Proc. R. Soc. Tasm.* **115**: 47–68 [65]. Type data: holotype, AM KS6997 M. adult, paratypes AM KS6998 F. adult, TMH J1890 2M.,2F. adult, from 145°56′E, 42°37′S, Gordon River Valley, SW Tas.

Distribution: Tas. Ecology: predator, litter; in moss.

MIMETIDAE

INTRODUCTION

The robber or pirate spiders, three-clawed spiders that enter the webs of other spiders to prey on them. A row of large curved spines interspersed with six to seven smaller spines along the prolateral margin of tibiae and metatarsi I and II. Chelicerae long, slender, usually fused at the base. Colulus present. One trichobothrium on metatarsi, none on tarsi. S. Heimer (pers. comm.) is currently revising the Australian Mimetidae; he believes that *Arcys* belongs in this family.

Ero C.L. Koch, 1836

Ero Koch, C.L. (1836). Arachniden *in* Panzer, G.W.F. *Faunae Insectorum Germaniae Initia.* Fortgesetzt von Herrich-Schäffer. Hefte 111–190. Regensburg [Heft 138 pl 4]. Publication date established from Bonnet, P. (1945–1961). *Bibliographia Araneorum.* Toulouse : Douladoure Vol. 2. Type species *Aranea tuberculata* De Geer, 1788 by monotypy.

This group is found worldwide, see Bonnet, P. (1945–1961). *Bibliographia Araneorum.* Toulouse : Douladoure Vol. 2.

Ero aphana (Walckenaer, 1802)

Aranea aphana Walckenaer, C.A. (1802). *Faune Parisienne.* Insectes. ou Histoire abrégée des insectes des environs de Paris. Paris Vol. 2 [Araignées, pp. 187–250] [206]. Publication date established from Bonnet, P. (1945–1961). *Bibliographia Araneorum.* Toulouse : Douladoure Vol. 2. Type data: syntypes, ?MNHP M.,F. adult, from Paris, France.

Distribution: NE coastal, Qld. Ecology: nomadic, predator.

Ero tasmaniensis Hickman, 1929

Ero tasmaniensis Hickman, V.V. (1929). Studies in Tasmanian spiders. Pt. III *Pap. Proc. R. Soc. Tasm.* **1928**: 96–118 [114] [emend. *tasmanensis* is regarded as an invalid emend. under Art. 32(a)(ii) of ICZN]. Type data: syntypes, QVM 39 *3M.,F. adult, from Trevallyn, Launceston, Tas.

Distribution: NE coastal, SE coastal, Qld., N.S.W., Tas. Ecology: nomadic, predator.

Mimetus Hentz, 1832

Mimetus Hentz, N.M. (1832). On North American spiders. *Am. J. Sci.* **21**: 99–109 [104]. Type species *Mimetus interfector* Hentz, 1850 by original designation.

This group is found worldwide, see Bonnet, P. (1945–1961). *Bibliographia Araneorum.* Toulouse : Douladoure Vol. 2.

Mimetus audax Hickman, 1929

Mimetus audax Hickman, V.V. (1929). Studies in Tasmanian spiders. Part III *Pap. Proc. R. Soc. Tasm.* **1928**: 96–118 [107]. Type data: syntypes, QVM 36 *F. adult, from Launceston, Tas.

Distribution: Tas. Ecology: nomadic, predator.

Mimetus aurioculatus Hickman, 1929

Mimetus aurioculatus Hickman, V.V. (1929). Studies in Tasmanian spiders. Part III *Pap. Proc. R. Soc. Tasm.* **1928**: 96–118 [110] [emendations *aureoculatus* by Bonnet and *auriculatus* by Roewer are considered invalid under Art. 32(ii) of ICZN]. Type data: syntypes, QVM 35 *M.,2F. adult, from Punch Bowl, Launceston, Tas.

Distribution: Tas. Ecology: nomadic, predator.

Mimetus maculosus Rainbow, 1904

Mimetus maculosus Rainbow, W.J. (1904). Studies in Australian Araneidae. No. 3. *Rec. Aust. Mus.* **5**: 326–336 [330]. Type data: syntypes, AM KS5821 M.,5F. adult, from Jenolan Caves District, N.S.W.

Distribution: SE coastal, NE coastal, N.S.W., Qld. Ecology: nomadic, predator.

MYSMENIDAE

INTRODUCTION

Minute, eight-eyed, three-clawed spiders with a more or less spherical abdomen, sometimes with a posterior extension and high carapace. Male with a clasping spur on metatarsus 1. Female usually has a ventral subdistal sclerotized spot on femur 1. Tiny denticles are present between cheliceral teeth. Cymbium of male with lobes or apophyses. Female palp without a claw. Colulus present. These spiders are found in litter and collected in pitfall traps but little is known of their habits. Some American and African mysmenids are kleptoparasites, living in webs of diplurid trap-door spiders.

Brignoli (1983) has recently transferred *Mysmena tasmaniae* to *Calodipoena*, the type species of which is *Calodipoena incredula* Gertsch & Davis from Texas. Further species have been described from Mexico, the West Indies and Panama. Brignoli's reason for this transfer was *M. tasmaniae*'s "alleged relationships with *Calodipoena incredula*". This seems a misinterpretation of Hickman's (1979) text which states only that *Mysmena tasmaniae* "has a strong resemblance to the American species *Mysmena incredula*". There are many undescribed mysmenids in Australia and until they are described it seems preferable to leave *tasmaniae* in *Mysmena* rather than transfer it to a Central American genus.

Mysmena Simon, 1894

Mysmena Simon, E. (1894). *Histoire Naturelle des Araignées.* 2nd edn. Paris : Roret Vol 1 (1892–1903) [588]. Type species *Theridion leucoplagiatum* Simon, 1879 by monotypy.

This group is also found in Algeria, Philippines, France, Italy, North & Central America and Sri Lanka, see Bonnet, P. (1945–1961). *Bibliographia Araneorum.* Toulouse : Douladoure Vol. 2.

Mysmena tasmaniae (Hickman, 1979)

Mysmena tasmaniae Hickman, V.V. (1979). Some Tasmanian spiders of the families Oonopidae, Anapidae and Mysmenidae. *Pap. Proc. R. Soc. Tasm.* **113**: 53–79 [74]. Type data: holotype, AM KS2711 *M. adult, paratypes AM KS9625 *F. adult, TMH J1437 *M.,F. adult, from Maxwell River Valley, Tas.

Distribution: Tas.

Ecology: sedentary, predator.

Biological references: Brignoli, P.M. (1983). *A Catalogue of Araneae described between 1940–1981.* Merrett, P. (ed.) Manchester : Manchester Univ. Press (taxonomy; Brignoli's transfer to *Calodipoena* is not accepted).

NICODAMIDAE

INTRODUCTION

Medium-sized, three-clawed spiders that have bright yellow to red-brown cephalothorax and black abdomen (occasionally with red posteriorly). Male palpal tibia with large, angulate apophysis. Cheliceral promargin with one tooth, retromargin smooth. One trichobothrium on metatarsi, none on tarsi. Wide posterior spiracle, tracheal tubes to abdomen only. *Nicodamus* lives in a small sheet web close to the ground; it hangs from the underside of the web.

Nicodamus Simon, 1887

Centropelma Koch, L. (1872). *in* Koch, L. & Keyserling, E. (1871–1883). *Die Arachniden Australiens.* Nürnberg : Bauer & Raspe Vol. 1 [246] [*non Centropelma* Sclater and Salvin, 1869]. Type species *Centropelma bicolor* L. Koch, 1872 by monotypy.

Nicodamus Simon, E. (1887). [Observations sur divers Arachnides: synonymies et descriptions.] *Ann. Soc. Entomol. Fr. (Bull.) (6)* **7**: 158–194 [194] [*nom. nov.* for *Centropelma* L. Koch, 1872].

Ozaleus Thorell, T. (1890). Studi sui ragni Malesi e Papuani. IV. *Ann. Mus. Civ. Stor. Nat. Genova* **8**: 1–419 [293]. Type species *Ozaleus tarandus* Thorell, 1890 by monotypy.

Taxonomic decision of Rainbow, W.J. (1911). A Census of Australian Araneidae. *Rec. Aust. Mus.* **9**: 107–391 [257].

This group is also found in Irian Jaya, Indonesia, see Bonnet, P. (1945–1961). *Bibliographia Araneorum.* Toulouse : Douladoure Vol. 2.

Nicodamus bicolor (L. Koch, 1872)

Centropelma bicolor Koch, L. (1872). *in* Koch, L. & Keyserling, E. (1871–1883). *Die Arachniden Australiens.* Nürnberg : Bauer & Raspe Vol. 1 [246 pl 20 figs 5–6]. Type data: syntypes, ?SMNS *M.,F. adult, from Australia (as New Holland).

Linyphia melanozantha Urquhart, A.T. (1893). On new species of Tasmanian Araneae. *Pap. Proc. R. Soc. Tasm.* **1892**: 94–130 [103]. Type data: syntypes, CMNZ M.,F. adult both lost, from Tas.

Taxonomic decision of Hickman, V.V. (1967). *Some Common Spiders of Tasmania.* Hobart : Tasmanian Museum and Art Gallery 112 pp. [74].

Distribution: NE coastal, SE coastal, SW coastal, NW coastal, Qld., N.S.W., Tas., W.A. Ecology: sedentary, terrestrial, predator. Biological references: McKeown, K.C. (1936). *Spider Wonders of Australia.* Sydney : Angus & Robertson 270 pp.; McKeown, K.C. (1952). *Australian Spiders.* Their lives and habits. Sydney : Angus & Robertson 274 pp.; Main, B.Y. (1976). *Spiders.* Sydney : William Collins 296 pp.; Mascord, R. (1970). *Australian Spiders in Colour.* Sydney : Reed 112 pp.; Clyne, D. (1969). *Australian Spiders.* Sydney : Nelson 168 pp.; Mascord, R. (1980). *Spiders of Australia.* A field guide. Sydney : Reed 128 pp.

Nicodamus bidens (Karsch, 1878)

Centropelma bidens Karsch, F. (1878). Exotisch-araneologisches. II. *Z. Ges. Naturw.* **51**: 771–826 [815]. Type data: holotype, ?ZMB *M. adult, from SE coastal N.S.W.

Distribution: SE coastal, N.S.W. Ecology: sedentary, predator.

Nicodamus dimidiatus Simon, 1897

Nicodamus dimidiatus Simon, E. (1897). Descriptions d'Arachnides nouveaux. *Ann. Soc. Entomol. Belg.* **41**: 8–17 [15]. Publication date established from Bonnet, P. (1945–1961). *Bibliographia Araneorum.* Toulouse : Douladoure Vol. 2. Type data: syntypes, ?MNHP *F.,M. adult, from eastern Australia, see Rainbow, W.J. (1911). A Census of Australian Araneidae. *Rec. Aust. Mus.* **9**: 107–391. Compiled from secondary source: Roewer, C.F. (1942, 1954). *Katalog der Araneae.* Bremen : Paul Budy Bd. 1 1–1040.

Distribution: Eastern Australia. Ecology: sedentary, predator. Biological references: Mascord, R. (1970). *Australian Spiders in Colour* Sydney : Reed 112 pp. (biology).

Nicodamus nodatus (Karsch, 1878)

Centropelma nodata Karsch, F. (1878) Exotisch-araneologisches. II. *Z. Ges. Naturw.* **51** 771–826 [815]. Type data: holotype, ?ZMB *M. adult from Tas.

Distribution: Tas. Ecology: sedentary, predator.

Nicodamus perigrinus (Walckenaer, 1841)

Theridion perigrinum Walckenaer, C.A. (1841). *Histoire Naturelle des Insectes Aptères.* Paris Vol. 2 549 pp. (1837–1847) [297]. Type data: none designated, from Australia. Ecology: sedentary, predator. Biological references: Keyserling, E. (1891). *Die Spinnen Amerikas. Brasilianische Spinnen.* Nürnberg 3: 1–278 (taxonomy).

Nicodamus semiflavus (L. Koch, 1865)

Theridium semiflavum Koch, L. (1865). Beschreibungen neuer Arachniden und Myriapoden. *Verh. Zool.-Bot. Ges. Wien* **15**: 857–892 (1865); **17**: 173–250 (1867) [858]. Type data: holotype (probable), ZMH *F. adult, from Wollongong, N.S.W.

Distribution: SE coastal, N.S.W. Ecology: sedentary, predator. Biological references: Simon, E. (1892–1903). *Histoire Naturelle des Araignées.* 2nd edn. Paris : Roret Vol. 2 (taxonomy); Mascord, R. (1970). *Australian Spiders in Colour.* Sydney : Reed 112 pp.

Nicodamus tarandus (Thorell, 1890)

Ozaleus tarandus Thorell, T. (1890). Studi sui ragni Malesi e Papuani, IV. *Ann. Mus. Civ. Stor. Nat. Genova* **8**: 1–419 [294]. Type data: syntypes, MCG *F.,M. adult, from ?northern Australia.

Distribution: ?northern Australia. Thorell stated that the origin of his specimens was uncertain.

OECOBIIDAE

INTRODUCTION

Small, three-clawed spiders with a large two-segmented anal tubercle with a fringe of long hairs. Terminal segments of posterior spinnerets long with spigots along inner edge. One trichobothrium on metatarsi, none on tarsi. The cribellate spider, *Oecobius annulipes*, has been introduced into Australia. It spins small sheet webs on walls and darts out from under the web when this is touched.

Oecobius Lucas, 1846

Oecobius Lucas, H. (1846). Histoire naturelle des animaux articulés. *Explor. Sci. Algér.* Zool. **1** (Arachn.): 89–271 [100]. Publication date established from Shear, W.A. (1970). The spider family Oecobiidae in North America, Mexico, and the West Indies. *Bull. Mus. Comp. Zool.* **140**: 129–164. Type species *Oecobius domesticus* Lucas, 1846 (= *Clotho cellariorum* Dugès, 1836) by subsequent designation, see Thorell, T. (1870). On European spiders. *N. Act. Reg. Soc. Sci. Upsal. (3)* **7**: 109–242 [112].

Thalamia Hentz, N.M. (1850). Descriptions and figures of the Araneides of the United States. *Bost. J. Nat. Hist.* **6**: 18–35, 271–295 [35]. Publication date established from Shear, W.A. (1870). The spider family Oecobiidae in North America, Mexico and the West Indies. *Bull. Mus. Comp. Zool.* **140**: 129–164. Type species *Thalamia parietalis* Hentz, 1850 (= *Oecobius annulipes* Lucas, 1846) by monotypy.

Omanus Thorell, T. (1870). On European spiders. *N. Act. Reg. Soc. Sci. Upsal. (3)* **7**: 109–242 [114]. Publication date established from Shear, W.A. (1970). The spider family Oecobiidae in North America, Mexico and the West Indies. *Bull. Mus. Comp. Zool.* **40**: 129–164. Type species *Oecobius navus* Blackwall, 1859 (= *Oecobius annulipes* Lucas, 1846) by monotypy.

Taxonomic decision of Shear, W.A. (1970). The spider family Oecobiidae in North America, Mexico and the West Indies. *Bull. Mus. Comp. Zool.* **140**: 129–164 [135].

This group is found worldwide.

Oecobius annulipes Lucas, 1846

Oecobius annulipes Lucas, H. (1846). Histoire naturelle des animaux articulés. *Explor. Sci. Algér.* Zool. **1** (Arachn.): 89–271 [102]. Type data: ?syntypes, ?MNHP *F. adult, from Algeria.

Oecobius navus Blackwall, J. (1859). Descriptions of newly discovered spiders captured by James Yate Johnson Esq., in the Island of Madeira. *Ann. Mag. Nat. Hist. (3)* **4**: 255–267 [266]. Type data: no types known, from Madeira, Spain, see Lehtinen, P.T. (1967). Classification of the cribellate spiders and some allied families, with notes on the evolution of the suborder Araneomorpha. *Ann. Zool. Fenn.* **4**: 199–468.

Taxonomic decision of Bonnet, P. (1945–1961). *Bibliographia Araneorum.* Toulouse : Douladoure Vol. 2 [3132].

Distribution: NE coastal, SW coastal, SE coastal, Qld., W.A., N.S.W., A.C.T. Ecology: sedentary, predator; introduced. Biological references: Main, B.Y. (1976). *Spiders.* Sydney : Collins 296 pp. (biology); Clyne, D. (1969). *Australian Spiders.* Sydney : Nelson 168 pp. (biology).

OONOPIDAE

INTRODUCTION

Small, two-clawed, haplogyne spiders that have a pair of tracheal spiracles just behind the book-lungs. Six or less eyes. Colulus absent, sometimes represented by two setae. Cheliceral margins smooth. Female palp without a claw. Tarsal claws pectinate in double row. One trichobothrium on metatarsi, none on tarsi. Most oonopids live in litter; some are well sclerotized with a shield covering, or partly covering, the abdomen; others are soft-bodied. *Orchestina* is found in foliage; it is soft bodied and has enlarged femora IV for jumping backwards. *Ilschnothyreus* spp. have been found throughout northeastern Qld. and on Murray Ils., Torres Strait but none has been described. *Oonopinus mollipes* Hickman, 1932 will be the type species for a monotypic new genus in the family Orsolobidae (Forster & Platnick, in prep.).

Gamasomorpha Karsch, 1881

Gamasomorpha Karsch, F. (1881). Diagnoses Arachnoidarum Japoniae. *Berl. Entomol. Z.* **25**: 35–40 [40]. Publication date established from Bonnet, P. (1945–1961). *Bibliographia Araneorum.* Toulouse : Douladoure Vol. 2. Type species *Gamasomorpha cataphracta* Karsch, 1881 by monotypy.

This group is found worldwide, see Bonnet, P. (1945–1961). *Bibliographia Araneorum.* Toulouse : Douladoure Vol. 2.

Gamasomorpha banksi Hickman, 1950

Gamasomorpha banksi Hickman, V.V. (1950). Araneae from Reevesby Island, South Australia. *Proc. R. Soc. Vic. (ns)* **60**: 1–16 [13]. Type data: syntypes, NMV K110 *M. adult, NMV K111 *F. adult, from Reevesby Is., Sir Joseph Banks Group, S.A.

Distribution: S Gulfs, S.A. Ecology: nomadic, predator.

Gamasomorpha clarki Hickman, 1950

Gamasomorpha clarki Hickman, V.V. (1950). Araneae from Reevesby Island, South Australia. *Proc. R. Soc. Vict. (ns)* **60**: 1–16 [15]. Type data: holotype, NMV K112 *M. adult, from Reevesby Is., Sir Joseph Banks Group, S.A.

Distribution: S Gulfs, S.A. Ecology: nomadic, predator.

Gamasomorpha loricata (L. Koch, 1873)

Oonops loricatus Koch, L. (1873). *in* Koch, L. & Keyserling, E. (1871–1883). *Die Arachniden Australiens.* Nürnberg : Bauer & Raspe Vol. 1 [449 pl 35 fig 3]. Type data: holotype, ZMH (MGH 8084) *F. adult, from Upolu, Samoa.

Distribution: SE coastal, NE coastal, Qld., N.S.W., Lord Howe Is. Ecology: nomadic, predator. Biological references: Simon, E. (1892–1903). *Histoire Naturelle des Araignées.* 2nd edn. Paris : Roret Vol. 1 (taxonomy).

Gamasomorpha servula Simon, 1908

Gamasomorpha servula Simon, E. (1908). Araneae. 1re partie pp. 359–446 *in* Michaelsen, W. & Hartneyer, R. (eds.) *Die Fauna Südwest-Australiens.* 1(12) Jena : Fischer [379]. Type data: holotype, ?ZMB M. adult, from Eradu, W.A.

Distribution: SW coastal, W.A. Ecology: nomadic, predator.

Oonopinus Simon, 1893

Oonopinus Simon, E. (1893). Arachnides. Voyage de M. E. Simon au Venezuela (décembre 1887-avril 1888). 21e Mémoire. *Ann. Soc. Entomol. Fr.* **61**: 423–462 [446]. Type species *Oonops angustatus* Simon, 1882 by original designation.

This group is also found in France, Spain, Sierra Leone, Venezuela, Puerto Rico and New Caledonia, see Bonnet, P. (1945–1961). *Bibliographia Araneorum.* Toulouse : Douladoure Vol. 2.

Oonopinus mollipes Hickman, 1932

Oonopinus mollipes Hickman, V.V. (1932). Studies in Tasmanian spiders. No. 5. *Pap. Proc. R. Soc. Tasm.* **1931**: 20–31 [23]. Type data: syntypes, QVM 43 *3F. adult, from Trevallyn, Launceston, Tas.

Distribution: Tas. Ecology: nomadic, predator; in grass. Biological references: Hickman, V.V. (1979). Some Tasmanian spiders of the families Oonopidae, Anapidae and Mysmenidae. *Pap. Proc.*

R. Soc. Tasm. **113**: 53–79 (taxonomy). This spider will be the type species for a new genus in the family Orsolobidae (Forster & Platnick, in prep.).

Oonops Templeton, 1835

Oonops Templeton, R. (1835). On the spiders of the genus *Dysdera* Latr. with the description of a new allied genus. *Zool. J.* **5**: 400–408 [404]. Type species *Oonops pulcher* Templeton, 1835 by monotypy.

This group is found worldwide, see Bonnet, P. (1945–1961). *Bibliographia Araneorum.* Toulouse : Douladoure Vol. 2.

Oonops leai Rainbow, 1920

Oonops leai Rainbow, W.J. (1920). Arachnida from Lord Howe and Norfolk Islands. *Rec. S. Aust. Mus.* **1**: 229–272 [232]. Type data: holotype, SAMA N1981241 F. ?adult, from Lord Howe Is.

Distribution: Lord Howe Is. Ecology: nomadic, predator.

Oonops pulcher Templeton, 1835

Oonops pulcher Templeton, R. (1835). On the spiders of the genus *Dysdera* Latr. with the description of a new allied genus. *Zool. J.* **5**: 400–408 [404]. Publication date established from Bonnet, P. (1945–1961). *Bibliographia Araneorum.* Toulouse : Douladoure Vol. 2. Type data: syntypes, F.,M. adult whereabouts unknown, from Belfast, Ireland.

Distribution: Tas. Ecology: nomadic, predator; introduced. Biological references: Hickman, V.V. (1979). Some Tasmanian spiders of the families Oonopidae, Anapidae and Mysmenidae. *Pap. Proc. R. Soc. Tasm.* **113**: 53–79.

Orchestina Simon, 1882

Orchestina Simon, E. (1882). Études arachnologiques. 13e Mémoire. Descriptions d'espèces et de genres nouveaux de la famille des Dysderidae. *Ann. Soc. Entomol. Fr.* **2**: 201–240 [237]. Publication date established from Bonnet, P. (1945–1961). *Bibliographia Araneorum.* Toulouse : Douladoure Vol. 2. Type species *Schaenobates pavesii* Simon, 1873 by monotypy.

This group is found worldwide, see Bonnet, P. (1945–1961). *Bibliographia Araneorum.* Toulouse : Douladoure Vol. 2.

Orchestina launcestoniensis Hickman, 1932

Orchestina launcestoniensis Hickman, V.V. (1932). Studies in Tasmanian spiders. No. 5. *Pap. Proc. R. Soc. Tasm.* **1931**: 20–31 [26]. Type data: syntypes, QVM 40 *M.,F. adult, from Punch Bowl, Launceston, Tas.

Distribution: Tas. Ecology: nomadic, predator; in grass.

ORSOLOBIDAE

INTRODUCTION

Pale, medium-sized, six-eyed spiders with a pair of posterior spiracles just behind the book-lungs. Vulva haplogyne, occasionally sclerotized. Onychium with spatulate hairs; superior claws large with two rows of teeth and expanded into wing-like projections on the outer side; inferior claw absent, sometimes represented by pad below claws. Colulus present. Cheliceral margins usually with one to two teeth. One trichobothrium on metatarsi, none on tarsi. Main synapomorphy is the elevated tarsal organ (Platnick, N.I., pers. comm.). The spiders are usually found in litter of rainforests. Forster and Platnick are revising the family and twenty six new species from Australia will be described.

Cornifalx Hickman, 1979

Cornifalx Hickman, V.V. (1979). Some Tasmanian spiders of the families Oonopidae, Anapidae and Mysmenidae. *Pap. Proc. R. Soc. Tasm.* **113**: 53–79 [54]. Type species *Cornifalx insignis* Hickman, 1979 by monotypy and original designation.

Cornifalx insignis Hickman, 1979

Cornifalx insignis Hickman, V.V. (1979). Some Tasmanian spiders of the families Oonopidae, Anapidae and Mysmenidae. *Pap. Proc. R. Soc. Tasm.* **113**: 53–79 [54]. Type data: holotype, AM KS2702 *M. adult, paratypes AM KS2703 *F. adult, TMH J1435 *F. adult, TMH J1436 *M. adult, from southwestern Tas. at 145°53'E, 42°38'S.

Distribution: Tas. Ecology: nomadic, predator, litter; in moss.

Tasmanoonops Hickman, 1930

Tasmanoonops Hickman, V.V. (1930). Studies in Tasmanian spiders. Part IV. *Pap. Proc. R. Soc. Tasm.* **1929**: 87–122 [97]. Type species *Tasmanoonops alipes* Hickman, 1930 by monotypy and original designation.

Tasmanoonops alipes Hickman, 1930

Tasmanoonops alipes Hickman, V.V. (1930). Studies in Tasmanian spiders. Part IV. *Pap. Proc. R. Soc. Tasm.* **1929**: 87–122 [98]. Type data: holotype, QVM 45 *F. adult, from Mt. Hobbs, Woodsdale, Tas.

Distribution: Tas., SW coastal, W.A. Ecology: nomadic, predator; under logs. Biological references: Main, B.Y. (1981). Australian spiders: diversity, distribution and ecology. pp. 807–852 *in* Keast, A. (ed.) *Ecological Biogeography of Australia.* The Hague : W. Junk.

Tasmanoonops fulvus Hickman, 1979

Tasmanoonops fulvus Hickman, V.V. (1979). Some Tasmanian spiders of the families Oonopidae, Anapidae and Mysmenidae. *Pap. Proc. R. Soc. Tasm.* **113**: 53–79 [61]. Type data: holotype, AM KS2704 *M. adult, paratype AM KS2705 *F. adult, from southwestern Tas. at 145°50'E, 42°56'S.

Distribution: Tas. Ecology: nomadic, predator, litter; in moss.

Tasmanoonops inornatus Hickman, 1979

Tasmanoonops inornatus Hickman, V.V. (1979). Some Tasmanian spiders of the families Oonopidae, Anapidae and Mysmenidae. *Pap. Proc. R. Soc. Tasm.* **113**: 53–79 [63]. Type data: holotype, AM KS13761 *M. adult, paratypes AM KS13762 *F. adult, TMH J1450 *F. adult, from southwestern Tas. at 145°53'E, 42°48'S.

Distribution: Tas. Ecology: nomadic, predator, litter; in moss.

Tasmanoonops magnus Hickman, 1979

Tasmanoonops magnus Hickman, V.V. (1979). Some Tasmanian spiders of the families Oonopidae, Anapidae and Mysmenidae. *Pap. Proc. R. Soc. Tasm.* **113**: 53–79 [66]. Type data: holotype, AM KS2706 *M. adult with paratype F. adult, from Western Creek, Tas.

Distribution: Tas. Ecology: nomadic, predator; in moss or logs.

PARARCHAEIDAE

INTRODUCTION

Tiny, three-clawed, entelegyne spiders with relatively long chelicerae, the bases of which are encircled by the carapace, thus separating them from the rest of the mouthparts. Viewed laterally, carapace square or rhomboid. Fang short. Row of peg teeth on distal promargin of chelicerae. Tarsi longer than metatarsi. Anterior book-lungs; median posterior spiracle just in front of spinnerets, tracheal tubes to abdomen. Conical colulus. One trichobothium on metatarsi, none on tarsi. No onychium. Female palp without claw. Male palp with paracymbium. *Pararchaea* is found in rainforest litter in temperate southeastern Australia and at cool, high altitudes in northeastern Australia.

Pararchaea Forster, 1955

Pararchaea Forster, R.R. (1955). Spiders of the family Archaeidae from Australia and New Zealand. *Trans. R. Soc. N.Z.* **83**: 391–403 [397]. Type species *Pararchaea alba* Forster, 1955 by original designation.

This group is also found in New Zealand, see Forster, R.R. & Platnick, N.I. (1984). A review of the archaeid spiders and their relatives, with notes on the limits of the Superfamily Palpimanoidea (Arachnida, Araneae). *Bull. Am. Mus. Nat. Hist.* **178**: 1–106.

Pararchaea binnaburra Forster, 1955

Pararchaea binnaburra Forster, R.R. (1955). Spiders of the family Archaeidae from Australia and New Zealand. *Trans. R. Soc. N.Z.* **83**: 391–403 [401]. Type data: holotype, QM *F. adult not received, paratype CMNZ CMA1078 F. adult, from Binna Burra, Lamington Natl. Park, Qld.

Distribution: NE coastal, Qld. Ecology: nomadic, predator, closed forest, litter.

Pararchaea bryophila Hickman, 1969

Pararchaea bryophila Hickman, V.V. (1969). New species of Toxopidae and Archaeidae (Araneida). *Pap. Proc. R. Soc. Tasm.* **103**: 1–11 [9]. Type data: holotype, AM KS6633 *M. adult, paratype AM KS6634 *F. adult, from Punch Bowl, Launceston, Tas.

Distribution: Tas. Ecology: nomadic, predator; in moss.

Pararchaea corticola Hickman, 1969

Pararchaea corticola Hickman, V.V. (1969). New species of Toxopidae and Archaeidae (Araneida). *Pap. Proc. R. Soc. Tasm.* **103**: 1–11 [3]. Type data: holotype, AM KS6635 *M. adult, paratype AM KS6636 *F. adult, from Hobart, Tas.

Distribution: Tas. Ecology: nomadic, predator; under eucalypt bark.

Pararchaea ornata Hickman, 1969

Pararchaea ornata Hickman, V.V. (1969). New species of Toxopidae and Archaeidae (Araneida). *Pap. Proc. R. Soc. Tasm.* **103**: 1–11 [7]. Type data: holotype, AM KS6638 *F. adult, from Hobart, Tas.

Distribution: Tas. Ecology: nomadic, predator; on gorse, *Ulex europeus*.

Pararchaea saxicola Hickman, 1969

Pararchaea saxicola Hickman, V.V. (1969). New species of Toxopidae and Archaeidae (Araneida). *Pap. Proc. R. Soc. Tasm.* **103**: 1–11 [5]. Type data: holotype, AM KS6639 *M. adult, paratype AM KS6640 *F. adult, from Hobart, Tas.

Distribution: Tas. Ecology: nomadic, predator; under stones.

PHOLCIDAE

INTRODUCTION

Haplogyne, three-clawed spiders with very long thin legs, hence the common name "daddy long legs". Chelicerae fused at base: median lamella ends distally with membraneous tooth-like structure. Eight or six eyes. When present, AME small; rest of eyes in two triads. Male palp complex with elaborate procursus (=paracymbium). Colulus present. One trichobothrium on metatarsi, none on tarsi. Female palp without a claw. Spiders hang from a lightly woven tangle web that often has an irregular horizontal sheet wihin it. *Belisana* sp. has been found on the Atherton Tableland and *Panjange* sp. at Iron Range, Queensland (Deeleman-Reinhold, C.L., pers. comm.).

Artema Walckenaer, 1837

Artema Walckenaer, C.A. (1837). *Histoire Naturelle des Insectes Aptères.* Paris Vol. 1 682 pp. (1837–1847) [656]. Type species *Artema mauriciana* Walckenaer, 1837 by subsequent designation, see Simon, E. (1892–1903). *Histoire Naturelle des Araignées.* 2nd edn. Paris : Roret Vol. 1 [466].

This group is found worldwide, see Bonnet, P. (1945–1961). *Bibliographia Araneorum.* Toulouse : Douladoure Vol. 2; Brignoli, P.M. (1981). Studies on the Pholcidae. 1. Notes on the genera *Artema* and *Physocyclus* (Araneae). *Bull. Am. Mus. Nat. Hist.* **170**: 90–100.

Artema atlanta Walckenaer, 1837

Artema atlanta Walckenaer, C.A. (1837). *Histoire Naturelle des Insectes Aptères.* Paris Vol. 1 (1837–1847) [656]. Type data: syntypes, M.,F. adult probably lost, from Brazil, S. America.

Artema mauriciana Walckenaer, C.A. (1837). *Histoire Naturelle des Insectes Aptères.* Paris Vol. 1 (1837–1847) [657]. Type data: syntypes, F.,M. adult probably lost, from Mauritius (Île de France).

Taxonomic decision of Chrysanthus, Fr. (1967). Spiders from South New Guinea. IX. *Tijdschr. Entomol.* **110**: 89–105 [92].

Distribution: NE coastal, N Gulf, Qld., N.T.; Moa Is., Torres Strait. Ecology: sedentary, predator; introduced.

Crossopriza Simon, 1893

Crossopriza Simon, E. (1893). *Histoire Naturelle des Araignées.* 2nd edn. Paris : Roret Vol. 1 [476]. Type species *Artema pristina* Simon, 1890 by original designation.

This group is also found in Egypt, Yemen, India, Burma, Vietnam, Brazil and Cuba, see Bonnet, P. (1945–1961). *Bibliographia Araneorum.* Toulouse : Douladoure Vol. 2.

Crossopriza lyoni (Blackwall, 1867)

Pholcus lyoni Blackwall, J. (1867). Descriptions of several species of East Indian spiders, apparently to be new or little known to arachnologists. *Ann. Mag. Nat. Hist. (3)* **19**: 387–394 [392]. Type data: syntypes, M.,F. adult whereabouts unknown, from Meerut, Agra and Delhi, India.

Distribution: N coastal, N Gulf, Lake Eyre basin, Murray-Darling basin, N.T., Qld. Ecology: sedentary, predator; introduced. Biological references: Simon, E. (1892–1903). *Histoire Naturelle des Araignées.* 2nd edn. Paris : Roret Vol. 1 (taxonomy).

Holocnemus Simon, 1873

Holocnemus Simon, E. (1873). Aranéides nouveaux un peu connus du midi de l'Europe. (2me Mémoire). *Mém. Soc. R. Sci. Liege (2)* **5**: 1–174 [48]. Publication date established from Bonnet, P. (1945–1961). *Bibliographia Araneorum.* Toulouse : Douladoure Vol. 2. Type species *Aranea rivulata* Forskol, 1775 (=*Aranea pluchii* Scopoli, 1763) by original designation.

This group is cosmopolitan.

Holocnemus pluchii (Scopoli, 1763)

Aranea pluchii Scopoli, J.A. (1763). *Entomologia Carniolica,* exhibens Insecta Cariolae indigena et distributa in ordines, genera, species, varietates. Methoda Linnaeana. Vindobonae 420 pp. [404]. Publication date established from Bonnet, P. (1945–1961). *Bibliographia Araneorum.* Toulouse : Douladoure Vol. 2. Type data: unknown, from Carniole, Austria.

Distribution: NE coastal, Qld. Ecology: sedentary, predator; introduced. Biological references: Hasselt, A.W.M. van (1886). *Catalogus Aranearum hucusque in Hollandiâ inventarum.* Hagae Comitis 145 pp. (taxonomy).

Micromerys Bradley, 1878

Micromerys Bradley, H.H.B. (1878). Araneides of the *Chevert* Expedition. Part II. *Proc. Linn. Soc. N.S.W.* **2**: 115–122 [118]. Type species *Micromerys gracilis* Bradley, 1878 by monotypy.

This group is also found in Indonesia (Deeleman-Reinhold, C.L., pers. comm.)

Micromerys gracilis Bradley, 1878

Micromerys gracilis Bradley, H.H.B. (1878). Araneides of the *Chevert* Expedition Part II. *Proc. Linn. Soc. N.S.W.* **2**: 115–122 [119]. Type data: syntypes, MMUS 2F. adult lost, from Cape York, Qld.

Distribution: NE coastal, Qld. Ecology: sedentary, predator.

Pholcus Walckenaer, 1805

Pholcus Walckenaer, C.A. (1805). *Tableau des Aranéides* ou caractères essentiels des tribus, genres, familles et races que renferme le genre Aranea de Linné, avec la désignation des espèces comprises dans chacune de ces divisions. Paris 88 pp. [80]. Type species *Aranea phalangioides* Fuesslin, 1775 by monotypy.

This group is found worldwide, see Bonnet, P. (1945–1961). *Bibliographia Araneorum.* Toulouse : Douladoure Vol. 2.

Pholcus ancoralis L. Koch, 1865

Pholcus ancoralis Koch, L. (1865). Beschreibungen neuer Arachniden und Myriapoden. *Verh. Zool.-Bot. Ges. Wien* **15**: 857–892 (1865); **17**: 173–250 (1867) [862]. Type data: syntypes, ZMH*, from Upolu, Samoa, see Rack, G. (1961). Die Entomologischen Sammlungen des Zoologischen Staatsinstituts und Zoologischen Museums Hamburg. II. *Mitt. Hamb. Zool. Mus. Inst.* **59**: 1–60.

Distribution: NE coastal, Qld., Great Barrier Reef; Heron Is., Prince of Wales Is., Torres Strait islands. Ecology: sedentary, predator.

Pholcus litoralis L. Koch, 1867

Pholcus litoralis Koch, L. (1867). Beschreibungen neuer Arachniden und Myriapoden. *Verh. Zool.-Bot. Ges. Wien* **15**: 857–892 (1865); **17**: 173–250 (1867) [193]. Type data: syntypes, ZMH *F. adult, from Brisbane, Qld., see Rack, G. (1961). Die Entomologischen Sammlungen des Zoologischen Staatsinstituts und Zoologischen Museums Hamburg. II. *Mitt. Hamb. Zool. Mus. Inst.* **59**: 1–60.

Distribution: NE coastal, Qld. Ecology: sedentary, predator.

Pholcus phalangioides (Fuesslin, 1775)

Aranea phalangioides Fuesslin, J.C. (1775). *Verzeichniss der ihm bekannten schweizerischen Insekten,* mit einer ausgemahlten Kupfertafel: nebst der Ankündigung eines neuen Insekten Werkes. Zurich und Winterthur 62 pp. [61]. Publication date established from Bonnet, P. (1945–1961). *Bibliographia Araneorum.* Toulouse : Douladoure Vol. 1. Type data: unknown, from Switzerland.

Distribution: NE coastal, SE coastal, S Gulfs, SW coastal, Qld., N.S.W., Vic., S.A., W.A., Cocos (Keeling) Is. Ecology: sedentary, predator; introduced. Biological references: Main, B.Y. (1976). *Spiders.* Sydney : William Collins 296 pp.; Mascord, R. (1970). *Australian Spiders in Colour.* Sydney : Reed 112 pp.; Clyne, D. (1969). *Australian Spiders.* Sydney : Nelson 168 pp.; Mascord, R. (1980). *Spiders of Australia.* A field guide. Sydney : Reed 128 pp. (biology); Walckenaer, C.A. (1805). *Tableau des Aranéides* ou caractères essentiels des tribus, genres, familles et races que renferme le genre Aranea de Linné, avec la désignation des espèces comprises dans chacune de ces divisions. Paris 88 pp. (taxonomy).

Physocyclus Simon, 1893

Physocyclus Simon, E. (1893). *Histoire Naturelle des Araignées.* 2nd edn. Paris : Roret Vol. 1 (1892–1903) [470]. Type species *Pholcus globosus* Taczanowski, 1873 by original designation.

This group is also found in North, Central & South America, Pacific islands and Malaysia, see Brignoli, P.M. (1981). Studies on the Pholcidae. 1. Notes on the genera *Artema* and *Physocyclus* (Araneae). *Bull. Am. Mus. Nat. Hist.* **170**: 90–100.

Physocyclus globosus (Taczanowski, 1873)

Pholcus globosus Taczanowski, W. (1873). Les aranéides de la Guyane Française. *Horae Soc. Entomol. Ross.* **10**: 56–115 [105]. Type data: syntypes, 2F. adult whereabouts unknown, from Cayenne, French Guyana. Compiled from secondary source: Brignoli, P.M. (1981). Studies on the Pholcidae. 1. Notes on the genera *Artema* and *Physocyclus* (Araneae). *Bull. Amer. Mus. Nat. Hist.* **170**: 90–100.

Distribution: NE coastal, N Gulf, Qld., Cocos (Keeling) Ils. Ecology: sedentary, predator; introduced. Biological references: Simon, E. (1892–1903). *Histoire Naturelle des Araignées.* 2nd edn. Paris : Roret Vol. 1 (taxonomy).

Psilochorus Simon, 1893

Psilochorus Simon, E. (1893). *Histoire Naturelle des Araignées.* 2nd edn. Paris : Roret Vol. 1 [482]. Type species *Theridion pullulum* Hentz, 1850 by monotypy.

This group is found worldwide, see Bonnet, P. (1945–1961). *Bibliographia Araneorum.* Toulouse : Douladoure Vol. 2.

Psilochorus sphaeroides (L. Koch, 1872)

Pholcus sphaeroides Koch, L. (1872). *in* Koch, L. & Keyserling, E. (1871–1883). *Die Arachniden Australiens.* Nürnberg : Bauer & Raspe Vol. 1 [283 pl 23 fig 6]. Type data: syntypes, ZMH M.,F. adult missing, from Rockhampton, Qld.

Distribution: NE coastal, Qld. Ecology: sedentary, predator. Biological references: Simon, E. (1892–1903). *Histoire Naturelle des Araignées.* 2nd edn. Paris : Roret Vol. 1 (taxonomy).

Smeringopus Simon, 1890

Smeringopus Simon, E. (1890). Études arachnologiques. 22ᵉ Mémoire. Étude sur les Arachnides de l'Yémen. *Ann. Soc. Entomol. Fr. (6)* **10**: 77–124 [94]. Publication date established from Bonnet, P. (1945–1961). *Bibliographia Araneorum.* Toulouse : Douladoure Vol. 2. Type species *Pholcus elongatus* Vinson, 1863 by monotypy.

This group is found worldwide, see Bonnet, P. (1945–1961). *Bibliographia Araneorum.* Toulouse : Douladoure Vol. 2.

Smeringopus pallidus (Blackwall, 1858)

Pholcus pallidus Blackwall, J. (1858). Descriptions of six newly discovered species and characters of a new genus of Araneidea. *Ann. Mag. Nat. Hist. (3)* **1**: 426–434 [433]. Type data: holotype, OUM *F. adult, from Pernambuco, Brazil.

Distribution: NE coastal, Qld., Cocos (Keeling) Is.; Torres Strait islands. Ecology: sedentary, predator; introduced. Biological references: Kraus, O. (1957). Araneenstudien 1. Pholcidae (Smeringopodinae, Ninetinae). *Senckenberg. Biol.* **38**: 217–243 (taxonomy).

Trichocyclus Simon, 1908

Trichocyclus Simon, E. (1908). Araneae. 1ʳᵉ partie pp. 359–446 *in* Michaelsen, W. & Hartmeyer, R. (eds.) *Die Fauna Südwest-Australiens.* 1(12) Jena : Fischer [407]. Type species *Trichocyclus nigropunctatus* Simon, 1908 by monotypy.

Trichocyclus nigropunctatus Simon, 1908

Trichocyclus nigropunctatus Simon, E. (1908). Araneae. 1ʳᵉ partie pp. 359–446 *in* Michaelsen, W. & Hartmeyer, R. (eds.) *Die Fauna Südwest-Australiens.* 1(12) Jena : Fischer [407]. Type data: syntypes, MNHP F. adult, ZMH juv., from Yalgoo, W.A.

Distribution: SW coastal, W.A. Ecology: sedentary, predator.

PISAURIDAE

INTRODUCTION

Medium to large, three-clawed spiders. Posterior row of eyes recurved. A row of trichobothria on metatarsi and tarsi. Male palpal tibia with apophysis (*cf.* Lycosidae). Trochanters deeply notched. Female pisaurids usually carry the egg sacs in the chelicerae and palps (*cf.* Lycosidae). *Dolomedes* spp., our commonest pisaurids, are hunting spiders and usually found near water. They rest by day under stones or stretched out on the underside of branches. Other pisaurids (*Dendrolycosa, Inola*) spin large sheet webs on which prey is trapped. The webs are usually built against tree trunks or rocks with retreats going back into crevices; the spiders run on the top of the sheet. A separate family, Dolomedidae, has been proposed but has not yet been well defined.

Dendrolycosa Doleschall, 1859

Dendrolycosa Doleschall, C.L. (1859). Tweede bijdrage tot de kennis der Arachniden van den Indischen Archipel. *Act. Soc. Sci. Ind.-Neerl.* **5**: 1–60 [51]. Publication date established from Bonnet, P. (1945–1961). *Bibliographia Araneorum.* Toulouse : Douladoure Vol. 2. Type species *Dendrolycosa fusca* Doleschall, 1859 by monotypy.

This group is also found in New Guinea, Malaysia, Nicobar Ils., Java, Sumatra, Burma and India, see Bonnet, P. (1945–1961). *Bibliographia Araneorum.* Toulouse : Douladoure Vol. 2. Species now known not to occur in Australia: *Dendrolycosa kochi* Simon, 1898 (*nom. nud.*), see Simon, E. (1892–1903). *Histoire Naturelle des Araignées.* 2nd edn. Paris : Roret Vol. 2 (285, 289); *D. kochi* mentioned twice without description or figure.

Dendrolycosa icadius (L. Koch, 1876)

Dolomedes icadius Koch, L. (1876). *in* Koch, L. & Keyserling, E. (1871–1883). *Die Arachniden Australiens.* Nürnberg : Bauer & Raspe Vol. 2 [859 pl 73 fig 5]. Type data: syntypes, ZMH (MGH 14628) M. juv. F. juv., from Rockhampton, Qld.

Distribution: NE coastal, Qld. Ecology: sedentary, predator. Biological references: Davies, V. Todd (1982). *Inola* nov. gen., a web-building pisaurid (Araneae : Pisauridae) from northern Australia with descriptions of three species. *Mem. Qd. Mus.* **20**: 479–487 (taxonomy).

Dolomedes Latreille, 1804

Dolomedes Latreille, P.A. (1804). Tableau méthodique des Insectes. *N. Dic. Hist. Nat.* **24**: 129–200 [135]. Publication date established from Bonnet, P. (1945–1961). *Bibliographia Araneorum.* Toulouse : Douladoure Vol. 2. Type species *Araneus fimbriatus* Clerck, 1758 by subsequent designation, see Walckenaer, C.A. (1805). *Tableau des Aranéides* ou caractères essentiels des tribus, genres, familles et races que renferme le genre Aranea de Linné, avec la désignation des espèces comprises dans chacune de ces divisions. Paris 88 pp. [16,17]. Compiled from secondary source: Carico, J.E. (1973). The Nearctic species of the genus *Dolomedes* (Araneae : Pisauridae). *Bull. Mus. Comp. Zool.* **144**: 435–488.

This group is found worldwide, see Bonnet, P. (1945–1961). *Bibliographia Araneorum.* Toulouse : Douladoure Vol. 2. Species now known not to occur in Australia: *Dolomedes mirificus* Walckenaer, 1837; this is a New Zealand species *vide* Walckenaer, C.A. (1837–1847). *Histoire Naturelle des Insectes Aptères.* Paris 4 vols+atlas (Australie-Nouvelle Zélande =New Zealand).

Dolomedes albicomus L. Koch, 1867

Dolomedes albicomus Koch, L. (1867). Beschreibungen neuer Arachniden und Myriapoden. *Verh. Zool.-Bot Ges. Wien* **15**: 857–892 (1865); **17**: 173–250 (1867) [199]. Type data: holotype, ZMH (MGH 3503) M. adult from Brisbane, Qld.

Distribution: NE coastal, Qld. Ecology: nomadic, predator.

Dolomedes chroesus Strand, 1911

Dolomedes chroesus Strand, E. (1911). Araneae von der Aru- und Kei-Inseln. *Abh. Senckenb. Naturf. Ges.* **34** 129–199 [166]. Type data: syntypes, ?SMF *juv., F adult, from Aru Is., Indonesia.

Distribution: N Gulf, Qld.; Torres Strait islands Ecology: nomadic, predator.

Dolomedes eberhardarum Strand, 1913

Dolomedes eberhardarum Strand, E. (1913). Drei neue Spinnen von Victoria in Australien. *Jb. Nassau. Ver. Naturk.* **66**: 204–209 [207] [as *Dolomedes Eberhardarum*]. Type data: holotype, NM Wiesbaden *M. adult, from Vic.

Distribution: SE coastal, Vic. Ecology: nomadic, predator.

Dolomedes facetus L. Koch, 1876

Dolomedes facetus Koch, L. (1876). *in* Koch, L. & Keyserling, E. (1871–1883). *Die Arachniden Australiens.* Nürnberg : Bauer & Raspe Vol. 2 [869 pl 74 fig 5–6]. Type data: lectotype, ZMH (MGH 6522) F. adult, paralectotypes ZMH juvs, designation by Chrysanthus, Fr. (1967). Spiders from South New Guinea. VIII. *Nova Guinea, Zool.* **37**: 401–426, from Rockhampton, Qld.

Distribution: SE coastal, NE coastal, N.S.W., Qld. Ecology: nomadic, predator. Biological references: McKeown, K.C. (1952). *Australian Spiders.* Their lives and habits. Sydney : Angus & Robertson 274 pp. (biology).

Dolomedes flaminius L. Koch, 1867

Dolomedes flaminius Koch, L. (1867). Beschreibungen neuer Arachniden und Myriapoden. *Verh. Zool.-Bot. Ges. Wien* **15**: 857–892 (1865); **17**: 173–250 (1867) [197]. Type data: holotype, ZMH M. adult, from Brisbane, Qld.

Distribution: NE coastal, Qld. Ecology: nomadic, predator.

Dolomedes instabilis L. Koch, 1876

Dolomedes instabilis Koch, L. (1876). *in* Koch, L. & Keyserling, E. (1871–1883). *Die Arachniden Australiens.* Nürnberg : Bauer & Raspe Vol. 2 [863 pl 74 fig 1]. Type data: holotype, whereabouts unknown F. adult, from Caigan, near Mudgee, N.S.W.

Distribution: SE coastal, N.S.W. Ecology: nomadic, predator.

Dolomedes stilatus Karsch, 1878

Dolomedes stilatus Karsch, F. (1878). Exotisch-araneologisches. II. *Z. Ges. Naturw.* **51**: 717–826 [814]. Type data: holotype, ?ZMB M. adult, from W.A. (as Austr. occid.).

Distribution: SW coastal, W.A. Ecology: nomadic, predator.

Hygropoda Thorell, 1895

Hygropoda Thorell, T. (1895). *Descriptive Catalogue of the Spiders of Burma.* London : British Museum 406 pp. [221]. Publication date established from Bonnet, P. (1945–1961). *Bibliographia Araneorum.* Toulouse : Douladoure Vol. 2. Type species *Tegenaria dolomedes* Doleschall, 1859 by subsequent designation, see Simon, E. (1892–1903). *Histoire Naturelle des Araignées.* 2nd edn. Paris : Roret Vol. 2 [316].

This group is also found in SE Asia, Venezuela, New Guinea, Madagascar and central Africa, see Bonnet, P. (1945–1961). *Bibliographia Araneorum.* Toulouse : Douladoure Vol. 2.

Hygropoda lineata (Thorell, 1881)

Dendrolycosa lineata Thorell, T. (1881). Studi sui ragni Malesi e Papuani. III. Ragni dell'Austro Malesia e del Capo York, conservati nel Museo Civico di Storia Naturale di Genova. *Ann. Mus. Civ. Stor. Nat. Genova* **17**: 1–720 [366]. Type data: holotype, MCG F. adult, from Somerset, Cape York, Qld.

Hygropoda spuripes Strand, E. (1911). Araneae von den Aru- und Kei-Inseln. *Abh. Senckenb. Naturf. Ges.* **34**: 129–199 [169]. Type data: holotype, SMF F. adult missing, paratype SMF M. adult missing, from Wokam, Aru Is., Indonesia.

Taxonomic decisions of Davies, V.T., this work.

Distribution: NE coastal, Qld.; Cape York Peninsula, Torres Strait Islands. Ecology: sedentary, predator.

Inola Davies, 1982

Inola Davies, V. Todd (1982). *Inola* nov. gen., a web-building pisaurid (Araneae : Pisauridae) from northern Australia with descriptions of three species. *Mem. Qd. Mus.* **20**: 479–487 [479]. Type species *Inola amicabilis* Davies, 1982 by original designation.

Inola amicabilis Davies, 1982

Inola amicabilis Davies, V. Todd (1982). *Inola* nov. gen., a web building pisaurid (Araneae : Pisauridae) from northern Australia with descriptions of three species. *Mem. Qd. Mus.* **20**: 479–87 [479]. Type data: holotype, QM S860 F. adult, paratypes QM S861–3 M. adult, QM S864–8 F. adult, QM S870–1 M. adult, QM S872 4M.,5F. adult, QM S873 M. adult, from The Granites Track near Home Rule, 35 km S Cooktown, Qld.

Distribution: NE coastal, Qld. Ecology: sedentary, predator.

Inola cracentis Davies, 1982

Inola cracentis Davies, V. Todd (1982). *Inola* nov. gen., a web-building pisaurid (Araneae : Pisauridae) from northern Australia with descriptions of three species. *Mem. Qd. Mus.* **20**: 479–487 [480]. Type data: holotype, QM S874 F. adult, paratypes QM S875–6 M. adult, QM S877 F. juv., from Boonjee, near Malanda, Atherton Tableland, Qld.

Distribution: NE coastal, Qld. Ecology: sedentary, predator.

Inola subtilis Davies, 1982

Inola subtilis Davies, V. Todd (1982). *Inola* nov. gen., a web-building pisaurid (Araneae : Pisauridae) from northern Australia with descriptions of three species. *Mem. Qd. Mus.* **20**: 479–487 [482]. Type data: holotype,

QM S878 F. adult, paratypes QM S879 F. adult & juv., QM S880 F. adult & juvs, from Redlynch, near Cairns, Qld.

Distribution: NE coastal, Qld. Ecology: sedentary, predator.

Megadolomedes Davies and Raven, 1980

Megadolomedes Davies, V. Todd & Raven, R.J. (1980). *Megadolomedes* nov. gen. (Araneae : Pisauridae) with a description of the male of the type species, *Dolomedes australianus* Koch, 1865. *Mem. Qd. Mus.* **20**: 135–141 [135]. Type species *Dolomedes australianus* L. Koch, 1865 by original designation.

Megadolomedes australianus (L. Koch, 1865)

Dolomedes australianus Koch, L. (1865). Beschreibungen neuer Arachniden und Myriapoden. *Verh. Zool.-Bot. Ges. Wien* **15**: 857–892 (1865); **17**: 173–250 (1867) [863]. Type data: holotype, ZMH (MGH 302) M. juv., from Wollongong, N.S.W.

Dolomedes cervinus Koch, L. (1876). *in* Koch, L. & Keyserling, E. (1871–1883). *Die Arachniden Australiens.* Nürnberg : Bauer & Raspe Vol. 2 [872 pl 75 fig 1]. Type data: holotype, juv. whereabouts unknown, from Nepean River, N.S.W.

Dolomedes trux Lamb, J. (1911). Descriptions of some new Queensland Araneidae. *Ann. Qd. Mus.* **10**: 169–174 [173]. Type data: holotype, QM G55 F. adult, from Brisbane, Qld.

Taxonomic decision of Davies, V. Todd & Raven, R.J. (1980). *Megadolomedes* nov. gen. (Araneae : Pisauridae) with a description of the male of the type species *Dolomedes australianus* Koch, 1865. *Mem. Qd. Mus.* **20**: 135–141 [136].

Distribution: NE coastal, SW coastal, SE coastal, SW coastal, Qld., N.S.W., Vic., Tas., W.A.; Prince Edward Is., Torres Strait. Ecology: nomadic, predator. Biological references: Main, B.Y. (1976). *Spiders.* Sydney : Collins 296 pp.

Perenethis L. Koch, 1878

Perenethis Koch, L. (1878). *in* Koch, L. & Keyserling, E. (1871–1883). *Die Arachniden Australiens.* Nürnberg : Bauer & Raspe Vol. 2 [980 pl 85 fig 7]. Type species *Perenethis venusta* L. Koch, 1878 by monotypy.

This group is also found in India, Burma, Papua New Guinea and Japan, see Roewer, C.F. (1942, 1954). *Katalog der Araneae*. Bruxelles : Inst. R. Sci. Nat. Belg. Bd. 2a 1–923. Simon, E. (1887). [Observations sur divers Arachnides: synonymies et descriptions]. *Ann. Soc. Entomol. Fr. (Bull.)* (6) **7**: 158–194 [186] placed *Perenethis* in synonymy with *Tetragonophthalma*.

Perenethis venusta L. Koch, 1878

Perenethis venusta Koch, L. (1878). *in* Koch, L. & Keyserling, E. (1871–1883). *Die Arachniden Australiens.* Nürnberg : Bauer & Raspe Vol. 2 [980 pl 85 fig 7]. Type data: syntypes, ZMH 2F. adult, from Rockhampton, Peak Downs, Qld.

Distribution: NE coastal, Qld. Ecology: nomadic, predator; in grass.

Incertae sedis

Dolomedes elegans Koch, L. (1876). *in* Koch, L. & Keyserling, E. (1871–1883). *Die Arachniden Australiens.* Nürnberg : Bauer & Raspe Vol. 2 [861 pl 73 fig 6] [*non Dolomedes elegans* Taczanowski, 1873]. Type data: holotype, ZMH M. adult (this specimen has been examined and is not *Dolomedes* and is unlikely to be *Nilus*; it is probably a new genus, from Port Mackay, Qld).

Nilus kochi Roewer, C.F. (1951). Neue Namen einiger Araneen-Arten. *Abh. Naturw. Verh. Bremen.* **32**: 437–456 [438] [*nom. nov.* and *n. comb.* for *Dolomedes elegans* L. Koch, 1876; little is known of *Nilus* as the type species, *N. curtus* O.P. Cambridge, is described from a juvenile specimen from Egypt and is yet to be adequately diagnosed].

Distribution: NE coastal, Qld. Biological references: Simon, E. (1892–1903). *Histoire Naturelle des Araignées.* 2nd edn. Paris : Roret Vol. 2 [290] (taxonomy).

PSECHRIDAE

INTRODUCTION

Large, three-clawed, cribellate spiders with claw tufts. Cribellum divided. Abdomen long. Leg I much longer than rest. Calamistrum more than one row of bristles. Row of trichobothria on metatarsi and tarsi. Juvenile *Fecenia ochracea* build a conical web with a central detritus-covered retreat; at a later stage this is replaced by a "planar pseudo-orb" in which a curled leaf is incorporated as a retreat.

Fecenia Simon, 1887

Mezentia Thorell, T. (1881). Studi sui ragni Malesi e Papuani. III. Ragni dell'Austro-Malesia e del Capo York, conservati nel Museo Civico di Storia Naturale di Genova. *Ann. Mus. Civ. Stor. Nat. Genova* **17**: 1–720 [203] [*non Mezentia* Stäl, 1878]. Type species *Mezentia angustata* Thorell, 1881 (= *Tegenaria ochracea* Doleschall, 1859) by monotypy.

Fecenia Simon, E. (1887). [Observations sur divers Arachnides: synonymies et descriptions.] *Ann. Soc. Entomol. Fr. (Bull.) (6)* **7**: 158–194 [194] [replacement name for *Mezentia* Thorell, 1881].

Taxonomic decision of Simon, E. (1887). [Observations sur divers Arachnides: synonymies et descriptions.] *Ann. Soc. Entomol. Fr. (Bull.) (6)* **7**: 158–194 [194].

This group is also found in Malaysia, Papua New Guinea, India, Burma and Indonesia, see Bonnet, P. (1945–1961). *Bibliographia Araneorum.* Toulouse : Douladoure Vol. 2.

Fecenia ochracea (Doleschall, 1859)

Tegenaria ochracea Doleschall, C.L. (1859). Tweede bijdrage tot de kennis der Arachniden van den Indischen Archipel. *Acta Soc. Sci. Ind.-Neerl.* **5**: 1–60 [50]. Publication date established from Bonnet, P. (1945–1961). *Bibliographia Araneorum.* Toulouse : Douladoure Vol. 2. Type data: holotype, RMNH *F. adult lost, from Ambon, Indonesia, see Lehtinen, P.T. (1967). Classification of the cribellate spiders and some allied families, with notes on the evolution of the suborder Araneomorpha. *Ann. Zool. Fenn.* **4**: 199–468; Levi, H.W. (1982). The spider genera *Psechrus* and *Fecenia* (Araneae : Psechridae). *Pac. Insects* **24**: 114–138.

Mezentia angustata Thorell, T. (1881). Studi sui ragni Malesi e Papuani. III. Ragni dell'Austro-Malesia e del Capo York, conservati nel Museo Civico di Storia Naturale di Genova. *Ann. Mus. Civ. Stor. Nat. Genova* **17**: 1–720 [204]. Type data: holotype, MCG *F. adult, from Ternaté Is., Indonesia.

Distribution: NE coastal, Qld.; Horn Is., Sabai Is., Torres Strait. Ecology: sedentary, predator. Biological references: Robinson, M.H. & Lubin, Y.D. (1979). Specialists and generalists. The ecology and behaviour of some web-building spiders from Papua New Guinea. *Pac. Insects* **21**: 133–164.

SCYTODIDAE

INTRODUCTION

The spitting spiders, medium to large, haplogyne spiders with median posterior spiracle. They ensnare prey by squirting a sticky secretion on to it from the fangs. Six eyes in three diads. Dome-shaped carapace and spindly legs. One trichobothrium on metatarsi, none on tarsi. Female palp without a claw. Three tarsal claws on onychium, two rows of teeth on superior claws. Chelicerae fused at base, membraneous median lamella ends distally in strong tooth. Colulus present. Two ventral pits situated below female genital groove receive fangs of male during copulation.

Scytodes Latreille, 1804

Scytodes Latreille, P.A. (1804). Tableau méthodique des Insectes. *N. Dic. Hist. Nat.* **24**: 129–200 [134]. Publication date established from Bonnet, P. (1945–1961). *Bibliographia Araneorum.* Toulouse : Douladoure Vol. 2. Type species *Aranea thoracica* Latreille, 1802 by monotypy, Compiled from secondary source: Valerio, C.E. (1981). Spitting spiders (Araneae, Scytodidae, *Scytodes*) from Central America. *Bull. Am. Mus. Nat. Hist.* **170**: 80–89.

Dictis Koch, L. (1872). *in* Koch, L. & Keyserling, E. (1871–1883). *Die Arachniden Australiens.* Nürnberg : Bauer & Raspe Vol. 1 [294]. Type species *Dictis striatipes* L. Koch, 1872 by monotypy, see Valerio, C.E. (1981). Spitting spiders (Araneae, Scytodidae, *Scytodes*) from Central America. *Bull. Am. Mus. Nat. Hist.* **170**: 80–89.

Taxonomic decision of Simon, E. (1893). Voyage de M.E. Simon au Venezuela (décembre 1887–avril 1888) 21e Mémoire. Arachnides. *Ann. Soc. Entomol. Fr.* **61**: 423–462 [427].

This group is found worldwide, see Bonnet, P. (1945–1961). *Bibliographia Araneorum.* Toulouse : Douladoure Vol. 2. Species now known not to occur in Australia: *Scytodes striatipes* (L. Koch, 1872), a Polynesian species mistakenly reported from Sydney, N.S.W. by Rainbow, W. (1902). Notes on the architecture, nesting habits, and life histories of Australian Araneidae based on specimens in the Australian Museum. Part iii. *Rec. Aust. Mus.* **4**: 312–316; *Scytodes thoracica* (Latreille, 1802), *Scytodes* sp. in houses in Sydney has been identified as *S. perfecta* by P.T. Lehtinen (Gray, M.R., pers. comm.); the same species occurs in Brisbane, Mt. Isa, Qld. and Adelaide, S.A.

Scytodes fusca Walckenaer, 1837

Scytodes fusca Walckenaer, C.A. (1837). *Histoire Naturelle des Insectes Aptères.* Paris Vol. 1 682 pp. [272]. Publication date established from Bonnet, P. (1945–1961). *Bibliographia Araneorum.* Toulouse : Douladoure Vol. 2. Type data: syntypes, MNHP F.,M. adult lost, from Cayenne, French Guiana, South America, see Valerio, C.E. (1981). Spitting spiders (Araneae, Scytodidae, *Scytodes*) from Central America. *Bull. Am. Mus. Nat. His.* **170**: 80–89.

Distribution: NE coastal, Qld.; Barrier Reef islands. Ecology: nomadic, predator.

Scytodes longipes Lucas, 1844

Scytodes longipes Lucas, H. (1844). Notice sur une nouvelle espèce d'Aranéide appartenant au genre *Scytodes* de M. Walckenaer. *Rev. Zool.* **1844**: 72. Type data: holotype, MNHP lost, from Mexico. Compiled from secondary source: Valerio, C.E. (1981). Spitting spiders (Araneae, Scytodidae, *Scytodes*) from Central America. *Bull. Am. Mus. Nat. Hist.* **170**: 80–89.

Scytodes marmorata Koch, L. (1872). *in* Koch, L. & Keyserling, E. (1871–1883). *Die Arachniden Australiens.* Nürnberg : Bauer & Raspe Vol. 1 [292 pl 24 fig 4]. Type data: syntypes (5), ZMH (MGH 6548), from Samoa.

Scytodes penicillata Rainbow, W.J. (1916). Arachnida from Northern Queensland. Part 1. *Rec. Aust. Mus.* **11**: 33–64 [38]. Type data: holotype, AM KS6678 F. adult, from Goondi, near Innisfail, Qld.

Taxonomic decision of Davies, V.T., this work.

Distribution: NE coastal, Qld. Ecology: nomadic, predator; in houses.

Scytodes pallida Doleschall, 1859

Scytodes pallida Doleschall, C.L. (1859). Tweede bijdrage tot de kennis der Arachniden van den Indischen Archipel. *Act. Soc. Sci. Ind.-Neerl.* **5**: 1–60 [48]. Type data: holotype, RMNH F. adult, from Amboina. Compiled from secondary source: Roewer, C.F. (1942, 1954). *Katalog der Araneae.* Bremen : Paul Budy Bd. 1 1–1040.

Distribution: NE coastal, Qld. Ecology: nomadic, predator.

Scytodes perfecta Banks, 1898

Scytodes perfecta Banks, N. (1898). Arachnida from Baja California, and other parts of Mexico. *Proc. Calif. Acad. Sci. (3)* **1**: 205–308 [210]. Type data: holotype, ?MCZ F. adult, from Mexico.

Distribution: NE coastal, N Gulf, SE coastal, S Gulfs, Qld., N.S.W., S.A. Ecology: nomadic, predator; introduced. Biological references: see *S. thoracica* in all these references: Clyne, D. (1969). *Australian Spiders.* Sydney : Nelson 168 pp.; Mascord, R. (1970). *Australian Spiders in Colour.* Sydney : Reed 112 pp.; Mascord, R. (1980). *Spiders of Australia.* A field guide. Sydney: Reed 128 pp. Southcott, R.V. (1976). Arachnidism and allied syndromes in the Australian region. *Rec. Adel. Child. Hosp.* **1**: 97–186.

Scytodes tardigrada Thorell, 1881

Scytodes tardigrada Thorell, T. (1881). Studi sui ragni Malesi e Papuani. III. Ragni dell'Austro Malesia e del Capo York, conservati nel Museo Civico di Storia Naturale di Genova. *Ann. Mus. Civ. Stor. Nat. Genova* **17**: 1–720 [181]. Type data: syntypes, MCG F.,M. adult, from Somerset, Cape York, Qld.

Distribution: NE coastal, Qld.; Cape York, Torres Strait islands. Ecology: nomadic, predator.

SEGESTRIIDAE

INTRODUCTION

Medium to large, three-clawed, haplogyne spiders with six eyes. Pair of tracheal spiracles just behind book-lungs (tracheae to cephalothorax). Colulus present. Long abdomen sometimes with chevron pattern. The spiders have a cylindrical shape and porrect chelicerae; both adaptations to life in a burrow. Unlike the lycosids and other burrowing three-clawed spiders the segestriids have the third pair of legs directed forwards. The common Australian genus, *Ariadna*, lives in insect holes in tree trunks and branches. The holes are lined with silk and about a dozen short trip-lines radiate from the opening.

Ariadna Savigny and Audouin, 1825

Ariadna Savigny, J.C. & Audouin, V. (1825). Explication sommaire des planches d'Arachnides de l'Égypte et de la Syrie. *Description de l'Égypte Histoire Naturelle.* **1**: 1–339 [109]. Publication date established from Bonnet, P. (1945–1961). *Bibliographia Araneorum.* Toulouse : Douladoure Vol. 2. Type species *Ariadna insidiatrix* Savigny and Audouin, 1825 by monotypy.

Macedonia Hogg, H.R. (1900). A contribution to our knowledge of the spiders of Victoria; including some new species and genera. *Proc. R. Soc. Vict.* **13**: 68–123 [86]. Type species *Macedonia burchelli* Hogg, 1900 by monotypy and original designation.

Taxonomic decision of Simon, E. (1908). Araneae. 1re partie pp. 359–446 *in* Michaelsen, R. & Hartmeyer, W. (eds.) *Die Fauna Südwest-Australiens. 1(12)* Jena : Fischer [380].

This group is found worldwide, see Bonnet, P. (1945–1961). *Bibliographia Araneorum.* Toulouse : Douladoure Vol. 2.

Ariadna burchelli (Hogg, 1900)

Macedonia burchelli Hogg, H.R. (1900). A contribution to our knowledge of the spiders of Victoria; including some new species and genera. *Proc. R. Soc. Vict.* **13**: 68–123 [86]. Type data: holotype, ?NMV *F. adult, from Macedon, Vic.

Distribution: SE coastal, Vic. Ecology: sedentary, predator.

Ariadna decatetracantha Main, 1954

Ariadna decatetracantha Main, B.Y. (1954). Spiders and Opiliones. *Aust. Geogr. Soc. Rep.* **1**: 37–53 [40]. Type data: holotype, WAM *F. adult, from Twin Peaks, Recherche Archipelago, W.A.

Distribution: SW coastal, W.A. Ecology: sedentary, predator.

Ariadna dysderina L. Koch, 1873

Ariadna dysderina Koch, L. (1873). *in* Koch, L. & Keyserling, E. (1871–1883). *Die Arachniden Australiens.* Nürnberg : Bauer & Raspe Vol. 1 [447 pl 35 fig 2] [as *Ariadne dysderina*]. Type data: holotype, ZMH (MGH 8083) *F. adult, from Bowen, Qld., see Rack, G. (1961). Die Entomologischen Sammlungen des Zoologischen Staatsinstituts und Zoologischen Museums Hamburg. II. *Mitt. Hamb. Zool. Mus. Inst.* **59**: 1–60.

Distribution: NE coastal, Qld. Ecology: sedentary, predator.

Ariadna major Hickman, 1929

Ariadna major Hickman, V.V. (1929). Studies in Tasmanian spiders. Part III. *Pap. Proc. R. Soc. Tasm.* **1928**: 96–118 [100]. Type data: holotype, QVM 41 *F. adult, from Mt. Hobbs, Woodsdale, Tas.

Distribution: Tas. Ecology: sedentary, predator; under stones.

Ariadna montana Rainbow, 1920

Ariadna montana Rainbow, W.J. (1920). Arachnida from Lord Howe and Norfolk Islands. *Rec. S. Aust. Mus.* **1**: 229–272 [233]. Type data: holotype, SAMA N1981244 F. ?adult, from Lord Howe Is.

Distribution: Lord Howe Is. Ecology: sedentary, predator.

Ariadna muscosa Hickman, 1929

Ariadna muscosa Hickman, V.V. (1929). Studies in Tasmanian spiders. Part III. *Pap. Proc. R. Soc. Tasm* **1928**: 96–118 [103]. Type data: syntypes, QVM 27 *M.,F. adult, from Punch Bowl Reserve, Launceston, Tas

Distribution: Tas. Ecology: sedentary, predator; in moss.

Ariadna octospinata (Lamb, 1911)

Macedonia octospinata Lamb, J. (1911). Descriptions of some new Queensland Araneidae. *Ann. Qd. Mus.* **10**: 169–174 [169]. Type data: holotype, QM W2119 F. adult, from Stafford-on-Kedron, Brisbane, Qld.

Distribution: NE coastal, Qld. Ecology: sedentary, predator. Biological references: Rainbow, W.J. (1916). Arachnida from Northern Queensland. Part. 1. *Rec. Aust. Mus.* **11**: 33–64 (taxonomy).

Ariadna segmentata Simon, 1893

Ariadna segmentata Simon, E. (1893). Études arachnologiques. 25ᵉ Mémoire. XL. Descriptions d'espèces et de genres nouveaux de l'ordre des Araneae. *Ann. Soc. Entomol. Fr.* **62**: 299–330 [306]. Type data: holotype, ?MNHP *F. adult, from Launceston, Tas.

Ariadna thyrianthina Simon, 1908

Ariadna thyrianthina Simon, E. (1908). Araneae. 1re partie pp. 359–446 *in* Michaelsen, R. & Hartmeyer, W. (eds.) *Die Fauna Südwest-Australiens.* 1(12) Jena : Fischer [380]. Type data: syntypes, ?MNHP *F. adult, from Edel Land, Baba Head, Tamala, Norseman, Albany, W.A.

Distribution: SW coastal, W.A. Ecology: sedentary, predator.

Gippsicola Hogg, 1900

Gippsicola Hogg, H.R. (1900). A contribution to our knowledge of the spiders of Victoria; including some new species and genera. *Proc. R. Soc. Vict.* **13**: 68–123 [85]. Type species *Gippsicola raleighi* Hogg, 1900 by monotypy and original designation. Simon (1908) *op. cit.* placed *Gippsicola* in synonymy with *Segestria*.

Gippsicola raleighi Hogg, 1900

Gippsicola raleighi Hogg, H.R. (1900). A contribution to our knowledge of the spiders of Victoria; including some new species and genera. *Proc. R. Soc. Vict.* **13**: 68–123 [88]. Type data: holotype, ?NMV F. adult, from Gembrook, Gippsland, Vic.

Distribution: SE coastal, Vic. Ecology: sedentary, predator.

SELENOPIDAE

INTRODUCTION

Very flat, medium-sized, laterigrade spiders found under bark and stones in warmer regions of the world. Two claws and claw tufts present. Carapace wider than long; sternum circular. Six eyes in anterior row and two (PLE) behind. PME between AME and ALE. PLE largest, situated behind ALE. Colulus absent. Trichobothria in two rows on metatarsi and tarsi. The Australian specimens examined have the third leg longest and the first leg shortest. Tarsal scopulae sparse or absent.

Selenops Dufour, 1817

Selenops Dufour, L. (1817). *in* Latreille, P.A. Les arachnides. pp. 73–125 *in* Cuvier, G. *Le Règne Animal,* distribué d'après son organisation, pour servir de base à l'histoire naturelle des animaux et d'introduction à l'anatomie comparée. Paris Vol. 3 [92]. Publication date established from Bonnet, P. (1945–1961). *Bibliographia Araneorum.* Toulouse : Douladoure Vol. 1. Type species *Selenops radiatus* Dufour, 1819 by subsequent designation, see Latreille, P.A. (1819). *N. Dic. Hist. Nat.* N. édn. 30 Paris [579]. Compiled from secondary source: Lehtinen, P.T. (1967). Classification of the cribellate spiders and some allied families, with notes on the evolution of the suborder Araneomorpha. *Ann. Zool. Fenn.* **4**: 199–468.

This group is found worldwide, see Bonnet, P. (1945–1961). *Bibliographia Araneorum.* Toulouse : Douladoure Vol. 2.

Selenops australiensis L. Koch, 1875

Selenops australiensis Koch, L. (1875). *in* Koch, L. & Keyserling, E. (1871–1883). *Die Arachniden Australiens.* Nürnberg : Bauer & Raspe Vol. 1 [615 pl 48 fig 6] [as *Selenops Australiensis*]. Type data: holotype, ZMH *juv. from Bowen, Qld., F. adult from Sydney was subsequently (1876) described by Koch *in* Koch, L. & Keyserling, E. (1871–1883). *Die Arachniden Australiens.* Nürnberg : Bauer & Raspe Vol. 2 (832 pl 71 fig 3) and is in ZMH.

Distribution: NE coastal, SE coastal, Qld., N.S.W. Ecology: nomadic, predator; under bark. Biological references: Mascord, R. (1970). *Australian Spiders in Colour.* Sydney : Reed 112 pp. (biology); Clyne, D. (1969). *Australian Spiders.* Sydney : Nelson 168 pp. (biology).

SYMPHYTOGNATHIDAE

INTRODUCTION

Minute to tiny, three-clawed spiders with book-lungs replaced by tracheae. Chelicerae fused, at least at their base. No scuta on abdomen. Six to four eyes in triads or diads. High clypeus. Female palp reduced to trochanter and coxa or just coxa (maxilla). Male palp without apophyses on femur, patella or tibia. Labium much wider than long, sternum broadly truncate posteriorly. Tarsi longer than metatarsi. Anterior spiracles (tracheae to cephalothorax and abdomen or to cephalothorax only). In *Anapistula* a pair of posterior spiracles (tracheae to cephalothorax) present. Webs are closely woven, horizontal orb webs or irregular threads in more or less horizontal plane. The family was first described (from Australia) by Hickman in 1931 and, until recently, served as a dumping ground for many minute spiders.

Anapistula Gertsch, 1941

Anapistula Gertsch, W.J. (1941). Report on some arachnids from Barro Colorado Island, Canal Zone. *Am. Mus. Novit.* 1146: 1–14 [2]. Type species *Anapistula secreta* Gertsch, 1941 by monotypy and original designation.

This group is also found on Barro Colorado Is., Panama Canal Zone, Bahama Ils., Columbia, Costa Rica, Mexico, Forida, U.S.A., Jamaica, and Zaire, see Gertsch, W.J. (1941). Report on some arachnids from Barro Colorado Island, Canal Zone. *Am. Mus. Novit.* 1146: 1–14; Forster, R.R & Platnick, N.I. (1977). A review of the family Symphytognathidae (Arachnida, Araneae). *Am. Mus. Novit.* 2619: 1–29.

Anapistula australia Forster, 1959

Anapistula australia Forster, R.R. (1959). Spiders of the family Symphytognathidae. *Trans. R. Soc. N.Z.* **86**: 269–329 [321]. Type data: holotype, QM S104 F. adult (slide), from Camp Mt., near Brisbane, Qld.

Distribution: NE coastal, Qld. Ecology: sedentary, predator.

Symphytognatha Hickman, 1931

Symphytognatha Hickman, V.V. (1931). A new family of spiders. *Proc. Zool. Soc. Lond.* **1931**: 1321–1328 [1322]. Type species *Symphytognatha globosa* Hickman, 1931 by monotypy.

This group is also found in Brazil, Mexico, Belize and Jamaica, see Forster, R.R. & Platnick, N.I. (1977). A review of the family Symphytognathidae (Arachnida, Araneae). *Am. Mus. Novit.* 2619: 1–29.

Symphytognatha blesti Forster and Platnick, 1977

Symphytognatha blesti Forster, R.R. & Platnick, N.I. (1977). A review of the spider family Symphytognathidae (Arachnida, Araneae). *Am. Mus. Novit.* 2619: 1–29 [7]. Type data: holotype, AM KS0657 *M. adult, paratype AM KS0664 *F. adult, from Berowra, N.S.W.

Distribution: SE coastal, N.S.W. Ecology: sedentary, predator.

Symphytognatha globosa Hickman, 1931

Symphytognatha globosa Hickman, V.V. (1931). A new family of spiders. *Proc. Zool. Soc. Lond.* **1931**: 1321–1328 [1322]. Type data: holotype, QVM 42 *M. adult, paratype QVM 46 *F. adult, from Punch Bowl, Launceston, Tas.

Distribution: Tas. Ecology: sedentary, predator.

TETRABLEMMIDAE

INTRODUCTION

Haplogyne spiders with characteristic pattern of abdominal sclerotization consisting of a large dorsal scutum, four ventral scuta, and three or four narrow, lateral and posterior scuta. Sexual dimorphism common. Usually six or four eyes, sometimes two, one or none. Onychium with three claws, single row of teeth on paired claws (*cf.* Oonopidae). Colulus present. Two rows of trichobothria on tibia, one trichobothrium on metatarsi, none on tarsi. Male palp without subtegulum and seldom with conductor. Posterior respiratory organs absent. In Victoria tetrablemmids are found in crevices of eucalyptus bark. Undescribed species have been found in litter in both sclerophyll and rainforest areas of Queensland and in monsoon forest of the Northern Territory. An undescribed *Ablemma* sp. has been collected in north Queensland (Harvey, M., pers. comm.).

Tetrablemma O.P. Cambridge, 1873

Tetrablemma Cambridge, O.P. (1873). On some new genera and species of Araneidea. *Proc. Zool. Soc. Lond.* **1873**: 112–129 [114]. Type species *Tetrablemma medioculatum* O.P. Cambridge, 1873 by monotypy.

This group is also found in India, Angola, Trinidad, Sri Lanka, Seychelle Ils., American Samoa, St. Helena Is., Indonesia, Vietnam, Nepal and South Mariana Is., see Lehtinen, P.T. (1981). Spiders of the Oriental-Australian region. III. Tetrablemmidae, with a world revision. *Acta Zool. Fenn.* **162**: 1–151. Shear, W.A. (1978). Taxonomic notes on the armored spiders of the family Tetrablemmidae and Pacullidae. *Am. Mus. Novit.* 2650: 1–46.

Tetrablemma okei Butler, 1932

Tetrablemma okei Butler, L.S.G. (1932). Studies in Australian spiders. No. 2. *Proc. R. Soc. Vict. (ns)* **44**: 103–117 [111]. Type data: holotype, NMV *F. adult, from Bairnsdale, Vic.

Distribution: SE coastal, Vic. Ecology: nomadic, predator.

TEXTRICELLIDAE

INTRODUCTION

Tiny, three-clawed spiders without book-lungs, having anterior spiracles (tracheae to abdomen only). Abdomen with or without a dorsal scutum. Eight eyes, AME small. Chelicera often with anterior prolateral boss; labium wider than long; sternum broadly obtuse. Colulus present. Female palp entire, without claw. Male palp with patellar apophysis. Tarsi longer than metatarsi. No tarsal trichobothria, one or none on metatarsi. Male with stridulating ridges on chelicerae and a long, stout, distal, prolateral bristle on tibia I. Small sheet web. This family was first described from Tasmania by Hickman in 1945; it is also found in New Zealand and New Guinea and a new genus, *Tricella* was described recently from Chile. *Eterosonycha* Butler, 1932 appears to be a textricellid. This name would have priority over *Textricella* Hickman, 1945 if, as seems likely, *Textricella parva* is found to be congeneric with *Eterosonycha alpina*.

Eterosonycha Butler, 1932

Eterosonycha Butler, L.S.G. (1932). Studies in Australian spiders. No. 2. *Proc. R. Soc. Vict. (ns)* **44**: 103–117 [114]. Type species *Eterosonycha alpina* Butler, 1932 by monotypy and original designation.

Eterosonycha alpina Butler, 1932

Eterosonycha alpina Butler, L.S.G. (1932). Studies in Australian spiders. No. 2. *Proc. R. Soc. Vict. (ns)* **44**: 103–117 [115]. Type data: syntypes, NMV K098 (on slide) *F. adult, NMV K099 *M. adult, from Mt. Kosciusko, N.S.W.

Distribution: SE coastal, N.S.W. Ecology: predator; in moss.

Textricella Hickman, 1945

Textricella Hickman, V.V. (1945). A new group of Apneumone spiders. *Trans. Conn. Acad. Arts Sci.* **36**: 135–148 [136]. Type species *Textricella parva* Hickman, 1945 by original designation.

This group is also found in New Zealand and Chile, see Forster, R.R. (1959). Spiders of the family Symphytognathidae. *Trans. R. Soc. N.Z.* **86**: 269–329; Forster, R.R. & Platnick, N. (1981). A textricellid spider from Chile (Araneae, Textricellidae). *Bull. Am. Mus. Nat. Hist.* **170**: 263–270.

Textricella complexa Forster, 1959

Textricella complexa Forster, R.R. (1959). Spiders of the family Symphytognathidae. *Trans. R. Soc. N.Z.* **86**: 269–329 [277]. Type data: holotype, QM S108 M. adult, paratype QM S108 F. adult, from Royal Natl. Park, Sydney, N.S.W.

Distribution: SE coastal, N.S.W. Ecology: sedentary, predator, litter.

Textricella fulva Hickman, 1945

Textricella fulva Hickman, V.V. (1945). A new group of Apneumone spiders. *Trans. Conn. Acad. Arts Sci.* **36**: 135–148 [140]. Type data: syntypes, AM KS6692 *M.,F. adult, from Mt. Wellington, Tas.

Distribution: Tas. Ecology: sedentary, predator; in moss.

Textricella hickmani Forster, 1959

Textricella hickmani Forster, R.R. (1959). Spiders of the family Symphytognathidae. *Trans. R. Soc. N.Z.* **86**: 269–329 [280]. Type data: holotype, QM S109 M. adult, paratype QM S109 F. adult, from Mt. Wellington, Hobart, Tas.

Distribution: Tas.; Mt. Wellington. Ecology: sedentary, predator; in moss.

Textricella lamingtonensis Forster, 1959

Textricella lamingtonensis Forster, R.R. (1959). Spiders of the family Symphytognathidae. *Trans. R. Soc. N.Z.* **86**: 269–329 [281]. Type data: holotype, M. adult not in QM, paratype QM S110 F. adult, from Lamington Natl. Park, Qld.

Distribution: NE coastal, Qld. Ecology: sedentary, predator, closed forest.

Textricella luteola Hickman, 1945

Textricella luteola Hickman, V.V. (1945). A new group of Apneumone spiders. *Trans. Conn. Acad. Arts Sci.* **36**: 135–148 [144]. Type data: syntypes, AM KS6693 *2M.,F. adult, from Mt. Wellington, Hobart, Tas.

Distribution: SE coastal, NE coastal, Qld., N.S.W., Tas. Ecology: sedentary, predator, litter; in moss.

Textricella parva Hickman, 1945

Textricella parva Hickman, V.V. (1945). A new group of Apneumone spiders. *Trans. Conn. Acad. Arts Sci.* **36**: 135–148 [137]. Type data: syntypes, AM KS6694 *2M.,F. adult, from Cascades, Hobart, Tas.

Distribution: Tas. Ecology: sedentary, predator; in moss.

THERIDIOSOMATIDAE

INTRODUCTION

The "ray" spiders, tiny to small, three-clawed spiders with spherical abdomen. Anterior sternal indentations on either side of labium. Sternum broadly truncate posteriorly. Female palp without a claw. Small orb-webs are made with few radials and spirals. From the centre of the orb a horizontal thread runs to a point of attachment. The spiders rest, back to the web, on this; as it shortens this thread the web is pulled into a cone shape and when prey touches the web the thread is released and the web springs forwards and straightens with the spider in its centre.

Theridiosoma O.P. Cambridge, 1879

Theridiosoma Cambridge, O.P. (1879). On some new and rare British spiders, with characters of a new genus. *Ann. Mag. Nat. Hist (5)* **4**: 190–215 [193]. Type species *Theridiosoma argentiolum* O.P. Cambridge, 1879 (= *Theridium gemmosum* L. Koch, 1877) by subsequent designation, see Simon, E. (1881). *Les Arachnides de France.* Paris Tom. 5 Pt. 1 179 pp. [26].

This group is found worldwide, see Bonnet, P. (1945–1961). *Bibliographia Araneorum.* Toulouse : Douladoure Vol. 2; Roewer, C.F. (1942,1954). *Katalog der Araneae.* Bremen : Paul Budy Bd. 1 1–1040.

Theridiosoma brauni Wunderlich, 1976

Theridiosoma brauni Wunderlich, J. (1976). Spinnen aus Australien. 1. Uloboridae, Theridiosomatidae und Symphytognathidae. *Senckenberg. Biol.* **57**: 113–124 [117]. Type data: holotype, SMF 29058 *M. adult, paratypes SMF 29059 *M.,F. adult, from Lamington Natl. Park, Qld.

Distribution: NE coastal, Qld. Ecology: sedentary, predator.

Theridiosoma circuloargenteum Wunderlich, 1976

Theridiosoma circuloargenteum Wunderlich, J. (1976). Spinnen aus Australien. 1. Uloboridae, Theridiosomatidae und Symphytognathidae. *Senckenberg. Biol.* **57**: 113–124 [118]. Type data: holotype, SMF 29060 *M. adult, from Ku-ring-gai Natl. Park, N.S.W.

Distribution: SE coastal, N.S.W. Ecology: sedentary, predator.

ULOBORIDAE

INTRODUCTION

Small to medium sized, three-clawed, cribellate spiders. Trichobothria in prolateral row on femora II–IV; two rows on tibiae, one trichobothrium on metatarsi, none on tarsi. Posterior tracheal spiracle with tracheae extending to cephalothorax. No venom glands. Ventral row of erect macrosetae on distal 1/2–1/3 of metatarsus IV and most of tarsus IV. Prolateral metatarsal concavity adjacent to calamistrum. Broad retrolateral patellar notches. Two apical cymbial setae on male palp. Uloborids are weavers of geometric orbs or webs derived from an orb.

Incertae sedis: *Uloborus canus* Macleay, 1827. The description, in Latin, occupies two lines and may be literally translated - "whitish *Uloborus* with convex thorax; of the legs the second pair is longer, with femur spotted". Even from this scanty description it is apparent that the spider does not belong in *Uloborus*. Walckenaer, C.A. (1837–1847). *Histoire Naturelle des Insectes Aptères.* Paris Vol. 2 [503] believed it to be *Philodromus* sp., but Lehtinen, P.T. (1967). Classification of the cribellate spiders and some allied families, with notes on the evolution of the suborder Araneomorpha. *Ann. Zool. Fenn.* **4**: 199–468 [273] believes it is a thomisid and this is probably correct. As the holotype is lost and the locality unknown, it is likely that the name will be suppressed eventually.

Miagrammopes O.P. Cambridge, 1869

Miagrammopes Cambridge, O.P. (1869). Descriptions and sketches of two new species of Araneidea, with characters of a new genus. *J. Linn. Soc. Lond.* **10**: 398–405 [400]. Type species *Miagrammopes thwaitesii* O.P. Cambridge, 1869 by subsequent designation, see Simon, E. (1892–1903). *Histoire Naturelle des Araignées.* 2nd edn. Paris : Roret Vol. 1 [220], see Opell, B.D. (1979). Revision of the genera and tropical American species of the spider family Uloboridae. *Bull. Mus. Comp. Zool.* **148**: 443–549.

This group is found worldwide, see Bonnet, P. (1945–1961). *Bibliographia Araneorum.* Toulouse : Douladoure Vol. 2.

Miagrammopes bradleyi O.P. Cambridge, 1874

Miagrammopes bradleyi Cambridge, O.P. (1874). On some new genera and species of Araneidea. *Ann. Mag. Nat. Hist. (4)* **14**: 169–183 [177] [as *Miagrammopes Bradleyi*]. Type data: holotype, OUM *F. adult, from "northern Australia".

Distribution: "northern Australia". Ecology: sedentary, predator. Biological references: Mascord, R. (1980). *Spiders of Australia.* A field guide. Sydney : Reed 128 pp. (biology).

Miagrammopes caudatus Keyserling, 1890

Miagrammopes caudata Keyserling, E. (1890). *Die Arachniden Australiens.* Supplement. Nürnberg : Bauer & Raspe (1884–1890) [253] [as *Miagrammopes candata*]. Type data: holotype, ?ZMB *juv., from Peak Downs, Qld.

Distribution: NE coastal, Qld. Ecology: sedentary, predator.

Miagrammopes fasciatus Rainbow, 1916

Miagrammopes fasciata Rainbow, W.J. (1916). Arachnida from Northern Queensland. Part 1. *Rec. Aust. Mus.* **11**:33–64 [36]. Type data: holotype, AM KS6637 *F. adult, from Gordonvale, Qld.

Distribution: NE coastal, Qld. Ecology: sedentary, predator.

Miagrammopsidis Wunderlich, 1976

Miagrammopsidis Wunderlich, J. (1976). Spinnen aus Australien. 1. Uloboridae, Theridiosomatidae und Symphytognathidae (Arachnida : Araneida). *Senckenberg. Biol.* **57**: 113–124 [115]. Type species *Miagrammopsidis flavus* Wunderlich, 1976 by monotypy and original designation.

Miagrammopsidis flavus Wunderlich, 1976

Miagrammopsidis flavus Wunderlich, J. (1976). Spinnen aus Australien 1. Uloboridae, Theridiosomatidae und Symphytognathidae. *Senckenberg. Biol.* **57**: 113–124 [116]. Type data: holotype, SMF 29087 *M. adult, from Hartley's Creek, Cairns, Qld.

Distribution: NE coastal, Qld. Ecology: sedentary, predator.

Philoponella Mello-Leitao, 1917

Philoponella Mello-Leitao, C.F. de (1917). Generos e especies novas de Araneidos. *Arch. Esc. Sup. Agr. Med. Veter.* **1**: 1–9 [8]. Publication date established from Lehtinen, P.T. (1967). Classification of the cribellate spiders and some allied families, with notes on the evolution of the suborder Araneomorpha. *Ann. Zool. Fenn.* **4**: 199–468. Type species *Uloborus republicanus* Simon, 1891 by monotypy and original designation. Compiled from secondary source: Opell, B.D. (1979). Revision of the genera and tropical American species of the spider family Uloboridae. *Bull. Mus. Comp. Zool.* **148**: 443–549.

This group is also found in U.S.A., West Indies and the Orient, see Opell, B.D. (1979). Revision of the genera and tropical American species of the spider family Uloboridae. *Bull. Mus. Comp. Zool.* **148**: 443–549.

Philoponella congregabilis (Rainbow, 1916)

Uloborus congregabilis Rainbow, W.J. (1916). Some new Araneidae from the county of Cumberland. *Aust. Zool.* **1**: 58–61 [59]. Type data: syntypes, AM KS6766 M.,2F. adult, from Parramatta, N.S.W.

Distribution: SE coastal, N.S.W. Ecology: sedentary, predator. Biological references: Lehtinen, P.T. (1967). Classification of the cribellate spiders and some allied families, with notes on the evolution of the suborder Araneomorpha. *Ann. Zool. Fenn.* **4**: 199–468 (taxonomy); Mascord, R. (1980). *Spiders of Australia.* A field guide. Sydney : Reed 128 pp. (biology); Main, B.Y. (1976). *Spiders.* Sydney : Collins 296 pp. (biology); Clyne, D. (1969). *Australian Spiders.* Sydney : Nelson 168 pp. (biology).

Philoponella pantherinus (Keyserling, 1890)

Uloborus pantherinus Keyserling, E. (1890). *Die Arachniden Australiens.* Supplement. Nürnberg : Bauer & Raspe (1884–1890) [234]. Type data: syntype (probable), ZMH *F. adult, from Sydney, N.S.W.

Taxonomic decision of Davies, V.T., this work.

Distribution: SE coastal, N.S.W. Ecology: sedentary, predator. Biological references: Mascord, R. (1980). *Spiders of Australia.* A field guide. Sydney : Reed 128 pp. (biology).

Philoponella variabilis (Keyserling, 1887)

Uloborus variabilis Keyserling, E. (1887). *Die Arachniden Australiens.* Supplement. Nürnberg : Bauer & Raspe (1884–1890) [229]. Type data: syntypes, ZMH*, BMNH *F. adult, from Rockhampton, Peak Downs, Gayndah, Qld. and Sydney, N.S.W., see Rack, G. (1961). Die Entomologischen Sammlungen des Zoologischen Staatsinstituts und Zoologischen Museums Hamburg. II. *Mitt. Hamb. Zool. Mus. Inst.* **59**: 1–60.

Distribution: NE coastal, SE coastal, Qld., N.S.W. Ecology: sedentary, predator. Biological references: Lehtinen, P.T. (1967). Classification of the cribellate spiders and some allied families, with notes on the evolution of the suborder Araneomorpha. *Ann. Zool. Fenn.* **4**: 199–468 (taxonomy).

Ranguma Lehtinen, 1967

Ranguma Lehtinen, P.T. (1967). Classification of the cribellate spiders and some allied families, with notes on the evolution of the suborder Araneomorpha. *Ann. Zool. Fenn.* **4**: 199–468 [395]. Type species *Miagrammopes similis* Kulczynski, 1908 by original designation.

This group is also found in Indochina, Malaysia, Burma, Japan, Singapore and New Guinea, see Lehtinen, P.T. (1967). Classification of the cribellate spiders and some allied families, with notes on the evolution of the suborder Araneomorpha. *Ann. Zool. Fenn.* **4**: 199–468.

Ranguma lehtineni Wunderlich, 1976

Ranguma lehtineni Wunderlich, J. (1976). Spinnen aus Australien. I. Uloboridae, Theridiosomatidae und Symphytognathidae. (Arachnida : Araneida). *Senckenberg. Biol.* **57**: 113–124 [113]. Type data: holotype, SMF 29085 *M. adult, from Lamington Natl. Park, Qld.

Distribution: NE coastal, Qld. Ecology: sedentary, predator.

Uloborus Latreille, 1806

Uloborus Latreille, P.A. (1806). *Genera Crustaceorum et Insectorum.* Aranéides. Tom. 1 : 82–127 [109]. Publication date established from Bonnet, P. (1945–1961). *Bibliographia Araneorum.* Toulouse : Douladoure Vol. 2. Type species *Uloborus walckenaerius* Latreille, 1806 by monotypy. Compiled from secondary source: Opell, B.D. (1979). Revision of the genera and tropical American species of the spider family Uloboridae. *Bull. Mus. Comp. Zool.* **148**: 443–549.

This group is found worldwide, see Bonnet, P. (1945–1961). *Bibliographia Araneorum.* Toulouse : Douladoure Vol. 2.

Uloborus barbipes L. Koch, 1872

Uloborus barbipes Koch, L. (1872) *in* Koch, L. & Keyserling, E. (1871–1883). *Die Arachniden Australiens.* Nürnberg : Bauer & Raspe Vol. 1 [229 pl 19 fig 9]. Type data: holotype, ZMH (MGH 7843) *F. adult, from Mackay, Qld., see Rack, G. (1961). Die Entomologischen Sammlungen des Zoologischen Staatsinstituts und Zoologischen Museums Hamburg. II. *Mitt. Hamb. Zool. Mus. Inst.* **59**: 1–60.

Distribution: NE coastal, Qld. Ecology: sedentary, predator.

Zosis Walckenaer, 1837

Zosis Walckenaer, C.A. (1837). *Histoire Naturelle des Insectes Aptères.* Paris 4 vols+atlas (1837–1847) [atlas pl 20 fig 2]. Publication date established from Opell, B.D. (1979). Revision of the genera and tropical American species of the spider family Uloboridae. *Bull. Mus. Comp. Zool.* **148**: 443–549. Type species *Zosis caraibe* Walckenaer, 1837 (=*Aranea geniculata* Olivier, 1789) by monotypy.

This group is cosmotropical, see Opell, B.D. (1979). Revision of the genera and tropical American species of the spider family Uloboridae. *Bull. Mus. Comp. Zool.* **148**: 443–549.

Zosis geniculatus (Olivier, 1789)

Aranea geniculata Olivier, G.A. (1789). Araignée, Aranea. *Encycl. Méth., Hist. Nat. Inst. Paris* **4**: 173–240 [214]. Type data: none designated, from "tropics", see Lehtinen, P.T. (1967). Classification of the cribellate spiders and some allied families, with notes on the evolution of the suborder Araneomorpha. *Ann. Zool. Fenn.* **4**: 199–468.

Uloborus spinitarsus Keyserling, E. (1887). *Die Arachniden Australiens.* Supplement. Nürnberg : Bauer & Raspe (1884–1890) [231]. Type data: holotype, ?ZMB *M. adult, from Middle Harbour, Sydney, N.S.W.

Taxonomic decision of Davies, V.T., this work.

Distribution: NE coastal, SE coastal, SW coastal, Qld., N.S.W., W.A. Ecology: sedentary, predator. Biological references: Mascord, R. (1980). *Spiders of Australia.* A field guide. Sydney : Reed 128 pp. (biology); Clyne, D. (1969). *Australian Spiders.* Sydney : Nelson 168 pp. (biology); Main, B.Y. (1976). *Spiders.* Sydney : Collins 296 pp.

ZODARIIDAE

INTRODUCTION

Medium to large, three-clawed hunting spiders, usually ground living. Six spinnerets, anterior long, median and posterior reduced. Serrula absent, or at least not seen at x50 magnification. Promargin of chelicera with one or two teeth, retromargin none. Eight eyes, both rows usually strongly procurved. High clypeus. One row trichobothria on metatarsi and tarsi. Zodariids are widespread in Australia, occurring in both dry desert areas and in the rainforests of eastern Australia. They are often active during the day. Many are dark brown with a pattern of white or yellow spots or stripes.

The following species are transferred to other families: *Storena albomaculata* Rainbow, 1902 =*Supunna albomaculata* comb. nov., *Storena variepes* Rainbow, 1912 =*Supunna picta* (L. Koch) synon. nov. and *Storena auripes* Rainbow, 1916 are transferred to the Clubionidae : Castianeirinae. The generic placement of the last is uncertain. *Storenosoma lycosoides* Hogg, 1900 belongs in Amaurobiidae. Forster & Wilton (1973) in The Spiders of New Zealand. Part IV *Otago Mus. Bull. No. 4* 309 pp. figured (figs 816, 820–3) but did not name or describe a *Storenosoma* sp. from N.S.W.

Incertae sedis: The following genera are regarded as not belonging in Zodariidae, and cannot at present be placed in a family: *Macedoniola* Strand, 1932 (*nom. nov.* for *Platycephala* Butler, 1932), the type of which is lost; *Sternodes* Butler, 1929 and *Perissopmeros* Butler, 1929.

Hetaerica Rainbow, 1916

Hetaerica Rainbow, W.J. (1916). Arachnida from Northern Queensland. Part 1. *Rec. Aust. Mus.* **11**: 33–64 [47]. Type species *Hetaerica aresca* Rainbow, 1916 by monotypy.

Hetaerica aresca Rainbow, 1916

Hetaerica aresca Rainbow, W.J. (1916). Arachnida from Northern Queensland. Part 1. *Rec. Aust. Mus.* **11**: 33–64 [48]. Type data: holotype, AM KS6744 juv., from Pentland, Qld.

Distribution: NE coastal, Qld. Ecology: arboreal, predator.

Neostorena Rainbow, 1914

Neostorena Rainbow, W.J. (1914). A new Victorian araneiad. *Aust. Zool.* **1**: 21–23 [21]. Type species *Neostorena venatoria* Rainbow, 1914 by original designation.

Neostorena venatoria Rainbow, 1914

Neostorena venatoria Rainbow, W.J. (1914). A new Victorian araneiad. *Aust. Zool.* **1**: 21–23 [22]. Type data: holotype, AM KS6627 F. juv., from Fern-Tree Gully, Vic.

Distribution: SE coastal, Vic. Ecology: terrestrial; predator.

Storena Walckenaer, 1805

Storena Walckenaer, C.A. (1805). *Tableau des Aranéides* ou caractères essentiels des tribus, genres, familles et races que renferme le genre Aranea de Linné, avec la désignation des espèces comprises dans chacune de ces divisions. Paris 88 pp. [83]. Publication date established from Bonnet, P. (1945–1961). *Bibliographia Araneorum.* Toulouse : Douladoure Vol. 2. Type species *Storena cyanea* Walckenaer, 1805 by monotypy.

Habronestes Koch, L. (1872). *in* Koch, L. & Keyserling, E. (1871–1883). *Die Arachniden Australiens.* Nürnberg : Bauer & Raspe Vol. 1 [299]. Type species *Habronestes striatipes* L. Koch, 1872 by subsequent designation, see Bonnet, P. (1945–1961). *Bibliographia Araneorum.* Toulouse : Douladoure Vol. 2 [2068].

Taxonomic decision of Simon, E. (1892–1903). *Histoire Naturelle des Araignées.* 2nd edn. Paris : Roret Vol. 1 [429].

This group is found worldwide excluding the northern Holarctic, see Bonnet, P. (1945–1961). *Bibliographia Araneorum.* Toulouse : Douladoure Vol. 2. Species now known not to occur in Australia: *Storena procera* Thorell, 1890 *vide* Thorell, T. (1890). Studi sui ragni Malesi e

Papuani. IV. *Ann. Mus. Civ. Stor. Nat. Genova* **17**: 1–419; Thorell says (p. 344) the place of capture is not known. It is either Dutch East Indies (Indonesia), Australia or the West Indies. As extensive field collecting has failed to re-collect the spider, it is now excluded.

Storena annulipes (L. Koch, 1867)

Enyo annulipes Koch, L. (1867). Beschreibungen neuer Arachniden und Myriapoden. *Verh. Zool.-Bot. Ges. Wien* **15**: 857–892 (1865); **17**: 173–250 (1867) [194]. Type data: holotype, ZMH juv., from Brisbane, Qld.

Distribution: NE coastal, Qld. Ecology: terrestrial, predator. Biological references: Simon, E. (1892–1903). *Histoire Naturelle des Araignées.* 2nd edn. Paris : Roret Vol. 1 (taxonomy).

Storena australiensis O.P. Cambridge, 1869

Storena australiensis Cambridge, O.P. (1869). Descriptions and sketches of some new species of Araneidea, with characters of a new genus. *Ann. Mag. Nat. Hist. (4)* **3**: 52–74 [58]. Type data: holotype, OUM *M. adult pinned, from unknown locality.

Distribution: unknown. Ecology: terrestrial, predator.

Storena braccata (L. Koch, 1865)

Enyo braccata Koch, L. (1865). Beschreibungen neuer Arachniden und Myriapoden. *Verh. Zool.-Bot. Ges. Wien* **15**: 857–892 (1865); **17**: 173–250 (1867) [859]. Type data: holotype(?), ZMH *F. juv., from Wollongong, N.S.W., see Rack, G. (1961). Die Entomologischen Sammlungen des Zoologischen Staatsinstituts und Zoologischen Museums Hamburg. II. *Mitt. Hamb. Zool. Mus. Inst.* **59**: 1–60.

Distribution: SE coastal, N.S.W. Ecology: terrestrial, predator. Biological references: Simon, E. (1892–1903). *Histoire Naturelle des Araignées.* 2nd edn. Paris : Roret Vol. 1 (taxonomy).

Storena bradleyi O.P. Cambridge, 1869

Storena bradleyi Cambridge, O.P. (1869). Descriptions and sketches of some new species of Araneidea, with characters of a new genus. *Ann. Mag. Nat. Hist. (4)* **3**: 52–74 [56] [as *Storena Bradleyi*]. Type data: syntypes, OUM *2M. adult, M. juv., from Sydney, N.S.W.

Distribution: SE coastal, N.S.W. Ecology: terrestrial, predator.

Storena colossea Rainbow, 1920

Storena colossea Rainbow, W.J. (1920). Arachnida from Lord Howe and Norfolk Islands. *Rec. S. Aust. Mus.* **1**: 229–272 [237]. Type data: syntypes, SAMA N1981256 F. adult, AM KS8878 *F. adult, from Lord Howe Is.

Distribution: Lord Howe Is. Ecology: terrestrial, predator.

Storena cyanea Walckenaer, 1805

Storena cyanea Walckenaer, C.A. (1805). *Tableau des Aranéides* ou caractères essentiels des tribus, genres, familles et races que renferme le genre Aranea de Linné, avec la désignation des espèces comprises dans chacune de ces divisions. Paris 88 pp. [83 pl 9 fig 85–86]. Type data: none designated, from unknown locality.

Distribution: unknown. Ecology: terrestrial, predator.

Storena eximia Simon, 1908

Storena eximia Simon, E. (1908). Araneae. 1re partie pp. 359–446 *in* Michaelsen, W. & Hartmeyer, R. (eds.) *Die Fauna Südwest-Australiens.* 1(12) Jena : Fischer [405]. Type data: syntypes, 2F. juv. whereabouts unknown, from Kalgoorlie and Boorabbin, W.A.

Distribution: W plateau, W.A. Ecology: terrestrial, predator.

Storena flavipedes (Urquhart, 1893)

Habronestes flavipedes Urquhart, A.T. (1893). On new species of Tasmanian Araneae. *Pap. Proc. R. Soc. Tasm.* **1892**: 94–130 [111]. Type data: holotype, CMNZ F. adult missing, from Tas.

Distribution: Tas. Ecology: terrestrial, predator. Biological references: Rainbow, W.J. (1911). A Census of Australian Araneidae. *Rec. Aust. Mus.* **9**: 107–391 (taxonomy).

Storena formosa Thorell, 1870

Storena formosa Thorell, T. (1870). Arancae nonnullae Novae Hollandiae, descriptae. *Öfvers. K. Vetensk Akad. Förh.* **27**: 367–389 [374]. Type data: holotype, ?NHRM *F. adult, from unknown locality.

Distribution: Murray-Darling basin, Lake Eyre basin, S Gulfs, W plateau, N.S.W., Qld., S.A., W.A.; semiarid inland areas of Australia. Ecology: terrestrial, predator. Biological references: Main, B.Y. (1976). *Spiders.* Sydney : Collins 296 pp.(biology); Mascord, R. (1970). *Australian Spiders in Colour.* Sydney : Reed 112 pp.

Storena graeffei L. Koch, 1866

Storena graeffei Koch, L. (1866). *Die Arachniden-Familie der Drassiden.* Nürnberg Vols. 1–6 304 pp. [192] [as *Storena Graeffei*]. Publication date established from Bonnet, P. (1945–1961). *Bibliographia Araneorum.* Toulouse : Douladoure Vol. 2. Type data: holotype, ZMH (MGH 301) juv., from N.S.W., see Roewer, C.F. (1942, 1954). *Katalog der Araneae.* Bremen : Paul Budy Bd. 1 1–1040; Rack, G. (1961). Die Entomologischen Sammlungen des Zoologischen Staatsinstituts und Zoologischen Museums Hamburg. II. *Mitt. Hamb. Zool. Mus. Inst.* **59**: 1–60.

Distribution: SE coastal, N.S.W. Ecology: terrestrial, predator. Biological references: Mascord, R. (1970). *Australian Spiders in Colour.* Sydney : Reed 112 pp.

Storena grimwadei Dunn, 1951

Storena grimwadei Dunn, R.A. (1951). Spiders of the Russell Grimwade Expedition. *Mem. Natl. Mus. Vict.* **17**: 9–18 [11]. Type data: holotype, NMV K113 *M. adult, paratypes NMV K114–5 *2F. adult, from 40 mi W of Eucla, W.A.

Distribution: W plateau, W.A. Ecology: terrestrial, predator.

Storena inornata Rainbow, 1916

Storena inornata Rainbow, W.J. (1916). Arachnida from Northern Queensland. Part 1. *Rec. Aust. Mus.* **11**: 33–64 [46]. Type data: holotype, F. adult missing, from Gordonvale, N Qld.

Distribution: NE coastal, Qld. Ecology: terrestrial, predator.

Storena leucosema Rainbow, 1920

Storena leucosema Rainbow, W.J. (1920). Arachnida from Lord Howe and Norfolk Islands. *Rec. S. Aust. Mus.* **1**: 229–272 [237]. Type data: syntypes, SAMA N1981254 F. adult, AM KS6685 *2F. adult, from Lord Howe Is.

Distribution: Lord Howe Is. Ecology: terrestrial, predator; in leaves and moss at high altitude.

Storena macedonensis Hogg, 1900

Storena macedonensis Hogg, H.R. (1900). A contribution to our knowledge of the spiders of Victoria; including some new species and genera. *Proc. R. Soc. Vict.* **13**: 68–123 [98]. Type data: holotype, BMNH 1907.2.24.38 M. adult, from Macedon, Vic.

Distribution: SE coastal, Vic. Ecology: terrestrial, predator.

Storena maculata O.P. Cambridge, 1869

Storena maculata Cambridge, O.P. (1869). Descriptions and sketches of some new species of Araneidea, with characters of a new genus. *Ann. Mag. Nat. Hist. (4)* **3**: 52–74 [59]. Type data: holotype, OUM *M. juv., from Swan River, W.A.

Distribution: SW coastal, W.A. Ecology: terrestrial, predator.

Storena ornata (Bradley, 1878)

Habronestes ornatus Bradley, H.H.B. (1878). Araneides of the *Chevert* Expedition. Part II. *Proc. Linn. Soc. N.S.W.* **2**: 115–122 [119]. Type data: holotype, MMUS M. juv. missing, from Cocoanut Is., Torres Strait.

Distribution: Qld.; Torres Strait islands. Ecology: terrestrial, predator. Biological references: Thorell, T. (1881). Studi sui ragni Malesi e Papuani. III. Ragni dell'Austro-Malesia e del Capo York, conservati nel Museo Civico di Storia Naturale di Genova. *Ann. Mus. Civ. Stor. Nat. Genova* **17**: 1–720 (taxonomy).

Storena picta (L. Koch, 1865)

Enyo picta Koch, L. (1865). Beschreibungen neuer Arachniden und Myriapoden. *Verh. Zool.-Bot. Ges. Wien* **15**: 857–892 (1865); **17**: 173–250 (1867) [861]. Type data: holotype, ?ZMH F. adult missing, from Wollongong, N.S.W.

Distribution: SE coastal, N.S.W. Ecology: terrestrial, predator. Biological references: Simon, E. (1892–1903). *Histoire Naturelle des Araignées.* 2nd edn. Paris : Roret Vol. 1 (taxonomy); Koch, L. & Keyserling, E. (1871–1883). *Die Arachniden Australiens.* Nürnberg : Bauer & Raspe Vol. 1 (taxonomy).

Storena rastellata Strand, 1913

Storena rastellata Strand, E. (1913). Über einige australische Spinnen des Senckenbergischen Museums. *Zool. Jb. (Syst.)* **35**: 599–624 [606]. Type data: holotype, SMF M. juv., from Central Australia.

Distribution: Lake Eyre basin, Qld., N.T., S.A. Ecology: terrestrial, predator.

Storena rufescens Thorell, 1881

Storena rufescens Thorell, T. (1881). Studi sui ragni Malesi e Papuani. III. Ragni dell'Austro Malesia e del Capo York, conservati nel Museo Civico di Storia Naturale di Genova. *Ann. Mus. Civ. Stor. Nat. Genova* **17**: 1–720 [188]. Type data: syntypes, MCG *F. adult, 2 juv., from Somerset, Cape York, Qld.

Distribution: N Gulf, Qld. Ecology: terrestrial, predator.

Storena scenica (L. Koch, 1872)

Habronestes scenicus Koch, L. (1872). *in* Koch, L. & Keyserling, E. (1871–1883). *Die Arachniden Australiens.* Nürnberg : Bauer & Raspe Vol. 1 [316 pl 25 fig 6]. Type data: holotype, ZMH (MGH 7876) M. adult, from Bowen, Qld.

Distribution: NE coastal, Qld. Ecology: terrestrial, predator. Biological references: Simon, E. (1892–1903). *Histoire Naturelle des Araignées.* 2nd edn. Paris : Roret Vol. 1 (taxonomy).

Storena scintillans O.P. Cambridge, 1869

Storena scintillans Cambridge, O.P. (1869). Descriptions and sketches of some new species of Araneidea, with characters of a new genus. *Ann. Mag. Nat. Hist. (4)* **3**: 52–74 [54]. Type data: holotype, OUM *F. adult, from Swan River, W.A.

Distribution: SW coastal, W.A. Ecology: terrestrial, predator.

Storena spirafera (L. Koch, 1872)

Habronestes spirafera Koch, L. (1872). *in* Koch, L. & Keyserling, E. (1871–1883). *Die Arachniden Australiens.* Nürnberg : Bauer & Raspe Vol. 1 [318 pl 25 fig 7] [as *Habronestes spirafer*]. Type data: holotype, M. adult missing from Port Mackay, Qld.

Distribution: NE coastal, Qld. Ecology: terrestrial, predator. Biological references: Simon, E. (1892–1903). *Histoire Naturelle des Araignées.* 2nd edn. Paris : Roret Vol. 1 (taxonomy).

Storena striatipes (L. Koch, 1872)

Habronestes striatipes Koch, L. (1872). *in* Koch, L. & Keyserling, E. (1871–1883). *Die Arachniden Australiens.* Nürnberg : Bauer & Raspe Vol. 1 [313 pl 25 fig 4]. Type data: syntypes, ZMH (MGH 6525) F. adult, F. adult whereabouts unknown, from Rockhampton and Bowen, Qld., see Rack, G. (1961). Die Entomologischen Sammlungen des Zoologischen Staatsinstituts und Zoologischen Museums Hamburg. II. *Mitt. Hamb. Zool. Mus. Inst.* **59**: 1–60.

Distribution: NE coastal, Qld. Ecology: terrestrial, predator. Biological references: Simon, E. (1892–1903). *Histoire Naturelle des Araignées.* 2nd edn. Paris : Roret Vol. 1 (taxonomy).

Storena tetrica Simon, 1908

Storena tetrica Simon, E. (1908). Araneae. 1re partie pp. 359–446 *in* Michaelsen, W. & Hartmeyer, R. (eds.) *Die Fauna Südwest-Australiens.* 1(12) Jena : Fischer [403]. Type data: holotype, ?ZMB *F. adult, from Albany, W.A.

Distribution: SW coastal, W.A. Ecology: terrestrial, predator.

Storena toddi Hickman, 1944

Storena toddi Hickman, V.V. (1944). The Simpson Desert Expedition, 1939. Scientific Report 1, Biology — Scorpions and spiders. *Trans. R. Soc. S. Aust.* **68**: 18–48 [38]. Type data: holotype, AM KS6686 *M. adult, from 6 mi N of junction of Todd & Hale Rivers, N.T.

Distribution: Lake Eyre basin, N.T. Ecology: terrestrial, predator.

Storena torosa Simon, 1908

Storena torosa Simon, E. (1908). Araneae. 1re partie pp. 359–446 *in* Michaelsen, W. & Hartmeyer, R. (eds.) *Die Fauna Südwest-Australiens.* 1(12) Jena : Fischer [404]. Type data: holotype, ?ZMB *F. adult, from Northampton, W.A.

Distribution: NW coastal, W.A. Ecology: terrestrial, predator.

Storena tricolor Simon 1908

Storena tricolor Simon, E. (1908). Araneae. 1re partie pp. 359–446 *in* Michaelsen, W. & Hartmeyer, R. (eds.) *Die Fauna Südwest-Australiens.* 1(12) Jena : Fischer [405]. Type data: syntypes, ZMH *M. juv., WAM, from Lion Mill, Collie, Boyanup, W.A.

Distribution: SW coastal, W.A. Ecology: terrestrial, predator.

Storena variegata O.P. Cambridge, 1869

Storena variegata Cambridge, O.P. (1869). Descriptions and sketches of some new species of Araneidea, with characters of a new genus. *Ann. Mag. Nat. Hist. (4)* **3**: 52–74 [53]. Type data: holotype, OUM *F. adult, from Swan River, W.A.

Distribution: SW coastal, S Gulfs, W.A., S.A. Ecology: terrestrial, predator.

Storena zebra Thorell, 1881

Storena zebra Thorell, T. (1881). Studi sui ragni Malesi e Papuani. III. Ragni dell'Austro-Malesia e del Capo York, conservati nel Museo Civico di Storia Naturale di Genova. *Ann. Mus. Civ. Stor. Nat. Genova* **17**: 1–720 [184]. Type data: syntypes, ?MCG *M. adult, F. juv., from Aru Is., Indonesia and Fly River, New Guinea.

Distribution: N Gulf, Qld.; Torres Strait islands. Ecology: terrestrial, predator.

ZORIDAE

INTRODUCTION

Small to large, two-clawed hunting spiders with claw-tufts. They usually have a narrow head region and long abdomen. Trochanters notched. Pairs (2–8) of ventral spines on tibiae I and II. Eyes 4.4, or 4.2.2; anterior row straight-recurved, posterior row recurved; usually ALE=AME (*Simonus* ALE smaller than AME). Low clypeus. Scopula present on anterior tarsi. Epigyne with anterior gonopores. Group of blunt or hooked spines on apex of male cymbium (exc. *Odo*) which is often grooved retrolaterally. Two rows trichobothria on metatarsi and tarsi. The strongly curved posterior row of eyes separates this family from Clubionidae and Miturgidae. The family is ill-defined and it is likely that some of the genera listed here will be transferred to other families. The Australian genera may even warrant a family of their own. *Horioctenoides*, the type of which is a penultimate female, may be a junior synonym of *Argoctenus*. *Horioctenoides* sp. from New Zealand [Forster, R.R. & Wilton, C.L. (1973). The Spiders of New Zealand. Part IV. *Otago Mus. Bull. No 4* (293–294)] was described and figured but not named. Unlike *Horioctenoides* Main it has tufts of hair near the eyes and on the abdomen. There are similar undescribed spiders in Australia. *Odo australiensis* , which is close to *Argoctenus*, has been left in the Central and South American genus *Odo*, pending revision of the family in Australia.

Argoctenus L. Koch, 1878

Argoctenus Koch, L. (1878). *in* Koch, L. & Keyserling, E. (1871–1883). *Die Arachniden Australiens.* Nürnberg : Bauer & Raspe Vol. 2 [990]. Type species *Argoctenus igneus* L. Koch, 1878 by subsequent designation, see Simon, E. (1892–1903). *Histoire Naturelle des Araignées.* 2nd edn. Paris : Roret Vol. 2 [132].

Aenigma Karsch, F. (1878). Exotisch-araneologisches II. *Z. Naturw.* **51**: 771–826 [825] [*non Aenigma* Strecker 1876]. Type species *Aenigma australiana* Karsch, 1878 by monotypy.

Aenigmaaranea Strand, E. (1929). Zoological and palaeontological nomenclatorical notes. *Act. Univ. Latv.* **20**: 1–29 [11] [*nom. nov.* for *Aenigma* Karsch, 1878].

Taxonomic decision of Davies, V.T., this work.

This group is also found in New Caledonia, see Lehtinen, P.T. (1967). Classification of the cribellate spiders and some allied families, with notes on the evolution of the suborder Araneomorpha. *Ann. Zool. Fenn.* **4**: 199–468; Forster, R.R. & Wilton, C.L. (1973). The spiders of New Zealand. Part IV. *Otago Mus. Bull. No. 4* 309 pp.

Argoctenus australianus (Karsch, 1878)

Aenigma australiana Karsch, F. (1878). Exotisch-araneologisches II. *Z. Naturw.* **51**: 771–826 [825]. Type data: holotype, ZMB F. adult (abdomen missing), from N.S.W.

Distribution: SE coastal, N.S.W. Ecology: nomadic, predator.

Argoctenus gracilis (Hickman, 1950)

Odo gracilis Hickman, V.V. (1950). Araneae from Reevesby Island, South Australia. *Proc. R. Soc. Vict. (ns)* **60**: 1–16 [8]. Type data: syntypes, NMV K146 M. adult, from Reevesby Is., S.A.

Taxonomic decision of Davies, V.T., this work.

Distribution: S Gulfs, S.A. Ecology: nomadic, predator.

Argoctenus hystriculus Simon, 1909

Argoctenus hystriculus Simon, E. (1909). Araneae. 2e partie pp. 155–212 *in* Michaelsen, W. & Hartmeyer, R. (eds.) *Die Fauna Südwest-Australiens.* 2(13) Jena : Fischer [164]. Type data: holotype, ?ZMB *F. adult, from Cannington, W.A.

Distribution: SW coastal, W.A. Ecology: nomadic, predator.

Argoctenus igneus L. Koch, 1878

Argoctenus igneus Koch, L. (1878). *in* Koch, L. & Keyserling, E. (1871–1883). *Die Arachniden Australiens.* Nürnberg : Bauer & Raspe Vol. 2 [990 pl 86 fig 4]. Type data: holotype, ?ZMH *F. adult lost, from King George Sound, W.A.

Distribution: SW coastal, W.A. Ecology: nomadic, predator.

Argoctenus nebulosus Simon, 1909

Argoctenus nebulosus Simon, E. (1909). Araneae. 2^e partie pp. 155–212 *in* Michaelsen, W. & Hartmeyer, R. (eds.) *Die Fauna Südwest-Australiens.* 2(13) Jena : Fischer [163]. Type data: holotype, ?ZMB *F. adult, from Cranbrook, W.A.

Distribution: SW coastal, W.A. Ecology: nomadic, predator.

Argoctenus pectinatus Hogg, 1900

Argoctenus pectinatus Hogg, H.R. (1900). A contribution to our knowledge of the spiders of Victoria; including some new species and genera. *Proc. R. Soc. Vict.* **13**: 68–123 [106]. Type data: holotype, *M. adult whereabouts unknown, from Frankston, Port Phillip Bay, Vic.

Distribution: SE coastal, Vic. Ecology: nomadic, predator.

Argoctenus pictus L. Koch, 1878

Argoctenus pictus Koch, L. (1878) *in* Koch, L. & Keyserling, E. (1871–1883). *Die Arachniden Australiens.* Nürnberg : Bauer & Raspe Vol. 2 [992 pl 86 fig 5]. Type data: syntypes, ZMH *F. adult ?lost, BMNH 1916.6.1.852 F. adult, from Sydney, N.S.W.

Distribution: SE coastal, N.S.W. Ecology: nomadic, predator.

Argoctenus vittatus (Rainbow, 1920)

Cycloctenus vittatus Rainbow, W.J. (1920). Arachnida from Lord Howe and Norfolk Islands. *Rec. S. Aust. Mus.* **1**: 229–272 [262] [as *Cyclostenus vittatus* Rainbow, 1920; inclusion in *Argoctenus* relegates *C. vittatus* to secondary junior homonymy of *Miturgina vittata* Simon, 1888, however Lehtinen, P.T. (1967). Classification of the cribellate spiders and some allied families, with notes on the evolution of the suborder Araneomorpha. *Ann. Zool. Fenn.* **4**: 199–468 says they may be synonymous]. Type data: syntypes, SAMA N1981340 *F. adult, AM KS6471 F. adult, from Lord Howe Is.

Distribution: Lord Howe Is. Ecology: nomadic, predator. Biological references: Lehtinen, P.T. (1967). Classification of the cribellate spiders and some allied families, with notes on the evolution of the suborder Araneomorpha. *Ann. Zool. Fenn.* **4**: 199–468 (taxonomy).

Elassoctenus Simon, 1909

Elassoctenus Simon, E. (1909). Araneae. 2^e partie pp. 155–212 *in* Michaelsen, W. & Hartmeyer, R. (eds.) *Die Fauna Südwest-Australiens.* 2(13) Jena : Fischer [164]. Type species *Elassoctenus harpax* Simon, 1909 by monotypy.

Elassoctenus harpax Simon, 1909

Elassoctenus harpax Simon, E. (1909). Araneae. 2^e partie pp. 155–212 *in* Michaelsen, W. & Hartmeyer, R. (eds.) *Die Fauna Südwest-Australiens.* 2(13) Jena : Fischer [165]. Type data: syntypes, ZMB F. adult, ZMH F. adult, from Geraldton, Wooroloo and Lion Mill, W.A.

Distribution: SW coastal, W.A. Ecology: nomadic, predator.

Hestimodema Simon, 1909

Hestimodema Simon, E. (1909). Araneae. 2^e partie pp. 155–212 *in* Michaelsen, W. & Hartmeyer, R. (eds.) *Die Fauna Südwest-Australiens.* 2(13) Jena : Fischer [165]. Type species *Hestimodema ambigua* Simon, 1909 by subsequent designation, see Rainbow, W.J. (1911). A Census of Australian Araneidae. *Rec. Aust. Mus.* **9**: 107–391 [251].

Hestimodema ambigua Simon, 1909

Hestimodema ambigua Simon, E. (1909). Araneae. 2^e partie pp. 155–212 *in* Michaelsen, W. & Hartmeyer, R. (eds.) *Die Fauna Südwest-Australiens.* 2(13) Jena : Fischer [166]. Type data: syntypes, ?ZMB *M. adult, ZMH F. adult, ?MNHP *F. adult, from Lion Mill and Cannington, W.A.

Distribution: SW coastal, W.A. Ecology: nomadic predator.

Hestimodema latevittata Simon, 1909

Hestimodema latevittata Simon, E. (1909). Araneae. 2^e partie pp. 155–212 *in* Michaelsen, W. & Hartmeyer, R. (eds.) *Die Fauna Südwest-Australiens.* 2(13) Jena : Fischer [167]. Type data: syntypes, MNHP F. adult, ZMB juv., from Fremantle and Donnybrook, W.A.

Distribution: SW coastal, W.A. Ecology: nomadic, predator.

Horioctenoides Main, 1954

Horioctenoides Main, B.Y. (1954). Spiders and Opiliones. *Aust. Geogr. Soc. Rep.* **1**: 37–53 [42]. Type species *Horioctenoides bidentatus* Main, 1954 by monotypy.

Horioctenoides bidentatus Main, 1954

Horioctenoides bidentatus Main, B.Y. (1954). Spiders and Opiliones. *Aust. Geogr. Soc. Rep.* **1**: 37–53 [42]. Type data: holotype, WAM F. juv., from Figure-of-Eight Is., Recherche Archipelago, W.A.

Distribution: SW coastal, W.A. Ecology: nomadic, predator.

Odo Keyserling, 1887

Odo Keyserling, E. (1887). Neue Spinnen aus Amerika. VII *Verh. Zool.-Bot.. Ges. Wien* **37**: 421–490 [454]. Publication date established from Bonnet, P. (1945–1961). *Bibliographia Araneorum.* Toulouse : Douladoure Vol. 2. Type species *Odo lenis* Keyserling, 1887 by monotypy.

This group is also found in South America, Virgin Ils. and Galapagos Ils., see Bonnet, P. (1945–1961). *Bibliographia Araneorum.* Toulouse : Douladoure Vol. 2.

Odo australiensis Hickman, 1944

Odo australiensis Hickman, V.V. (1944). The Simpson Desert Expedition, 1939. Scientific Report 1. Biology — Scorpions and spiders. *Trans. R. Soc. S. Aust.* **68**: 18–48 [40]. Type data: syntypes, AM KS6629 F. adult, from 11 mi E of Hale River, Simpson Desert, N.T. AM KS6628 M. adult, from centre of Simpson Desert, N.T.

Distribution: Lake Eyre basin, N.T. Ecology: nomadic, predator.

Odomasta Simon, 1909

Odomasta Simon, E. (1909). Araneae. 2e partie pp. 155–212 *in* Michaelsen, W. & Hartmeyer, R. (eds.) *Die Fauna Südwest-Australiens.* 2(13) Jena : Fischer [167]. Type species *Odo guttipes* Simon, 1903 by monotypy and original designation.

Odomasta guttipes (Simon, 1903)

Odo guttipes Simon, E. (1903). Descriptions d'Arachnides nouveaux. *Ann. Soc. Entomol. Belg.* **47**: 21–39 [29]. Type data: holotype, MNHP M. adult, from Hobart, Tas.

Distribution: Tas. Ecology: nomadic, predator.

Simonus Ritsema, 1881

Ctenophthalmus Simon, E. (1880). Matériaux pour servir à une faune arachnologique de la Nouvelle-Calédonie. *Ann. Soc. Entomol. Belg.* **23** C.R.: 164–175 [173] [*non Ctenophthalmus* Kolenati, 1857]. Publication date established from Lehtinen, P.T. (1967). Classification of the cribellate spiders and some allied families, with notes on the evolution of the suborder Araneomorpha. *Ann. Zool. Fenn.* **4**: 199–468. Type species *Ctenophthalmus lineatus* Simon, 1880 by monotypy.

Simonus Ritsema, C. (1881). Naamsverandering van het arachnidengenus *Ctenophthalmus* Simon. *Tijdschr. Entomol.* **24**: 91 [*nom. nov.* for *Ctenophthalmus* Simon, 1880]. Publication date established from Lehtinen, P.T. (1967). Classification of the cribellate spiders and some allied families, with notes on the evolution of the suborder Araneomorpha. *Ann. Zool. Fenn.* **4**: 199–468.

Ctenomma Thorell, T. (1892). Studi sui ragni Malesi e Papuani. IV. *Ann. Mus. Civ. Stor. Nat. Genova (2)* **11**: 1–490 [131] [second *nom. nov.* for *Ctenophthalmus* Simon, 1880].

This group is also found in New Caledonia, see Bonnet, P. (1945–1961). *Bibliographia Araneorum.* Toulouse : Douladoure Vol. 2.

Simonus lineatus (Simon, 1880)

Ctenophthalmus lineatus Simon, E. (1880). Matériaux pour servir à une faune arachnologique de la Nouvelle Calédonie. *Ann. Soc. Entomol. Belg.* **23** C.R.: 164–175 [173]. Type data: holotype, MNHP F. adult, from Noumea, New Caledonia.

Distribution: SW coastal, W.A. Ecology: nomadic, predator. Biological references: Lehtinen, P.T. (1967). Classification of the cribellate spiders and some allied families, with notes on the evolution of the suborder Araneomorpha. *Ann. Zool. Fenn.* **4**: 199–468 (taxonomy).

Thasyraea L. Koch, 1878

Thasyraea Koch, L. (1878). *in* Koch, L. & Keyserling, E. (1871–1883). *Die Arachniden Australiens.* Nürnberg : Bauer & Raspe Vol. 2 [982]. Type species *Thasyraea ornata* L. Koch, 1878 by subsequent designation, see Simon, E. (1892–1903). *Histoire Naturelle des Araignées.* 2nd edn. Paris : Roret Vol. 2 [124].

Thasyraea lepida L. Koch, 1878

Thasyraea lepida Koch, L. (1878). *in* Koch, L. & Keyserling, E. (1871–1883). *Die Arachniden Australiens.* Nürnberg : Bauer & Raspe Vol. 2 [984 pl 86 fig 2]. Type data: holotype, ZMH (MGH 16488) F. juv., from Sydney, N.S.W.

Distribution: SE coastal, N.S.W. Ecology: nomadic, predator.

Thasyraea ornata L. Koch, 1878

Thasyraea ornata Koch, L. (1878). *in* Koch, L. & Keyserling, E. (1871–1883). *Die Arachniden Australiens.* Nürnberg : Bauer & Raspe Vol. 2 [983 pl 86 fig 1]. Type data: holotype, ZMH (MGH 16487) F. adult, from Gayndah, Qld.

Distribution: NE coastal, Qld. Ecology: nomadic, predator.

PSEUDOSCORPIONIDA

M.S. Harvey

INTRODUCTION

The arachnid order Pseudoscorpionida (or Chelonethida) is an important group of predators that occupies virtually every terrestrial habitat, even colonizing such inhospitable places as the intertidal zone of the seashore. They attract little attention outside scientific circles because of their small size, inoffensive nature and secretive habits, but several are known to be synanthropic, and some have been found in bee hives (Weygoldt, 1969). Their role in the food cycle is probably quite significant since they are often found in high densities, especially in leaf litter. For example, Gabbutt (1967) recorded *Chthonius ischnocheles* (Hermann), *Roncus lubricus* L. Koch and *Neobisium muscorum* (Leach) in densities of up to 900 per square metre in beech leaf litter.

Approximately 3,000 species and subspecies of pseudoscorpions are recognized from around the world and, because only 99 valid names are applied to the Australian fauna, it is obvious that many more forms are yet to be described. This is borne out by the large number of undescribed species that are present in collections. Yet before new taxa can be named, an adequate knowledge of previously described species is needed. This is often only achieved through the examination of type material, and with the desire to document the type depositories of Australian Pseudoscorpionida, I published a checklist containing full bibliographic synonymies with a key to the known Australian genera (Harvey, 1981). This catalogue expands on that work by (a) providing type localities, (b) adding those species described from Australia's dependencies and territories, and (c) listing species described since 1979.

Two species which have been recorded from Australia were excluded from the 1981 faunal list: *Chelifer cancroides* (Linnaeus) (Cheliferidae) and *Ideobisium antipodum* (Simon) (Syarinidae). The former was probably based upon a misidentification, while the latter was an incorrect transcription of locality data. Unless otherwise stated, numbers of genera and species are exclusive of fossil taxa, a summary of which may be found in Schawaller (1978).

CLASSIFICATION

The familial classification adopted in this catalogue follows that proposed by Muchmore (1982), which is an adaptation of Chamberlin's (1931) system. The order of genera within each family is simply alphabetical and no attempt has been made to place them into subfamilies.

MORPHOLOGY AND BIOLOGY

An excellent account of the morphology and general biology of the pseudoscorpions is given by Chamberlin (1931), and a further account of the biology is given by Weygoldt (1969). The terms used in the familial diagnoses are taken from Chamberlin (1931). The following size ranges are utilized for body length: small—0.5–2.0 mm, medium—2.0–5.0 mm, large—5.0–10.0 mm.

ACKNOWLEDGEMENTS

I wish to thank the following curators for placing type material at my disposal or for confirming the presence of certain types in collections in their care (a list of curators who were contacted in an effort to locate the whereabouts of some types, but whose collections did not contain any Australian type material may be found in Harvey (1981)): Dr E.M. Benedict and Dr D.R. Malcolm (PUO), Dr L. Cederholm (LU), Mr M.R. Gray (AM), Ms A. Green (TM), Dr J. Gruber (NHMW), Dr L.E. Koch (WAM), Mr D.C. Lee (SAMA), Prof. H.W. Levi (MCZ), Dr N.I. Platnick (AMNH), Dr G. Rack (ZMH), Mr F.R. Wanless, Miss B. Brewster and Mr P.D. Hillyard (BMNH). This work was funded by a grant from the Australian Biological Resources Study, and was completed at CSIRO, Division of Entomology, to whom I am grateful for research facilities.

M.S.H.

References

Beier, M. (1932). Pseudoscorpionidea I. Suborder Chthoniinea et Neobisiinea. *Das Tierreich* **57**: i–xx 1–258

Beier, M. (1932). Pseudoscorpionidea II. Suborder Cheliferinea. *Das Tierreich* **58**: i–xxi 1–294

Chamberlin, J.C. (1929). A synoptic classification of the false scorpions or chela-spinners, with a report on a cosmopolitan collection of the same. Part I. The Heterosphyronida (Chthoniidae) (Arachnida-Chelonethida). *Ann. Mag. Nat. Hist. (10)* **4**: 50–80

Chamberlin, J.C. (1930). A synoptic classification of the false scorpions or chela-spinners, with a report on a cosmopolitan collection of the same. Part II. The Diplosphyronida (Arachnida-Chelonethida). *Ann. Mag. Nat. Hist. (10)* **5**: 1–48, 585–620

Chamberlin, J.C. (1931). The arachnid order Chelonethida. *Stanford Univ. Publ. Biol. Ser.* 7: 1–284

Gabbutt, P.D. (1967). Quantitative sampling of the pseudoscorpion *Chthonius ischnocheles* from beech litter. *J. Zool. Lond.* **151**: 469–478

Harvey, M.S. (1981). A checklist of the Australian Pseudoscorpionida. *Bull. Br. Arachnol. Soc.* **5**: 237–252

Muchmore, W.B. (1982). Pseudoscorpionida. pp. 96–102 *in* Parker, S.P. (ed.) *Synopsis and Classification of Living Organisms.* New York : McGraw-Hill Vol. 2

Schawaller, W. (1978). Neue Pseudoskorpione aus dem Baltischen Bernstein der Stuttgarter Bernsteinsammlung (Arachnida : Pseudoscorpionidea). *Stuttgarter Beitr. Naturk.* Series B No. 42: 1–22

Weygoldt, P. (1969). *The Biology of Pseudoscorpions.* Cambridge : Harvard Univ. Press xiv 145 pp.

ATEMNIDAE

INTRODUCTION

This is a family of medium to large pseudoscorpions that are usually corticolous. Currently 19 genera and more than 170 species are recognized, with the majority being described from Asia and Africa. Two subfamilies have been described, Atemninae and Miratemninae, but only the former has been recorded from Australia. The Miratemninae is sometimes treated as a family, but I believe that it belongs within the Atemnidae.

Characterized by: monotarsate; heterofemorate; venom apparatus present in only the fixed chelal finger.

References

Beier, M. (1932). Pseudoscorpionidea II. Suborder Cheliferinea. *Das Tierreich* **58**: i–xxi 1–294

Oratemnus Beier, 1932

Oratemnus Beier, M. (1932). Revision der Atemnidae (Pseudoscorpionidae). *Zool. Jb. Abt. Syst.* **62**: 547–610 [588]. Type species *Chelifer (Lamprochernes) articulosus* Simon, 1889 by original designation.

Steiratemnus Beier, M. (1948). Über Pseudoscorpione der australischen region. *Eos, Madrid* **24**: 525–562 [527]. Type species *Chelifer punctatus* L. Koch, 1885 by original designation.

Taxonomic decision of Beier, M. (1966). On the Pseudoscorpionidea of Australia. *Aust. J. Zool.* **14**: 275–303 [296].

This group is also found in Asia, see Beier, M. (1932). Revision der Atemnidae (Pseudoscorpionidae). *Zool. Jb. Abt. Syst.* **62**: 547–610.

Oratemnus cavernicola Beier, 1976

Oratemnus cavernicola Beier, M. (1976). A cavernicolous atemnid pseudoscorpion from New South Wales. *J. Aust. Entomol. Soc.* **15**: 271–272 [271]. Type data: holotype, AM KS0066 F., paratype AM KS0067 tritonymph, from Jump-Up Cave, Grey Range, N.S.W.

Distribution: Murray-Darling basin, N.S.W.; known only from type locality. Ecology: terrestrial, (predator), caves.

Oratemnus curtus (Beier, 1954)

Steiratemnus curtus Beier, M. (1954). Report from Prof. T. Gislén's expedition to Australia in 1951–1952. 7. Pseudoscorpionidea. *Lunds Univ. Arsskr., Avd 2* **50**(3): 1–26 [15]. Type data: holotype, ZMLU *F., paratype NHMW *F., from Pemberton, W.A.

Distribution: SW coastal, W.A.; known only from type locality. Ecology: terrestrial, (predator), under bark, woodland.

Oratemnus distinctus (Beier, 1948)

Steiratemnus distinctus Beier, M. (1948). Über Pseudoscorpione der australischen region. *Eos, Madrid* **24**: 525–562 [531]. Type data: holotype, BMNH 1979.8.23.1 *F., from N.S.W.

Distribution: N.S.W., W plateau, W.A.; N.S.W. type locality as State only. Ecology: terrestrial, (predator), caves. Biological references: Richards, A.M. (1971). An ecological study of the cavernicolous fauna of the Nullarbor Plain, southern Australia. *J. Zool. Lond.* **164**: 1–60 (distribution).

Oratemnus punctatus (L. Koch, 1885)

Chelifer punctatus Koch, L. (1885). Ordo Chelonethi. pp. 44–51 pls 4, 6 *in* Koch, L. & Keyserling, E. *Die Arachniden Australiens nach der Natur beschrieben und abgebildet.* Nürnberg : Bauer & Raspe Vol. 2 [45 pl 4]. Type data: syntypes, BMNH 1891.8.1.1076–1077*, BMNH 1913.9.1.352–356*, from Gayndah, Qld.

Chelifer brevidigitatus Koch, L. (1885). Ordo Chelonethi. pp. 44–51 pls 4,6 *in* Koch, L. & Keyserling, E. *Die Arachniden Australiens nach der Natur beschrieben und abgebildet.* Nürnberg : Bauer & Raspe Vol. 2 [48 pl 4]. Type data: syntypes, BMNH 1913.9.1.218–219*, from Rockhampton, Qld.

Taxonomic decision of Beier, M. (1948). Über Pseudoscorpione der australischen region. *Eos, Madrid* **24**: 525–562 [529].

Distribution: NE coastal, SE coastal, Murray-Darling basin, S Gulfs, W plateau, W.A., N.S.W., A.C.T., S.A., Qld., Vic., Lord Howe Is. Ecology: terrestrial, (predator), under bark; also in litter and under stones. Biological references: Beier, M. (1948). Über Pseudoscorpione der australischen region. *Eos, Madrid* **24**: 525–562 (distribution, redescription); Beier, M. (1966). On the Pseudoscorpionidea of Australia. *Aust. J. Zool.* **14**: 275–303 (distribution); Beier, M. (1969). Neue Pseudoskorpione aus Australien. *Ann. Naturh. Mus. Wien* **73**: 171–187 (distribution); Beier, M. (1974). Neue Pseudoskorpione aus Australien und Neu-Guinea. *Ann. Naturh. Mus. Wien* **78**: 203–213 (distribution); Beier, M. (1976). The pseudoscorpions of New Zealand, Norfolk and Lord Howe. *N.Z. J. Zool.* **3**: 199–246 (distribution).

Paratemnus Beier, 1932

Paratemnus Beier, M. (1932). Revision der Atemnidae (Pseudoscorpionidea). *Zool. Jb. Abt. Syst.* **62**: 547–610 [562]. Type species *Chelifer (Atemnus) guineensis* Ellingsen, 1906 by original designation.

This group is also found in Africa, southeast Asia, North and South America, see Beier, M. (1932). Revision der Atemnidae (Pseudoscorpionidea). *Zool. Jb. Abt. Syst.* **62**: 547–610.

Paratemnus pococki (With, 1907)

Chelifer pococki With, C.J. (1907). On some new species of the Cheliferidae, Hans., and Garypidae, Hans., in the British Museum. *J. Linn. Soc. Zool.* **30**: 49–85 [63]. Type data: holotype, BMNH 1913.9.1.352–356 *F. + paratypes, from Christmas Is.

Distribution: Christmas Is. Ecology: terrestrial, (predator).

CHEIRIDIIDAE

INTRODUCTION

Cheiridiids are very small pseudoscorpions that have been recorded from most areas of the world, usually in litter, but some are corticolous or nidicolous, and one, *Cheiridium museorum* (Leach), is often synanthropic. Six genera and approximately 60 species currently are recognized. These have been placed in two subfamilies, Cheiridiinae and Pycnocheiridiinae, but only the former is recorded from Australia. The latter only contains *Pycnocheiridium mirum* Beier from South Africa.

Characterized by: monotarsate; homofemorate; pedal femora often not divided; carapace sub-triangular; body sub-ovate; one pair of eyes present.

References

Beier, M. (1932). Pseudoscorpionidea II. Suborder Cheliferinea. *Das Tierreich* **58**: i–xxi 1–294

Cryptocheiridium Chamberlin, 1931

Cryptocheiridium (Cryptocheiridium) Chamberlin, 1931

Cryptocheiridium Chamberlin, J.C. (1931). The arachnid order Chelonethida. *Stanford Univ. Publ. Biol. Ser.* 7: 1–284 [238]. Type species *Cheiridium subtropicum* Tullgren, 1907 by original designation.

This group is also found in Africa and southeast Asia, see Mahnert, V. (1982). Die Pseudoskorpione (Arachnida) Kenyas II. Feaellidae; Cheiridiidae. *Rev. Suisse Zool.* **89**: 115–134.

Cryptocheiridium (Cryptocheiridium) australicum Beier, 1969

Cryptocheiridium australicum Beier, M. (1969). Neue Pseudoskorpione aus Australien. *Ann. Naturh. Mus. Wien* **73**: 171–187 [183]. Type data: holotype, SAMA N1980189 *M., from Murra-el-elevyn Cave, W.A.

Distribution: W plateau, W.A. Ecology: terrestrial, (predator), caves. Biological references: Richards, A.M. (1971). An ecological study of the cavernicolous fauna of the Nullarbor Plain, southern Australia. *J. Zool. Lond.* **164**: 1–60 (ecology).

CHELIFERIDAE

INTRODUCTION

The Cheliferidae is a large, cosmopolitan family of pseudoscorpions that contains small to large individuals. Many are corticolous, but some are found under rocks or in leaf litter. Two subfamilies are recognized: Cheliferinae and Dactylocheliferinae, but only the latter are recorded from Australia. Approximately 60 genera and 260 species currently are recognized.

Characterized by: monotarsate; heterofemorate; venom apparatus present in both chelal fingers; coxal sacs and ramshorn organs usually present in males; junction between femur I and femur II of the first pair of legs oblique.

References

Beier, M. (1932). Pseudoscorpionidea II. Suborder Cheliferinea. *Das Tierreich* **58**: i–xxi 1–294

Australochelifer Beier, 1974

Australochelifer Beier, M. (1974). Neue Pseudoskorpione aus Australien und Neu-Guinea. *Ann. Naturh. Mus. Wien* **78**: 203–213 [209]. Type species *Australochelifer pygmaeus* Beier, 1974 by original designation.

Australochelifer pygmaeus Beier, 1974

Australochelifer pygmaeus Beier, M. (1974). Neue Pseudoskorpione aus Australien und Neu-Guinea. *Ann. Naturh. Mus. Wien* **78**: 203–213 [209]. Type data: holotype, SAMA 197758 M., paratypes SAMA 197759–60 2 nymphs, from Waste Point, N.S.W.

Distribution: Murray-Darling basin, N.S.W.; known only from type locality. Ecology: terrestrial, (predator), litter, open forest.

Protochelifer Beier, 1948

Protochelifer Beier, M. (1948). Über Pseudoscorpione der australischen region. *Eos, Madrid* **24**: 525–562 [552]. Type species *Protochelifer novaezealandiae* Beier, 1948 by original designation.

This group is also found in New Zealand, see Beier, M. (1976). The pseudoscorpions of New Zealand, Norfolk and Lord Howe. *N.Z. J. Zool.* **3**: 199–246.

Protochelifer australis (Tubb, 1937)

Idiochelifer australis Tubb, J.A. (1937). Reports of the expedition of the McCoy Society for field investigation and research (Lady Julia Percy Island). 19. Arachnida. *Proc. R. Soc. Vict. (ns)* **49**: 412–421 [414]. Type data: syntypes, AMNH *M., 2 adults, 2 nymphs whereabouts unknown, from Lady Julia Percy Is., Vic.

Distribution: SE coastal, Vic.; known only from type locality. Ecology: terrestrial, (predator), under stones. Biological references: Chamberlin, J.C. (1949). New and little-known false scorpions from various parts of the world (Arachnida, Chelonethida), with notes on structural abnormalities in two species. *Am. Mus. Novit.* 1430: 1–57 (redescription).

Protochelifer brevidigitatus (Tubb, 1937)

Idiochelifer brevidigitatus Tubb, J.A. (1937). Reports of the expedition of the McCoy Society for field investigation and research (Lady Julia Percy Island). 19. Arachnida. *Proc. R. Soc. Vict. (ns)* **49**: 412–421 [414]. Type data: holotype, M. whereabouts unknown, from Lady Julia Percy Is., Vic.

Distribution: SE coastal, Vic.; known only from type locality. Ecology: terrestrial, (predator), under stones.

Protochelifer cavernarum Beier, 1967

Protochelifer cavernarum Beier, M. (1967). Some Pseudoscorpionidea from Australia, chiefly from caves. *Aust. J. Zool.* **14**: 199–205 [203]. Type data: syntypes, SAMA N1977167–71 3M.,2F., AM KS5318 1 tritonymph, 1 protonymph, AM KS5317 1 tritonymph, NHMW *3F.,2M., from Murder Cave, Cliefden, N.S.W., Belfry Cave, Timor, N.S.W. and Timor Caves, Timor, N.S.W.

Protochelifer cavernarum aitkeni Beier, M. (1968). Some cave-dwelling Pseudoscorpionidea from Australia and New Caledonia. *Rec. S. Aust. Mus.* **15**: 757–765

[764]. Type data: holotype, SAMA N196838 *F., paratypes SAMA *2F., 1 tritonymph, from Abrakurrie Cave, W.A.

Taxonomic decision of Harvey, M.S. (1981). A checklist of the Australian Pseudoscorpionida. *Bull. Br. Arachnol. Soc.* **5**: 237–252 [238].

Distribution: SE coastal, Murray-Darling basin, W plateau, SW coastal, N.S.W., Vic., S.A., W.A. Ecology: terrestrial, (predator), caves. Biological references: Richards, A.M. (1971). An ecological study of the cavernicolous fauna of the Nullarbor Plain, southern Australia. *J. Zool. Lond.* **164**: 1–60 (biology, distribution); Beier, M. (1968). Some cave-dwelling Pseudoscorpionidea from Australia and New Caledonia. *Rec. S. Aust. Mus.* **15**: 757–765 (distribution); Beier, M. (1969). Neue Pseudoskorpione aus Australien. *Ann. Naturh. Mus. Wien* **73**: 171–187 (distribution).

Protochelifer naracoortensis Beier, 1968

Protochelifer naracoortensis Beier, M. (1968). Some cave-dwelling Pseudoscorpionidea from Australia and New Caledonia. *Rec. S. Aust. Mus.* **15**: 757–765 [763]. Type data: holotype, SAMA N196831 M., paratypes SAMA N196833–7, N197711a-d, N197912a-c, N197913a-c, N197914a-b, N197915 9M.,6F., 1 tritonymph, 2 deutonymphs, 4 protonymphs, NHMW *3M.,3F., from Bat Cave, Naracoorte, S.A.

Distribution: SE coastal, S.A.; known only from type locality. Ecology: terrestrial, (predator), caves.

Protochelifer victorianus Beier, 1966

Protochelifer victorianus Beier, M. (1966). On the Pseudoscorpionidea of Australia. *Aust. J. Zool.* **14**: 275–303 [301]. Type data: holotype, AM KS5863 M., from Olsen's Road, Gippsland, Vic.

Distribution: SE coastal, Vic.; known only from type locality. Ecology: terrestrial, (predator), under bark.

CHERNETIDAE

INTRODUCTION

The Chernetidae is the largest family of pseudoscorpions and contains small to large species. Many are corticolous, but others are found in a variety of habitats including leaf litter, under rocks, and several are known to be phoretic on insects. This family currently includes nearly 100 genera and over 550 species and is distributed over much of the world. Several subfamilial and tribal names have been proposed in the literature but, as discussed by Muchmore (1972), the validity of these categories is in serious doubt.

Characterized by: monotarsate; venom apparatus present in moveable chelal finger; accessory teeth present on chelal fingers; elevated slit sensillum present on all pedal tarsi.

References

Beier, M. (1932). Pseudoscorpionidea II. Suborder Cheliferinea. *Das Tierreich* **58**: i–xxi 1–294

Muchmore, W.B. (1972). Observations on the classification of some European pseudoscorpions. *Bull. Br. Arachnol. Soc.* **2**: 112–115

Apatochernes Beier, 1948

Apatochernes Beier, M. (1948). Über Pseudoscorpione der australischen Region. *Eos, Madrid* **24**: 525–562 [540]. Type species *Apatochernes cheliferoides* Beier, 1948 by original designation.

This group is also found in New Zealand, see Beier, M. (1976). The pseudoscorpions of New Zealand, Norfolk and Lord Howe. *N.Z. J. Zool.* **3**: 199–246.

Apatochernes posticus Beier, 1976

Apatochernes posticus Beier, M. (1976). The pseudoscorpions of New Zealand, Norfolk and Lord Howe. *N.Z. J. Zool.* **3**: 199–246 [227]. Type data: holotype, ANIC 10001 M., paratype 1 deutonymph whereabouts unknown, from Mt. Pitt, Norfolk Is.

Distribution: Norfolk Is.; known only from type locality. Ecology: terrestrial, (predator).

Austrochernes Beier, 1932

Austrochernes Beier, M. (1932). Pseudoscorpionidea II. Suborder Cheliferinea. *Das Tierreich* **58**: i–xxi 1–294 [170]. Type species *Chelifer australiensis* With, 1905 by original designation.

Austrochernes australiensis (With, 1905)

Chelifer australiensis With, C.J. (1905). On Chelonethi, chiefly from the Australian region, in the collection of the British Museum, with observations on the "coxal sac" and on some cases of abnormal segmentation. *Ann. Mag. Nat. Hist. (7)* **15**: 94–143, 328 [101]. Type data: syntypes, BMNH 1893.12.17 *2M., from Qld.

Distribution: Qld. Ecology: terrestrial, (predator).

Calymmachernes Beier, 1954

Calymmachernes Beier, M. (1954). Report from Prof. T. Gislén's expedition to Australia in 1951–1952. 7. Pseudoscorpionidea. *Lunds Univ. Arsskr., Avd 2* **50**(3): 1–26 [24]. Type species *Calymmachernes angulatus* Beier, 1954 by original designation.

Calymmachernes angulatus Beier, 1954

Calymmachernes angulatus Beier, M. (1954). Report from Prof. T. Gislén's expedition to Australia in 1951–1952. 7. Pseudoscorpionidea. *Lunds Univ. Arsskr., Avd 2* **50**(3): 1–26 [25]. Type data: holotype, ZMLU *F., from Walpole, W.A.

Distribution: SW coastal, W.A.; known only from type locality. Ecology: terrestrial, (predator).

Conicochernes Beier, 1948

Conicochernes Beier, M. (1948). Über Pseudoscorpione der australischen Region. *Eos, Madrid* **24**: 525–562 [532]. Type species *Chelifer brevispinosus* L. Koch, 1885 by original designation.

Conicochernes brevispinosus (L. Koch, 1885)

Chelifer brevispinosus Koch, L. (1885). Ordo Chelonethi. pp. 44–51 pls 4, 6 *in* Koch, L. & Keyserling, E. *Die Arachniden Australiens nach der Natur beschrieben und abgebildet.* Nürnberg : Bauer & Raspe Vol. 2 [46 pl]. Type data: syntypes (19), BMNH 1891.8.1.1085-6*, BMNH 1913.9.1.220–37*, from Gayndah, Rockhampton and Peak Downs (as Peack Downs), Qld.

Chelifer keyserlingi With, C.J. (1907). On some new species of the Cheliferidae, Hans., and Garypidae, Hans., in the British Museum. *J. Linn. Soc. Zool.* **30**: 49–85 [53]. Type data: holotype, BMNH 1879.55 *F., from Rockhampton, Qld.

Chelifer (Lamprochernes) silvestrii Beier, M. (1930). Alcuni pseudoscorpioni esotici raccolti dal Prof. F. Silvestri. *Boll. Lab. Zool. Gen. Agr. R. Scuola Agric. Portici* **23**: 197–209 [198]. Type data: syntypes (probable), 1 adult, 1 semi-adult whereabouts unknown, from Mt. Lofty, S.A.

Taxonomic decision of Beier, M. (1948). Über Pseudoscorpione der australischen region. *Eos, Madrid* **24**: 525–562 [533].

Distribution: NE coastal, Murray-Darling basin, SE coastal, S Gulfs, Qld., A.C.T., (N.S.W.), Vic., S.A. Ecology: terrestrial, (predator), under bark. Biological references: Beier, M. (1948). Über Pseudoscorpione der australischen region. *Eos, Madrid* **24**: 525–562 (redescription); Beier, M. (1966). On the Pseudoscorpionidea of Australia. *Aust. J. Zool.* **14**: 275–303 (distribution).

Conicochernes crassus Beier, 1954

Conicochernes crassus Beier, M. (1954). Report from Prof. T. Gislén's expedition to Australia in 1951–1952. 7. Pseudoscorpionidea. *Lunds Univ. Arsskr., Avd 2* **50**(3): 1–26 [21]. Type data: syntypes, ZMLU *1M.,2F.,1 tritonymph, NHMW *1M.,1F., from Denmark, W.A.

Distribution: SW coastal, W.A.; known only from type locality. Ecology: terrestrial, (predator), under bark.

Conicochernes globosus Beier, 1954

Conicochernes globosus Beier, M. (1954). Report from Prof. T. Gislén's expedition to Australia in 1951–1952. 7. Pseudoscorpionidea. *Lunds Univ. Arsskr., Avd 2* **50**(3): 1–26 [19]. Type data: syntypes, ZMLU *4M.,1F.,1 nymph, NHMW *2M.,1F., from Porongorup, W.A.

Distribution: SW coastal, W.A.; known only from type locality. Ecology: terrestrial, (predator), under bark.

Conicochernes incrassatus (Beier, 1933)

Haplochernes incrassatus Beier, M. (1933). Two new species of Cheliferinea. *Ann. Mag. Nat. Hist. (10)* **11**: 644–647 [644]. Type data: holotype, BMNH 1933.1.27.17 *F., from Upper Ferntree Gully (as Upper Fern Tree Gulley), Vic.

Distribution: SE coastal, Vic.; known only from type locality. Ecology: terrestrial, (predator).

Haplochernes Beier, 1932

Haplochernes Beier, M. (1932). Pseudoscorpionidea II. Suborder Cheliferinea. *Das Tierreich* **58**: i–xxi 1–294 [108]. Type species *Chelifer boncicus* Karsch, 1881 by original designation.

This group is also found in southeast Asia and Madagascar, see Beier, M. (1932). Pseudoscorpionidea II. Suborder Cheliferinea. *Das Tierreich* **58**: i–xxi 1–294.

Haplochernes norfolkensis Beier, 1976

Haplochernes norfolkensis Beier, M. (1976). The pseudoscorpions of New Zealand, Norfolk and Lord Howe. *N.Z. J. Zool.* **3**: 199–246 [214]. Type data: holotype, ANIC 10000 M, paratypes ANIC 2F.,1 tritonymph, paratypes *DSIR 2F.,2 tritonymphs, from Steels Point, Norfolk Is.

Distribution: Norfolk Is. Ecology: terrestrial, (predator); beaten from foliage.

Haplochernes pygmaeus (L. Koch, 1885)

Chelifer pygmaeus Koch, L. (1885). Ordo Chelonethi. pp. 44–51 pls 4, 6 *in* Koch, L. & Keyserling, E. *Die Arachniden Australiens nach der Natur beschrieben und abgebildet.* Nürnberg : Bauer & Raspe Vol. 2 [49 pl 6]. Type data: syntypes, BMNH 1891.8.1.1078–1084*, BMNH 1913.9.1.357–396*, from Gayndah, Qld.

Distribution: NE coastal, Qld.; known only from type locality. Ecology: terrestrial, (predator).

Haplochernes ramosus (L. Koch, 1885)

Chelifer ramosus Koch, L. (1885). Ordo Chelonethi. pp. 44–51 pls 4, 6 *in* Koch, L. & Keyserling, E. *Die Arachniden Australiens nach der Natur beschrieben und abgebildet.* Nürnberg : Bauer & Raspe Vol. 2 [47 pl 4]. Type data: syntype (1), BMNH 1913.9.1.391*, from Rockhampton, Qld.

Distribution: NE coastal, SE coastal, Lake Eyre basin, Qld., N.S.W., N.T. Ecology: terrestrial, (predator), under bark. Biological references: Beier, M. (1966). On the Pseudoscorpionidea of Australia. *Aust. J. Zool.* **14**: 275–303 (redescription and distribution).

Megachernes Beier, 1932

Megachernes Beier, M. (1932). Pseudoscorpionidea II. Suborder Cheliferinea. *Das Tierreich* **58**: i–xxi 1–294 [128]. Type species *Chernes grandis* Beier, 1930 by original designation.

This group is also found in Asia, see Beier, M. (1932). Pseudoscorpionidea II. Suborder Cheliferinea. *Das Tierreich* **58**: i–xxi 1–294.

Megachernes penicillatus Beier, 1948

Megachernes penicillatus Beier, M. (1948). Phoresie und Phagophile bei Pseudoscorpioniden. *Österr. Zool. Z.* **1**: 441–497 [478]. Type data: syntypes, BMNH 1924.10.27.24–26 *3F., from Dinner Creek, Ravenshoe (as Ravenshor), Qld. (*ex Phascogale* sp.=?*Antechinus*).

Distribution: NE coastal, Qld.; known only from type locality. Ecology: terrestrial, (predator).

Megachernes queenslandicus Beier, 1948

Megachernes queenslandicus Beier, M. (1948). Phoresie und Phagophile bei Pseudoscorpioniden. *Österr. Zool. Z.* **1**: 441–497 [480]. Type data: holotype, BMNH 1924.10.27.23 *F., paratype BMNH 41 *F., from Dinner Creek, Ravenshoe (as Ravenshor), Qld. (*ex Rattus fuscipes assimilis*)(holotype only).

Distribution: NE coastal, Qld.; known only from Ravenshoe and Kuranda, Qld. Ecology: terrestrial, (predator).

Nesidiochernes Beier, 1957

Nesidiochernes Beier, M. (1957). Pseudoscorpionidea. *Insects Micronesia* **3**: 1–64 [50]. Type species *Nesidiochernes maculatus* Beier, 1957 by original designation.

This group is also found in southeast Asia and New Zealand, see Beier, M. (1957). Pseudoscorpionidea. *Insects Micronesia* **3**: 1–64; Beier, M. (1976). The pseudoscorpions of New Zealand, Norfolk and Lord Howe. *N.Z. J. Zool.* **3**: 199–246.

Nesidiochernes australicus Beier, 1966

Nesidiochernes australicus Beier, M. (1966). On the Pseudoscorpionidea of Australia. *Aust. J. Zool.* **14**: 275–303 [298]. Type data: holotype, AM KS5864 M., paratypes AM KS5865 1F., NHMW *1F., from Cobar, N.S.W.

Distribution: Murray-Darling basin, S Gulfs, N.S.W., S.A.; known only from Cobar, N.S.W. and Yorke Peninsula, S.A. Ecology: terrestrial, (predator), litter, low open woodland. Biological references: Beier, M. (1974). Neue Pseudoskorpione aus Australien und Neu-Guinea. *Ann. Naturh. Mus. Wien* **78**: 203–213 (distribution).

Nesidiochernes slateri Beier, 1974

Nesidiochernes slateri Beier, M. (1974). Neue Pseudoskorpione aus Australien und Neu-Guinea. *Ann. Naturh. Mus. Wien* **78**: 203–213 [208]. Type data: holotype, AM KS0014 M., paratype AM KS001 1F., from Peak Head, S of Albany, W.A.

Distribution: SW coastal, W.A.; known only from type locality. Ecology: terrestrial, (predator).

Nesochernes Beier, 1932

Nesochernes Beier, M. (1932). Pseudoscorpionidea II. Suborder Cheliferinea. *Das Tierreich* **58**: i–xxi 1–294 [184]. Type species *Nesochernes gracilis* Beier, 1932 by original designation.

This group is also found in New Zealand, see Beier, M. (1976). The pseudoscorpions of New Zealand, Norfolk and Lord Howe. *N.Z. J. Zool.* **3**: 199–246.

Nesochernes gracilis norfolkensis Beier, 1976

Nesochernes gracilis norfolkensis Beier, M. (1976). The pseudoscorpions of New Zealand, Norfolk and Lord Howe. *N.Z. J. Zool.* **3**: 199–246 [239]. Type data: holotype, ANIC 10002 M, paratypes ANIC 7M., 4F., 5 tritonymphs, 7 deutonymphs, 1 protonymph, *(DSIR and CMC), from Mt. Bates, Norfolk Is. (holotype and some paratypes only).

Distribution: Norfolk Is. Ecology: terrestrial, (predator), litter.

Parachernes Chamberlin, 1931

Parachernes Chamberlin, J.C. (1931). *Parachernes ronnaii*, a new genus and species of false scorpions from Brazil (Arachnida-Chelonethida). *Entomol. News* **42**: 192–195 [192]. Type species *Parachernes ronnaii* Chamberlin, 1931 by original designation.

This group is found world-wide, see Beier, M. (1932). Pseudoscorpionidea II. Suborder Cheliferinea. *Das Tierreich* **58**: i–xxi 1–294; Muchmore, W.B. & Alteri, C. (1974). The genus *Parachernes* (Pseudoscorpionida, Chernetidae) in the United States, with descriptions of new species. *Trans. Am. Entomol. Soc.* **99**: 477–506.

Parachernes sabulosus (Tullgren, 1909)

Chelifer sabulosus Tullgren, A. (1909). Chelonethi. pp. 411–415 *in* Michaelsen, W. & Hartmeyer, R. (eds.) *Die Fauna Südwest-Australiens.* Jena : Gustav Fischer Vol. 2 [411]. Type data: holotype, ZMH *F., from Brown Station, Dirk Hartog, W.A.

Distribution: NW coastal, W.A.; known only from type locality. Ecology: terrestrial, (predator).

Paraustrochernes Beier, 1966

Paraustrochernes Beier, M. (1966). On the Pseudoscorpionidea of Australia. *Aust. J. Zool.* **14**: 275–303 [299]. Type species *Paraustrochernes victorianus* Beier, 1966 by original designation.

This group is also found in Papua New Guinea, see Beier, M. (1974). Neue Pseudoskorpione aus Australien und Neu-Guinea. *Ann. Naturh. Mus. Wien* **78**: 203–213.

Paraustrochernes victorianus Beier, 1966

Paraustrochernes victorianus Beier, M. (1966). On the Pseudoscorpionidea of Australia. *Aust. J. Zool.* **14**: 275–303 [300]. Type data: syntypes, AM KS5867–8 3M.,1F.,1 deutonymph, NHMW *2M.,2F., from Boolarra, Morwell River Road, Vic.

Distribution: SE coastal, Vic.; known only from type locality. Ecology: terrestrial, (predator), under bark.

Sundochernes Beier, 1932

Sundochernes Beier, M. (1932). Pseudoscorpionidea II. Suborder Cheliferinea. *Das Tierreich* **58**: i–xxi 1–294 [162]. Type species *Chelifer modiglianii* Ellingsen, 1911 by original designation.

This group is also found in southeast Asia (and South America), see Beier, M. (1932). Pseudoscorpionidea II. Suborder Cheliferinea. *Das Tierreich* **58**: i–xxi 1–294; Beier, M. (1974). Brasilianische Pseudoscropione aus dem Museum in Genf. *Rev. Suisse Zool.* **81**: 899–909.

Sundochernes australiensis Beier, 1954

Sundochernes australiensis Beier, M. (1954). Report from Prof. T. Gislén's expedition to Australia in 1951–1952. 7. Pseudoscorpionidea. *Lunds Univ. Arsskr., Avd 2* **50**(3): 1–26 [16]. Type data: holotype, ZMLU *F., from Denmark, W.A.

Distribution: SW coastal, W.A.; known only from type locality. Ecology: terrestrial, (predator), under bark.

Sundochernes dewae Beier, 1967

Sundochernes dewae Beier, M. (1967). Some Pseudoscorpionidea from Australia, chiefly from caves. *Aust. Zool.* **14**: 199–205 [200]. Type data: holotype, AM KS5867 M, paratypes AM 1F.,2 tritonymphs, 2 deutonymphs, 1 protonymph, NHMW *1M.,2F.,1 nymph, from Brewarrina, N.S.W.

Distribution: Murray-Darling basin, N.S.W.; known only from type locality. Ecology: terrestrial, (predator); from nest of Galah (*Cacatua roseicapilla*).

Sundochernes dubius Beier, 1954

Sundochernes dubius Beier, M. (1954). Report from Prof. T. Gislén's expedition to Australia in 1951–1952. 7. Pseudoscorpionidea. *Lunds Univ. Arsskr., Avd 2* **50**(3): 1–26 [18]. Type data: holotype, ZMLU *F., from Augusta, W.A.

Distribution: SW coastal, W.A.; known only from type locality. Ecology: terrestrial, (predator).

Sundochernes grayi Beier, 1976

Sundochernes grayi Beier, M. (1976). The pseudoscorpions of New Zealand, Norfolk and Lord Howe. *N.Z. J. Zool.* **3**: 199–246 [225]. Type data: holotype, AM KS20 *F., from near Old Settlement, Lord Howe Is.

Distribution: Lord Howe Is.; known only from type locality. Ecology: terrestrial, (predator).

Sundochernes guanophilus Beier, 1967

Sundochernes guanophilus Beier, M. (1967). Some Pseudoscorpionidea from Australia, chiefly from caves. *Aust. Zool.* **14**: 199–205 [202]. Type data: holotype, SAMA N1966163 M., paratypes SAMA N1966164 1F., NHMW *1F., from Fig Tree Cave, Wombeyan, N.S.W.

Distribution: SE coastal, N.S.W.; known only from type locality. Ecology: terrestrial, (predator), caves.

Sundochernes queenslandicus Beier, 1974

Sundochernes queenslandicus Beier, M. (1974). Neue Pseudoskorpione aus Australien und Neu-Guinea. *Ann. Naturh. Mus. Wien* **78**: 203–213 [207]. Type data: holotype, SAMA N197761 M., from Marburg, Qld.

Distribution: NE coastal, Qld.; known only from type locality. Ecology: terrestrial, (predator), litter.

Troglochernes Beier, 1969

Troglochernes Beier, M. (1969). Neue Pseudoskorpione aus Australien. *Ann. Naturh. Mus. Wien* **73**: 171–187 [185]. Type species *Troglochernes imitans* Beier, 1969 by original designation.

Troglochernes imitans Beier, 1969

Troglochernes imitans Beier, M. (1969). Neue Pseudoskorpione aus Australien. *Ann. Naturh. Mus. Wien* **73**: 171–187 [185]. Type data: holotype, SAMA N1908192 *M., paratypes SAMA N1980193 *F., NHMW 2F., from Dingo Cave, W.A.

Distribution: W plateau, W.A. Ecology: terrestrial, (predator), caves. Biological references: Beier, M. (1974). Neue Pseudoskorpione aus Australien und Neu-Guinea. *Ann. Naturh. Mus. Wien* **78**: 203–213 (distribution).

CHTHONIIDAE

INTRODUCTION

Members of this family are small to medium sized pseudoscorpions usually confined to leaf litter and soil. They are abundant in most areas of the world and 34 genera and over 450 species are currently recognized. Several tribes have been erected (mostly when the Chthoniidae was regarded as a subfamily (Chamberlin, 1929, Beier, 1932)), and all except the Lechytiini have been recorded from Australia.

Characterized by: heterotarsate; spiracles transversely orientated, and usually without an associated plate; genital area of female non-sclerotic; pedipalpal tibia pyriform; carapace usually trapezoidal and with 30 or fewer setae.

References

Beier, M. (1932). Pseudoscorpionidea I. Suborder Chthoniinea et Neobisiinea. *Das Tierreich* **57**: i–xx 1–258

Chamberlin, J.C. (1929). A synoptic classification of the false scorpions or chela-spinners, with a report on a cosmopolitan collection of the same. Part I. The Heterosphyronida (Chthoniidae) (Arachnida-Chelonethida). *Ann. Mag. Nat. Hist. (10)* **4**: 50–80

Austrochthonius Chamberlin, 1929

Austrochthonius Chamberlin, J.C. (1929). A synoptic classification of the false scorpions or chela-spinners, with a report on a cosmopolitan collection of the same. Part II. The Diplosphyronida (Arachnida-Chelonethida). *Ann. Mag. Nat. Hist. (10)* **5**: 1–48, 585–620 [68]. Type species *Chthonius chilensis* Chamberlin, 1923 by original designation.

Paraustrochthonius Beier, M. (1931). Zur Kenntnis der Chthoniiden (Pseudoskorpione). *Zool. Anz.* **93**: 49–56 [52]. Type species *Paraustrochthonius tullgreni* Beier, 1930 by original designation.

Taxonomic decision of Beier, M. (1976). The pseudoscorpions of New Zealand, Norfolk and Lord Howe. *N.Z. J. Zool.* **3**: 199–246 [203].

This group is also found in South America and South Africa, see Vitali-di Castri, V. (1975). Nuevos *Austrochthonius* sud-americanos (Pseudoscorpionida, Chthoniidae). *Physis, B. Aires (C)* **34**: 117–127.

Austrochthonius australis Hoff, 1951

Austrochthonius australis Hoff, C.C. (1951). New species and records of chthoniid pseudoscorpions. *Am. Mus. Novit.* 1483: 1–13 [5]. Type data: holotype, AMNH *F., from Mt. Slide, Vic.

Distribution: SE coastal, Murray-Darling basin, SW coastal, Tas., N.S.W., A.C.T., Vic., W.A. Ecology: terrestrial, (predator), litter. Biological references: Beier, M. (1966). On the Pseudoscorpionidea of Australia. *Aust. J. Zool.* **14**: 275–303 (distribution); Beier, M. (1969). Neue Pseudoskorpione aus Australien. *Ann. Naturh. Mus. Wien* **73**: 171–187 (distribution); Beier, M. (1974). Neue Pseudoskorpione aus Australien und Neu-Guinea. *Ann. Naturh. Mus. Wien* **78**: 203–213 (distribution).

Austrochthonius cavicola Beier, 1968

Austrochthonius cavicola Beier, M. (1968). Some cave-dwelling Pseudoscorpionidea from Australia and New Caledonia. *Rec. S. Aust. Mus.* **15**: 757–765 [757]. Type data: holotype, SAMA N196828 M., from Cathedral Cave, Naracoorte, S.A.

Distribution: SE coastal, S.A.; known only from type locality. Ecology: terrestrial, (predator), caves.

Lagynochthonius Beier, 1951

Lagynochthonius Beier, M. (1951). Die Pseudoscorpione Indochinas. *Mém. Mus. Natl. Hist. Nat. Paris (ns)* **1**: 47–123 [61] [originally described as a subgenus of *Tyrannochthonius* Chamberlin, 1929]. Type species *Chthonius johni* Redikorzev, 1922 by original designation.

This group is also found in Asia and Pacific Region, see Chamberlin, J.C. (1962). New and little-known false scorpions, principally from caves, belonging to the families Chthoniidae and Neobisiidae (Arachnida, Chelonethida). *Bull. Am. Mus. Nat. Hist.* **123**: 303–352; Murthy, V.A. & Ananthakrishnan, T.N. (1977). Indian Chelonethi. *Orient. Insect Monogr.* **4**: 1–210.

Lagynochthonius australicus (Beier, 1966)

Tyrannochthonius (Lagynochthonius) australicus Beier, M. (1966). On the Pseudoscorpionidea of Australia. *Aust. J. Zool.* **14**: 275–303 [284]. Type data: holotype, AM KS5853 tritonymph, from Denmark, W.A.

Distribution: SW coastal, N coastal, W.A., N.T. Ecology: terrestrial, (predator), litter, tall forest. Biological references: Beier, M. (1974). Neue Pseudoskorpione aus Australien und Neu-Guinea. *Ann. Naturh. Mus. Wien* **78**: 203–213 (distribution).

Paraliochthonius Beier, 1956

Paraliochthonius (Paraliochthonius) Beier, 1956

Paraliochthonius Beier, M. (1956). Ein neuer *Blothrus* (Pseudoscorp.) aus Sardinien, und ueber zwei Pseudoscorpione des West Mediterranean litorals. *Fragm. Entomol.* **2**: 55–63 [58]. Type species *Chthonius singularis* Menozii, 1924 by original designation.

Morikawia Chamberlin, J.C. (1962). New and little-known false scorpions, principally from caves, belonging to the families Chthoniidae and Neobisiidae (Arachnida, Chelonethida). *Bull. Am. Mus. Nat. Hist.* **123**: 303–352 [311]. Type species *Chthonius johnstoni* Chamberlin, 1923 by original designation.

This group is found world-wide, see Beier, M. (1976). The pseudoscorpions of New Zealand, Norfolk and Lord Howe. *N.Z. J. Zool.* **3**: 199–246.

Paraliochthonius (Paraliochthonius) cavicola (Beier, 1967)

Morikawia cavicola Beier, M. (1967). Some Pseudoscorpionidea from Australia, chiefly from caves. *Aust. Zool.* **14**: 199–205 [199]. Type data: holotype, SAMA N1966165 M., paratype SAMA N1966166 1F., NHMW *1M., from The Grill Cave, Bungonia, N.S.W.

Distribution: SE coastal, N.S.W.; only known from type locality. Ecology: terrestrial, (predator), caves.

Paraliochthonius (Paraliochthonius) kermadecensis Beier, 1976

Paraliochthonius (Paraliochthonius) kermadecensis Beier, M. (1976). The pseudoscorpions of New Zealand, Norfolk and Lord Howe. *N.Z. J. Zool.* **3**: 199–246 [205]. Type data: holotype, DSIR *F., paratype AM KS12743 1F., from Kermadec Is. (holotype only).

Distribution: Lord Howe Is.; Kermadec Is. Ecology: terrestrial, (predator); found in deserted nest of *Turdus merula*, on foliage.

Paraliochthonius (Paraliochthonius) norfolkensis Beier, 1976

Paraliochthonius (Paraliochthonius) norfolkensis Beier, M. (1976). The pseudoscorpions of New Zealand, Norfolk and Lord Howe. *N.Z. J. Zool.* **3**: 199–246 [207]. Type data: holotype, ANIC 9998 M., paratypes ANIC 2M.,3F., 1 tritonymph, DSIR*, OM*, AIM*, from Mt. Pitt, Norfolk Is. (holotype and some paratypes only).

Distribution: Norfolk Is.; New Zealand. Ecology: terrestrial, (predator), litter.

Paraliochthonius (Paraliochthonius) queenslandicus (Beier, 1969)

Morikawia queenslandica Beier, M. (1969). Neue Pseudoskorpione aus Australien. *Ann. Naturh. Mus. Wien* **73**: 171–187 [174]. Type data: holotype, M. whereabouts unknown, paratypes NHMW *4M.,2F., 4M.,3F. whereabouts unknown, from Tamborine Mt., Joalah Natl. Park (as Jamborine Mt., Toalah National Park), Qld.

Distribution: NE coastal, Qld.; only known from type locality. Ecology: terrestrial, (predator), litter.

Paraliochthonius (Paraliochthonius) semihorridus (Beier, 1969)

Morikawia semihorrida Beier, M. (1969). Neue Pseudoskorpione aus Australien. *Ann. Naturh. Mus. Wien* **73**: 171–187 [176]. Type data: holotype, F. whereabouts unknown, from Mt. Nebo, Qld.

Distribution: NE coastal, Qld; only known from type locality. Ecology: terrestrial, (predator), litter.

Paraliochthonius (Pholeochthonius) Beier, 1976

Pholeochthonius Beier, M. (1976). The pseudoscropions of New Zealand, Norfolk and Lord Howe. *N.Z. J. Zool.* **3**: 199–246 [209]. Type species *Paraliochthonius (Pholeochthonius) cavernicola* Beier, 1976 by monotypy.

Paraliochthonius (Pholeochthonius) cavernicola Beier, 1976

Paraliochthonius (Pholeochthonius) cavernicola Beier, M. (1976). The pseudoscorpions of New Zealand, Norfolk and Lord Howe. *N.Z. J. Zool.* **3**: 199–246 [209]. Type data: holotype, AM KS21 M., from Lord Howe Is.

Distribution: Lord Howe Is.; only known from type locality. Ecology: terrestrial, (predator), caves.

Pseudotyrannochthonius Beier, 1930

Pseudotyrannochthonius Beier, M. (1930). Alcuni pseudoscorpioni esotici raccolti dal Prof. F. Silvestri. *Boll. Lab. Zool. Gen. Agr. R. Scuola Agric. Portici* **23**: 197–209 [207]. Type species *Chthonius silvestrii* Ellingsen, 1905 by original designation.

Tubbichthonius Hoff, C.C. (1951). New species and records of chthoniid pseudoscorpions. *Am. Mus. Novit.* 1483: 1–13 [10]. Type species *Tubbichthonius solitarius* Hoff, 1951 by original designation.

Spelaeochthonius Morikawa, K. (1954). On some pseudoscorpions in Japanese lime-grottoes. *Mem. Ehime Univ. (2B)* **2**: 79–87 [83]. Type species *Spelaeochthonius kubotai* Morikawa, 1954 by original designation.

Taxonomic decision of Beier, M. (1966). On the Pseudoscorpionidea of Australia. *Aust. J. Zool.* **14**: 275–303 [285]; Muchmore, W.B. (1967). Pseudotyrannochthoniine pseudoscorpions from the western United States. *Trans. Am. Microscop. Soc.* **86**: 132–139 [134].

This group is also found in North America, South America and Japan, see Beier, M. (1932). Pseudoscorpionidea I. Suborder Chthoniinea et Neobisiinea. *Das Tierreich* **57**: i–xx 1–258.

Pseudotyrannochthonius australiensis Beier, 1966

Pseudotyrannochthonius australiensis Beier, M. (1966). On the Pseudoscorpionidea of Australia. *Aust. J. Zool.* **14**: 275–303 [287]. Type data: holotype, AM KS5323 F., from Edith, N.S.W.

Distribution: Murray-Darling basin, N.S.W.; only known from type locality. Ecology: terrestrial, (predator), litter.

Pseudotyrannochthonius bornemisszai Beier, 1966

Pseudotyrannochthonius bornemisszai Beier, M. (1966). On the Pseudoscorpionidea of Australia. *Aust. J. Zool.* **14**: 275–303 [286]. Type data: holotype, AM KS5854 F., paratype AM deutonymph lost, from Boolarra, Vic.

Distribution: SE coastal, Vic.; only known from type locality. Ecology: terrestrial, (predator), litter, tall forest.

Pseudotyrannochthonius giganteus Beier, 1971

Pseudotyrannochthonius giganteus Beier, M. (1971). A new chthoniid pseudoscorpion from Western Australia. *J. Aust. Entomol. Soc.* **10**: 233–234 [233]. Type data: holotype, AM KS5324 F., paratypes AM KS5324 1M., NHMW *1M., from Calgadup Cave, Augusta, W.A.

Distribution: SW coastal, W.A.; only known from type locality. Ecology: terrestrial, (predator), caves.

Pseudotyrannochthonius gigas Beier, 1969

Pseudotyrannochthonius gigas Beier, M. (1969). Neue Pseudoskorpione aus Australien. *Ann. Naturh. Mus. Wien* **73**: 171–187 [178]. Type data: holotype, SAMA N1980191 F., paratype AM KS5326 F., from Harmon One Cave, Byaduk, Vic.

Distribution: SE coastal, Vic.; only known from type locality. Ecology: terrestrial, (predator), caves.

Pseudotyrannochthonius hamiltonsmithi Beier, 1968

Pseudotyrannochthonius hamiltonsmithi Beier, M. (1968). Some cave-dwelling Pseudoscorpionidea from Australia and New Caledonia. *Rec. S. Aust. Mus.* **15**: 757–765 [759]. Type data: holotype, SAMA N196830 F., from Mt. Widderin Cave, Skipton, Vic.

Distribution: SE coastal, Vic.; only known from type locality. Ecology: terrestrial, (predator), caves.

Pseudotyrannochthonius jonesi (Chamberlin, 1962)

Tubbichthonius jonesi Chamberlin, J.C. (1962). New and little-known false scorpions, principally from caves, belonging to the families Chthoniidae and Neobisiidae (Arachnida, Chelonethida). *Bull. Am. Mus. Nat. Hist.* **123**: 303–352 [317]. Type data: holotype, AMNH JC-204.01001 *F., from "probably in Blue Mountains near Sydney".

Distribution: SE coastal, N.S.W.; only known from type locality. Ecology: terrestrial, (predator), caves.

Pseudotyrannochthonius queenslandicus Beier, 1969

Pseudotyrannochthonius queenslandicus Beier, M. (1969). Neue Pseudoskorpione aus Australien. *Ann. Naturh. Mus. Wien* **73**: 171–187 [177]. Type data: holotype, M. whereabouts unknown, paratypes NHMW *2M., 1 tritonymph whereabouts unknown, from Tamborine Mt., Joalah Natl. Park (as Jamborine Mt., Toalah National Park), Qld.

Distribution: NE coastal, Qld; only known from type locality. Ecology: terrestrial, (predator), litter.

Pseudotyrannochthonius solitarius (Hoff, 1951)

Tubbichthonius solitarius Hoff, C.C. (1951). New species and records of chthoniid pseudoscorpions. *Am. Mus. Novit.* 1483: 1–13 [9]. Type data: holotype, AMNH *M., from Mt. Slide, Vic.

Distribution: SE coastal, Murray-Darling basin, SW coastal, W.A., N.S.W., A.C.T., Tas. Ecology: terrestrial, (predator), litter. Biological references: Beier, M. (1966). On the Pseudoscorpionidea of Australia. *Aust. J. Zool.* **14**: 275–303; Dartnall, A.J. (1970). Some Tasmanian chthoniid pseudoscorpions. *Pap. Proc. R. Soc. Tasm.* **104**: 65–68; Beier, M. (1974). Neue Pseudoskorpione aus Australien und Neu-Guinea. *Ann. Naturh. Mus. Wien* **78**: 203–213.

Pseudotyrannochthonius tasmanicus Dartnall, 1970

Pseudotyrannochthonius tasmanicus Dartnall, A.J. (1970). Some Tasmanian chthoniid pseudoscorpions. *Pap. Proc. R. Soc. Tasm.* **104**: 65–68 [66]. Type data: holotype, TMH J613 *M., paratypes TMH J614–616 *3F., from King George V Cave, Tas.

Distribution: Tas. Ecology: terrestrial, (predator), caves.

Pseudotyrannochthonius typhlus Dartnall, 1970

Pseudotyrannochthonius typhlus Dartnall, A.J. (1970). Some Tasmanian chthoniid pseudoscorpions. *Pap. Proc. R. Soc. Tasm.* **104**: 65–68 [67]. Type data: holotype, TMH J602 *M., paratypes TMH *J603, J611 1 tritonymph, 1 deutonymph, from Sennacheribs Passage, Georgies Hall Cave, Mole Creek, Tas. (holotype and some paratypes only).

Distribution: Tas. Ecology: terrestrial, (predator), caves.

Sathrochthonius Chamberlin, 1962

Sathrochthonius Chamberlin, J.C. (1962). New and little-known false scorpions, principally from caves, belonging to the families Chthoniidae and Neobisiidae (Arachnida, Chelonethida). *Bull. Am. Mus. Nat. Hist.* **123**: 303–352 [303]. Type species *Sathrochthonius tuena* Chamberlin, 1962 by original designation.

This group is also found in South America and the Pacific, see Vitali-di Castri, V. (1974). Presencia en America sur del genero *Sathrochthonius* (Pseudoscorpionida) con descripcion de una nueva especie. *Physis, B. Aires (C)* **33**: 193–201.

Sathrochthonius crassidens Beier, 1966

Sathrochthonius crassidens Beier, M. (1966). On the Pseudoscorpionidea of Australia. *Aust. J. Zool.* **14**: 275–303 [280]. Type data: holotype, AM KS5851 M., from Edith, N.S.W.

Distribution: Murray-Darling basin, N.S.W.; only known from type locality. Ecology: terrestrial, (predator), litter.

Sathrochthonius insulanus Beier, 1976

Sathrochthonius insulanus Beier, M. (1976). The pseudoscorpions of New Zealand, Norfolk and Lord Howe. *N.Z. J. Zool.* **3**: 199–246 [202]. Type data: holotype, AM KS22 M., paratypes AM KS12742 F., NHMW *1F., from Lord Howe Is.

Distribution: Lord Howe Is.; only known from type locality. Ecology: terrestrial, (predator), litter.

Sathrochthonius tuena Chamberlin, 1962

Sathrochthonius tuena Chamberlin, J.C. (1962). New and little-known false scorpions, principally from caves, belonging to the families Chthoniidae and Neobisiidae (Arachnida, Chelonethida). *Bull. Am. Mus. Nat. Hist.* **123**: 303–352 [304]. Type data: holotype, AMNH JC-2013.02001 *M., paratypes AMNH JC-2014.02002–4 *1M.,2F., from "probably in the Blue Mountains near Sydney".

Distribution: SE coastal, Murray-Darling basin, N.S.W. Ecology: terrestrial, (predator), caves. Biological references: Beier, M. (1967). Some Pseudoscorpionidea from Australia, chiefly from caves. *Aust. Zool.* **14**: 199–205 (distribution); Beier, M. (1968). Some cave-dwelling Pseudoscorpionidea from Australia and New Caledonia. *Rec. S. Aust. Mus.* **15**: 757–765 (distribution).

Sathrochthonius webbi Muchmore, 1982

Sathrochthonius webbi Muchmore, W.B. (1982). A new cavernicolous *Sathrochthonius* from Australia (Pseudoscorpionida : Chthoniidae). *Pac. Insects* **24**: 156–158 [156]. Type data: holotype, BPBM 12592 *F., paratypes SAMA*, BPBM *3 specimens, *FSM, from Holly Jump Lava Cave, 25 km E of Warwick, Qld.

Distribution: Murray-Darling basin, Qld.; only known from type locality. Ecology: terrestrial, (predator), caves.

Tyrannochthonius Chamberlin, 1929

Tyrannochthonius Chamberlin, J.C. (1929). A synoptic classification of the false scorpions or chela-spinners, with a report on a cosmopolitan collection of the same. Part I. The Heterosphyronida (Chthoniidae) (Arachnida-Chelonethida). *Ann. Mag. Nat. Hist. (10)* **4**: 50–80 [74]. Type species *Chthonius terribilis* With, 1906 by original designation.

This group is found world-wide, see Beier, M. (1932). Pseudoscorpionidea I. Suborder Chthoniinea et Neobisiinea. *Das Tierreich* **57**: i–xx 1–258; Beier, M. (1951). Die Pseudoscorpione Indochinas. *Mém. Mus. Natl. Hist. Nat. Paris (ns)* **1**: 47–123; Hoff, C.C. (1959). The pseudoscorpions of Jamaica. Part I. The genus *Tyrannochthonius* (Heterosphyronida : Chthoniidae). *Bull. Inst. Jamaica, Sci. Ser.* No. 10(1): 1–39; Mahnert, V. (1978). Pseudoskorpione (ausgenommen Olpiidae, Garypidae) aus Congo-Brazzaville (Arachnida, Pseudoscorpiones). *Folia Entomol. Hung. (ns)* **31**: 69–133; Mahnert, V. (1979). Pseudoskorpione (Arachnida) aus dem Amazonas-Gebiet (Brasilien). *Rev. Suisse Zool.* **86**: 719–810.

Tyrannochthonius laevis Beier, 1966

Tyrannochthonius (Tyrannochthonius) laevis Beier, M. (1966). On the Pseudoscorpionidea of Australia. *Aust. J. Zool.* **14**: 275–303 [283]. Type data: holotype, AM KS5852 M., paratypes AM KS5852 7 specimens, NHMW *3 specimens, from Kimberley Research Station, W.A.

Distribution: N coastal, W.A.; only known from type locality. Ecology: terrestrial, (predator), litter.

Incertae sedis

Chthonius caecus Tullgren, A. (1909). Chelonethi. pp. 411–415 *in* Michaelsen, W. & Hartmeyer, R. (eds.) *Die Fauna Südwest-Australiens.* Jena : Gustav Fischer Vol. 2 [414] [junior homonym of *Chthonius coecus* Packard, 1884; the male holotype has been lost, see Weidner, H. (1959). Die Entomologischen Sammlungen des

Zoologischen Staatsinstituts und Zoologischen Museums Hamburg. I. Teil. Pararthropoda und Chelicerata I. *Mitt. Hamb. Zool. Mus. Inst.* **57**: 89–142 and the original description is insufficient to recognize the species or the genus to which it belongs]. Type data: holotype, ZMH M. lost, from Brunswick, W.A.

Sathrochthonius tullgreni Chamberlin, J.C. (1962). New and little-known false scorpions, principally from caves, belonging to the families Chthoniidae and Neobisiidae (Arachnida, Chelonethida). *Bull. Am. Mus. Nat. Hist.* **123**: 303–352 [306] [replacement name for *Chthonius caecus* Tullgren, 1909].

GARYPIDAE

INTRODUCTION

Garypids are small to large pseudoscorpions found in habitats such as leaf litter, under rocks and under bark of trees. The cosmopolitan genus *Garypus* L. Koch is confined to the littoral zone and contains some of the world's largest pseudoscorpions. Currently 15 genera and approximately 150 species are recognized and have been placed in two subfamilies, both of which have been recorded from Australia: Garypinae and Geogarypinae. The latter is monogeneric.

Characterized by: carapace sub-triangular; four corneate eyes on ocular tubercles, situated away from the anterior margin of the carapace; usually diplotarsate; venom apparatus present in both chelal fingers; pleural membrane granulo-plicate.

References

Beier, M. (1932). Pseudoscorpionidea I. Suborder Chthoniinea et Neobisiinea. *Das Tierreich* **57**: i–xx 1–258

Anagarypus Chamberlin, 1930

Anagarypus Chamberlin, J.C. (1930). A synoptic classification of the false scorpions or chela-spinners, with a report on a cosmopolitan collection of the same. Part II. The Diplosphyronida (Arachnida-Chelonethida). *Ann. Mag. Nat. Hist. (10)* **5**: 1–48, 585–620 [615]. Type species *Anagarypus oceanusindicus* Chamberlin, 1930 by original designation.

This group is also found in the Indian Ocean Region, see Muchmore, W.B. (1982). The genus *Anagarypus* (Pseudoscorpionida : Garypidae). *Pac. Insects* **24**: 159–163.

Anagarypus australianus Muchmore, 1982

Anagarypus australianus Muchmore, W.B. (1982). The genus *Anagarypus* (Pseudoscorpionida : Garypidae). *Pac. Insects* **24**: 159–163 [160]. Type data: holotype, BPBM 12593 *M., paratypes BPBM*, FSM*, from Nymph Is., Qld. (holotype and some paratypes only).

Distribution: NE coastal, N Gulf, Qld., N.T.; coastal islands. Ecology: terrestrial, (predator); under stones.

Anagarypus heatwolei Muchmore, 1982

Anagarypus heatwolei Muchmore, W.B. (1982). The genus *Anagarypus* (Pseudoscorpionida : Garypidae). *Pac. Insects* **24**: 159–163 [162]. Type data: holotype, BPBM 12594 *F., paratypes BPBM *2F., 1 nymph, from Barrow Is., W.A.

Distribution: NW coastal, W.A.; known only from type locality. Ecology: terrestrial (predator); under stones.

Garypus L. Koch, 1873

Garypus Koch, L. (1873). *Uebersichtliche Darstellung der Europaischen Chernetiden (Pseudoscorpione).* Nürnberg : Bauer & Raspe vi 68 pp. [38]. Type species *Garypus litoralis* L. Koch, 1873 (= *Chelifer beauvoisii* Audouin, 1826) by subsequent designation, see Simon, E. (1879). *Les Arachnides de France, Chernetes.* Vol. 7 Paris : Roret vii 78 pp.

This group is found world-wide, see Beier, M. (1932). Pseudoscorpionidea I. Suborder Chthoniinea et Neobisiinea. *Das Tierreich* **57**: i–xx 1–258.

Garypus longidigitus Hoff, 1947

Garypus longidigitus Hoff, C.C. (1947). New species of diplosphyronid pseudoscorpions from Australia. *Psyche Camb.* **54**: 36–56 [41]. Type data: holotype, MCZ F., from Weir, Murray Ils., Qld.

Distribution: NE coastal, Qld.; known only from type locality. Ecology: terrestrial, (predator), littoral.

Geogarypus Chamberlin, 1930

Geogarypus (Geogarypus) Chamberlin, 1930

Geogarypus Chamberlin, J.C. (1930). A synoptic classification of the false scorpions or chela-spinners, with a report on a cosmopolitan collection of the same. Part

II. The Diplosphyronida (Arachnida-Chelonethida). *Ann. Mag. Nat. Hist. (10)* **5**: 1–48, 585–620 [609]. Type species *Garypus minor* L. Koch, 1873 by original designation.

This group is found world-wide, see Beier, M. (1932). Pseudoscorpionidea I. Suborder Chthoniinea et Neobisiinea. *Das Tierreich* **57**: i–xx 1–258.

Geogarypus (Geogarypus) rhantus Harvey, 1981

Geogarypus (Geogarypus) rhantus Harvey, M.S. (1981). *Geogarypus rhantus* sp. nov. (Pseudoscorpionida : Garypidae : Geogarypinae); a generic addition to the Australian fauna. *Mem. Qd. Mus.* **20**: 279–283 [279]. Type data: holotype, QM S352 M., paratypes QM S350–S351, S353, S647–S667 8M.,6F., 10 tritonymphs, from Lamond Hill, Qld. (holotype and some paratypes only).

Distribution: NE coastal, Qld.; known only from Lamond Hill, Mt Cook, Home Rule, Qld. Ecology: terrestrial (predator), litter, closed forest.

Synsphyronus Chamberlin, 1930

Taxonomic decision of Chamberlin, J.C. (1943). The taxonomy of the false scorpion genus *Synsphyronus*, with remarks on the sporadic loss of stability in generally constant morphological characters. *Ann. Entomol. Soc. Am.* **36**: 486–500 [488].

Synsphyronus (Synsphyronus) Chamberlin, 1930

Synsphyronus Chamberlin, J.C. (1930). A synoptic classification of the false scorpions or chela-spinners, with a report on a cosmopolitan collection of the same. Part II. The Diplosphyronida (Arachnida-Chelonethida). *Ann. Mag. Nat. Hist. (10)* **5**: 1–48, 585–620 [616]. Type species *Synsphyronus paradoxus* Chamberlin, 1930 by original designation.

Synsphyronus (Synsphyronus) callus Hoff, 1947

Synsphyronus (Synsphyronus) callus Hoff, C.C. (1947). New species of diplosphyronid pseudoscorpions from Australia. *Psyche Camb.* **54**: 36–56 [50]. Type data: holotype, MCZ F., paratypes MCZ 1F., 1 tritonymph, from Rottnest Is., W.A. (holotype and one paratype only).

Distribution: SW coastal, W.A.; known only from Rottnest Is. and Cottesloe, W.A. Ecology: terrestrial, (predator), woodland.

Synsphyronus (Synsphyronus) mimetus Chamberlin, 1943

Synsphyronus (Synsphyronus) mimetus Chamberlin, J.C. (1943). The taxonomy of the false scorpion genus *Synsphyronus*, with remarks on the sporadic loss of stability in generally constant morphological characters. *Ann. Entomol. Soc. Am.* **36**: 486–500 [492]. Type data: holotype, PUO JC-619.02001 M., paratypes WAM 83/589–90 1M.,1F., PUO JC-619.02002–9 2M.,3F., 1 tritonymph, MCZ JC-619.002013–14 1M.,1F., JC-619.02015–40 whereabouts unknown, from Corny Point, S.A.

Distribution: S Gulfs, SE coastal, S.A., Vic.; known only from Corny Point, S.A. and Lady Julia Percy Is., Vic. Ecology: terrestrial, (predator), woodland; under stones. Biological references: Chamberlin, J.C. (1949). New and little-known false scorpions from various parts of the world (Arachnida, Chelonethida), with notes on structural abnormalities in two species. *Am. Mus. Novit.* 1430: 1–57 (distribution).

Synsphyronus (Synsphyronus) paradoxus Chamberlin, 1930

Synsphyronus (Synsphyronus) paradoxus Chamberlin, J.C. (1930). A synoptic classification of the false scorpions or chela-spinners, with a report on a cosmopolitan collection of the same. Part II. The Diplosphyronida (Arachnida-Chelonethida). *Ann. Mag. Nat. Hist. (10)* **5**: 1–48, 585–620 [617]. Type data: holotype, PUO JC-481.01001 M., paratypes PUO JC-481.01002–4 3F., from Menindee, N.S.W.

Distribution: Murray-Darling basin, N.S.W.; known only from type locality. Ecology: terrestrial, (predator), woodland.

Synsphyronus (Maorigarypus) Chamberlin, 1930

Maorigarypus Chamberlin, J.C. (1930). A synoptic classification of the false scorpions or chela-spinners, with a report on a cosmopolitan collection of the same. Part II. The Diplosphyronida (Arachnida-Chelonethida). *Ann. Mag. Nat. Hist. (10)* **5**: 1–48, 585–620 [617]. Type species *Maorigarypus melanochelatus* Chamberlin, 1930 by original designation.

Idiogarypus Chamberlin, J.C. (1943). The taxonomy of the false scorpion genus *Synsphyronus*, with remarks on the sporadic loss of stability in generally constant morphological characters. *Ann. Entomol. Soc. Am.* **36**: 486–500 [499]. Type species *Garypus hansenii* With, 1909 by original designation.

Taxonomic decision of Morris, J.C.H. (1948). The taxonomic position of *Idiogarypus hansenii* (With). *Pap. Proc. R. Soc. Tasm.* **1947**: 37–41 [37].

This group is also found in New Zealand, see Beier, M. (1976). The pseudoscorpions of New Zealand, Norfolk and Lord Howe. *N.Z. J. Zool.* **3**: 199–246.

Synsphyronus (Maorigarypus) dewae Beier, 1969

Synsphyronus (Maorigarypus) dewae Beier, M. (1969). Neue Pseudoskorpione aus Australien. *Ann. Naturh. Mus. Wien* **73**: 171–187 [181]. Type data: holotype, NHMW F., paratypes AM KS5855 2 deutonymphs, from Truro, S.A.

Distribution: S Gulfs, S.A.; known only from type locality. Ecology: terrestrial, (predator), woodland; from nest of Galah (*Cacatua roseicapilla*).

Synsphyronus (Maorigarypus) elegans Beier, 1954

Synsphyronus (Maorigarypus) elegans Beier, M. (1954). Report from Prof. T. Gislén's expedition to Australia in 1951–1952. 7. Pseudoscorpionidea. *Acta Univ. Lund (N.F.) (2)* **50**(3): 1–26 [13]. Type data: syntypes, ZMLU 8M.,11F., 1 tritonymph, 1 protonymph, NHMW 3M.,3F., from Yorkrakine Hill (as York Rakine Tammin), W.A.

Distribution: SW coastal, W.A.; known only from type locality. Ecology: terrestrial, (predator), woodland; under stones.

Synsphyronus (Maorigarypus) fallaciosus Beier, 1966

Synsphyronus (Maorigarypus) fallaciosus Beier, M. (1966). On the Pseudoscorpionidea of Australia. *Aust. J. Zool.* **14**: 275–303 [294]. Type data: holotype, AM KS5856 M., paratypes AM KS5857–8 2F., 1 tritonymph, 1 deutonymph, NHMW 2F., from Gunyah Gunyah (as Gunyah), Vic.

Distribution: SE coastal, Vic., N.S.W.; known only from Gunyah Gunyah, Vic. and Wilsons Valley, N.S.W. Ecology: terrestrial, (predator), under bark, woodland. Biological references: Beier, M. (1974). Neue Pseudoskorpione aus Australien und Neu-Guinea. *Ann. Naturh. Mus. Wien* **78**: 203–213 (distribution).

Synsphyronus (Maorigarypus) gigas Beier, 1971

Synsphyronus (Maorigarypus) gigas Beier, M. (1971). A new *Synsphyronus* Chamberlin (Pseudoscorpiones) from the Great Victoria desert. *J. Aust. Entomol. Soc.* **10**: 161–162 [161]. Type data: syntypes, AM KS5198 1M.,3F., 2 deutonymphs, 1 protonymph, NHMW 2F., from Milne Rock [as grid reference 553623 Lennis 1:250 000 sheet], W.A.

Distribution: W plateau, W.A.; known only from type locality. Ecology: terrestrial, (predator), desert; under stones.

Synsphyronus (Maorigarypus) gisleni Beier, 1954

Synsphyronus (Maorigarypus) gisleni Beier, M. (1954). Report from Prof. T. Gislén's expedition to Australia in 1951–1952. 7. Pseudoscorpionidea. *Lunds Univ. Arsskr., Avd 2* **50**(3): 1–26 [9]. Type data: holotype, ZMLU F., from Denmark, W.A.

Distribution: SW coastal, W.A.; known only from type locality. Ecology: terrestrial, (predator), under bark, woodland.

Synsphyronus (Maorigarypus) grayi Beier, 1974

Synsphyronus (Maorigarypus) grayi Beier, M. (1974). Neue Pseudoskorpione aus Australien und Neu-Guinea. *Ann. Naturh. Mus. Wien* **78**: 203–213 [205]. Type data: syntypes, AM KS0015 1M.,2F., NHMW 1M.,1F., from Mullamullang Cave, W.A.

Distribution: W plateau, W.A.; known only from type locality. Ecology: terrestrial, (predator), caves.

Synsphyronus (Maorigarypus) hansenii (With, 1908)

Garypus hansenii With, C.J. (1908). Remarks on the Chelonethi. *Vidensk. Meddr. Dansk Naturh. Foren. (6)* **10**: 1–25 [12]. Type data: holotype, BMNH 1901.6.4.29 M., from Tas.

Distribution: Tas. Ecology: terrestrial, (predator), under bark, woodland; in house, in pelage of bandicoot. Biological references: Morris, J.C.H. (1948). The taxonomic position of *Idiogarypus hansenii* (With). *Pap. Proc. R. Soc. Tasm.* **1947**: 37–41 (redescription and distribution); Beier, M. (1954). Report from Prof. T. Gislén's expedition to Australia in 1951–1952. 7. Pseudoscorpionidea. *Lunds Univ. Arsskr., Avd 2* **50**(3): 1–26 (redescription and distribution); Dartnall, A.J. (1969). Pseudoscorpions. *Tasm. Nat.* No. 17: 2 (distribution); Dartnall, A.J. (1975). A note on the natural history of a pseudoscorpion. *Tasm. Nat.* No. 40: 5–6 (distribution).

Synsphyronus (Maorigarypus) magnus Hoff, 1947

Synsphyronus (Maorigarypus) magnus Hoff, C.C. (1947). New species of diplosphyronid pseudoscorpions from Australia. *Psyche Camb.* **54**: 36–56 [47]. Type data: holotype, MCZ F., from Margaret River, W.A.

Distribution: SW coastal, W.A.; known only from type locality. Ecology: terrestrial, (predator), under bark, woodland.

Synsphyronus (Maorigarypus) mimulus Chamberlin, 1943

Synsphyronus (Maorigarypus) mimulus Chamberlin, J.C. (1943). The taxonomy of the false scorpion genus *Synsphyronus*, with remarks on the sporadic loss of stability in generally constant morphological characters. *Ann. Entomol. Soc. Am.* **36**: 486–500 [496]. Type data: holotype, PUO JC-619.03001 M., paratypes WAM 81/329–330 1M.,1F., PUO JC-619.03002–8 3M.,4F., MCZ JC-619.03013–14 1M.,1F., from Corny Point, S.A.

Distribution: S Gulfs, S.A.; known only from type locality. Ecology: terrestrial, (predator), woodland; under stones.

Synsphyronus (Maorigarypus) niger Hoff, 1947

Synsphyronus (Maorigarypus) niger Hoff, C.C. (1947). New species of diplosphyronid pseudoscorpions from Australia. *Psyche Camb.* **54**: 36–56 [45]. Type data: holotype, MCZ M., from Adelaide, S.A.

Distribution: S Gulfs, S.A.; known only from type locality. Ecology: terrestrial, (predator), woodland.

Synsphyronus (Maorigarypus) nullarborensis Beier, 1969

Synsphyronus (Maorigarypus) nullarborensis Beier, M. (1969). Neue Pseudoskorpione aus Australien. *Ann.*

Naturh. Mus. Wien **73**: 171–187 [179]. Type data: holotype, WAM 78/601 F., from Toolinna Rock Hole, W.A.

Distribution: W plateau, S. Gulfs, W.A., S.A.; known only from Toolinna Rock Hole, W.A. and Morphett Vale, S.A. Ecology: terrestrial, (predator), caves. Biological references: Beier, M. (1974). Neue Pseudoskorpione aus Australien und Neu-Guinea. *Ann. Naturh. Mus. Wien* **78**: 203–213 (distribution).

Synsphyronus (Maorigarypus) viridis (Tubb, 1937)

Maorigarypus viridis Tubb, J.A. (1937). Reports of the expedition of the McCoy Society for field investigation and research (Lady Julia Percy Island). 19. Arachnida. *Proc. R. Soc. Vict. (ns)* **49**: 412–421 [412]. Type data: holotype, F. whereabouts unknown, from Lady Julia Percy Is., Vic.

Distribution: SE coastal, Vic,; known only from type locality. Ecology: terrestrial, (predator); under stones.

OLPIIDAE

INTRODUCTION

This is a large family containing small to large pseudoscorpions distributed in most parts of the world and found in a variety of habitats. Currently approximately 50 genera and over 300 species are recognized and are placed in two subfamilies: Olpiinae and Garypininae, both of which are represented in Australia.

Characterized by: diplotarsate; four corneate eyes not on ocular tubercles, placed near the anterior margin of the carapace; carapace sub-rectangular; venom apparatus present in both chelal fingers; pleural membrane smoothly plicate.

References

Beier, M. (1932). Pseudoscorpionidea I. Suborder Chthoniinea et Neobisiinea. *Das Tierreich* **57**: i–xx 1–258

Austrohorus Beier, 1966

Austrohorus Beier, M. (1966). On the Pseudoscorpionidea of Australia. *Aust. J. Zool.* **14**: 275–303 [288]. Type species *Austrohorus exsul* Beier, 1966 by original designation.

Austrohorus exsul Beier, 1966

Austrohorus exsul Beier, M. (1966). On the Pseudoscorpionidea of Australia. *Aust. J. Zool.* **14**: 275–303 [289]. Type data: holotype, AM KS5862 M., from Morawa, W.A.

Distribution: NW coastal; known only from type locality. Ecology: terrestrial, (predator), salt lake; under log.

Euryolpium Redikorzev, 1938

Euryolpium Redikorzev, V. (1938). Les pseudoscorpions des l'Indochine française receullis par M.C. Dawydoff. *Mém. Mus. Natl. Hist. Nat. Paris (ns)* **10**: 69–116 [82]. Type species *Euryolpium agniae* Redikorzev, 1938 by original designation.

This group is also found in southeast Asia, see Murthy, V.A. & Ananthakrishnan, T.N. (1977). Indian Chelonethi. *Orient. Insect Monogr.* **4**: 1–210; Heurtault, J. (1980). Données nouvelles sur les genres *Xenolpium, Antiolpium, Indolpium* et *Euryolpium* (Arachnides, Pseudoscorpions). *Rev. Suisse Zool.* **87**: 143–154.

Euryolpium granulosum (Hoff, 1947)

Xenolpium granulosum Hoff, C.C. (1947). New species of diplosphyronid pseudoscorpions from Australia. *Psyche Camb.* **54**: 36–56 [39]. Type data: holotype, MCZ *F., from Mullewa, W.A.

Xenolpium robustum Beier, M. (1948). Über Pseudoscorpione der australischen region. *Eos, Madrid* **24**: 525–562 [525]. Type data: holotype, BMNH 1931.12.14.5 *F., from Pender Bay, W.A.

Taxonomic decision of Beier, M. (1966). On the Pseudoscorpionidea of Australia. *Aust. J. Zool.* **14**: 275–303 [291].

Distribution: NW coastal, N coastal, W.A.; known only from Mullewa and Pender Bay, W.A. Ecology: terrestrial, (predator).

Euryolpium michaelseni (Tullgren, 1909)

Olpium michaelseni Tullgren, A. (1909). Chelonethi. pp. 411–415 *in* Michaelsen, W. & Hartmeyer, R. (eds.) *Die Fauna Südwest-Australiens.* Jena : Gustav Fischer Vol. 2 [412]. Type data: holotype, ZMH *F., from Yalgoo, W.A.

Distribution: NW coastal, N coastal, Murray-Darling basin, W.A., N.S.W. Ecology: terrestrial, (predator), litter. Biological references: Chamberlin, J.C. (1930). A synoptic classification of the false scorpions or chela-spinners, with a report on a cosmopolitan collection of the same. Part II. The Diplosphyronida (Arachnida-Chelonethida). *Ann. Mag. Nat. Hist. (10)* **5**: 1–48, 585–620 (distribution); Beier, M. (1966). On the Pseudoscorpionidea of Australia. *Aust. J. Zool.* **14**: 275–303 (distribution).

Olpium L. Koch, 1873

Olpium Koch, L. (1873). *Uebersichtliche Darstellung der Europaischen Chernetiden (Pseudoscorpione).* Nürnberg vi 68 pp. [33]. Type species *Obisium pallipes* Lucas, 1849 by current usage; an application will be made to the International Commission of Zoological Nomenclature for the validation of the acceptance of *Obisium pallipes* Lucas, 1849 as type species.

This group is also found in Africa, Asia and Europe, see Beier, M. (1932). Pseudoscorpionidea I. Suborder Chthoniinea et Neobisiinea. *Das Tierreich* **57**: i–xx 1–258; Murthy, V.A. & Ananthakrishnan, T.N. (1977). Indian Chelonethi. *Orient. Insect Monogr.* **4**: 1–210.

Olpium australicum Beier, 1969

Olpium australicum Beier, M. (1969). Neue Pseudoskorpione aus Australien. *Ann. Naturh. Mus. Wien* **73**: 171–187 [182]. Type data: holotype, SAMA N1980190 *M., from near Madura, W.A.

Distribution: W plateau, W.A., S.A.; known only from Madura, W.A. and Koonalda, S.A. Ecology: terrestrial, (predator), litter. Biological references: Beier, M. (1974). Neue Pseudoskorpione aus Australien und Neu-Guinea. *Ann. Naturh. Mus. Wien* **78**: 203–213 (distrubiton).

Protogarypinus Beier, 1954

Protogarypinus Beier, M. (1954). Report from Prof. T. Gislén's expedition to Australia in 1951–1952. 7. Pseudoscorpionidea. *Lunds Univ. Arsskr., Avd 2* **50**(3): 1–26 [3]. Type species *Protogarypinus giganteus* Beier, 1954 by original designation.

Protogarypinus dissimilis Beier, 1974

Protogarypinus dissimilis Beier, M. (1974). Neue Pseudoskorpione aus Australien und Neu-Guinea. *Ann. Naturh. Mus. Wien* **78**: 203–213 [204]. Type data: syntypes, SAMA N197762a–t 1M.,3F., 16 nymphs, NHMW *1M.,2F., 5 nymphs, from Alligator Gorge, S.A.

Distribution: S Gulfs, S.A.; known only from type locality. Ecology: terrestrial, (predator), litter.

Protogarypinus giganteus Beier, 1954

Protogarypinus giganteus Beier, M. (1954). Report from Prof. T. Gislén's expedition to Australia in 1951–1952. 7. Pseudoscorpionidea. *Lunds Univ. Arsskr., Avd 2* **50**(3): 1–26 [4]. Type data: holotype, ZMLU *F., from Denmark, W.A.

Distribution: SW coastal, W.A.; known only from type locality. Ecology: terrestrial, (predator), under bark.

Solinus Chamberlin, 1930

Solinus Chamberlin, J.C. (1930). A synoptic classification of the false scorpions or chela-spinners, with a report on a cosmopolitan collection of the same. Part II. The Diplosphyronida (Arachnida-Chelonethida). *Ann. Mag. Nat. Hist. (10)* **5**: 1–48, 585–620 [596]. Type species *Garypinus corticolus* Chamberlin, 1923 by original designation.

This group is found circumtropical, see Chamberlin, J.C. (1930). A synoptic classification of the false scorpions or chela-spinners, with a report on a cosmopolitan collection of the same. Part II. The Diplosphyronida (Arachnida-Chelonethida). *Ann. Mag. Nat. Hist. (10)* **5**: 1–48, 585–620; Beier, M. (1963). *Bestimmungsbucher zur Bodenfauna Europas.* Ordnung Pseudoscorpionidea (Afterskorpione). Berlin : Akademie-Verl. Vol. 1 vi 313 pp.; Beier, M. (1967). Pseudoskorpione aus dem tropischen Ostafrika (Kenya, Tanzania, Uganda, etc.). *Ann. Naturh. Mus. Wien* **70**: 73–93; Beier, M. (1971). Pseudoskorpione unter Araucarien Rinde in Neu-Guinea. *Ann. Naturh. Mus. Wien* **75**: 367–373.

Solinus australiensis Chamberlin, 1930

Solinus australiensis Chamberlin, J.C. (1930). A synoptic classification of the false scorpions or chela-spinners, with a report on a cosmopolitan collection of the same. Part II. The Diplosphyronida (Arachnida-Chelonethida). *Ann. Mag. Nat. Hist. (10)* **5**: 1–48, 585–620 [597]. Type data: holotype, PUO JC-480.03001 *F., from Barringun, N.S.W.

Distribution: Murray-Darling basin, SE coastal, N.S.W., Vic.; known only from Barringun, N.S.W. and Lady Julia Percy Is., Vic. Ecology: terrestrial, (predator); under rocks. Biological references: Tubb, J.A. (1937). Reports of the expedition of the McCoy Society for field investigation and research (Lady Julia Percy Island). 19. Arachnida. *Proc. R. Soc. Vict. (ns)* **49**: 412–421 (distribution).

Xenolpium Chamberlin, 1930

Xenolpium Chamberlin, J.C. (1930). A synoptic classification of the false scorpions or chela-spinners, with a report on a cosmopolitan collection of the same. Part II. The Diplosphyronida (Arachnida-Chelonethida). *Ann. Mag. Nat. Hist. (10)* **5**: 1–48, 585–620 [600]. Type species *Olpium pacificum* With, 1907 by original designation.

Antiolpium Beier, M. (1962). On some Pseudoscorpionidea from New Zealand. *Rec. Canterbury Mus.* **7**: 399–402 [399]. Type species *Olpium zealandiensis* Hoff, 1947 (= *Olpium pacificum* With, 1907) by original designation.

Taxonomic decision of Beier, M. (1976). The pseudoscorpions of New Zealand, Norfolk and Lord Howe. *N.Z. J. Zool.* **3**: 199–246 [210].

This group is also found in Africa, New Zealand and southeast Asia, see Beier, M. (1965). Die Pseudoscorpioniden Neu-Guineas und der benachbarten Inseln. *Pac. Insects* **7**: 749–796; Beier, M. (1965). Pseudoskorpione aus dem Tschad-Gebiet. *Ann. Naturh. Mus. Wien* **68**: 365–374.

Xenolpium bornemisszai Beier, 1966

Xenolpium bornemisszai Beier, M. (1966). On the Pseudoscorpionidea of Australia. *Aust. J. Zool.* **14**: 275–303 [292]. Type data: holotype, AM KS5859 M., paratypes AM KS5860 F., NHMW *F., from Gnangara Lake (as Gnangara), W.A.

Distribution: SW coastal, W.A.; known only from Gnangara Lake and Moora, W.A. Ecology: terrestrial, (predator), litter.

Xenolpium longiventer (L. Koch, 1885)

Olpium longiventer Koch, L. (1885). Ordo Chelonethi. pp. 44–51 pls 4, 6 *in* Koch, L. & Keyserling, E. *Die Arachniden Australiens nach der Natur beschrieben und abgebildet.* Vol. 2 Nürnberg : Bauer & Raspe [50]. Type data: syntype (1), BMNH 1913.9.1.455*, from Peak Downs (as Peack Downs), Qld.

Distribution: NE coastal, SE coastal, Tas., Qld., Vic. Ecology: terrestrial, (predator), litter. Biological references: Beier, M. (1966). On the Pseudoscorpionidea of Australia. *Aust. J. Zool.* **14**: 275–303 (distribution); Beier, M. (1969). Neue Pseudoskorpione aus Australien. *Ann. Naturh. Mus. Wien* **73**: 171–187 (distribution); Heurtault, J. (1980). Données nouvelles sur les genres *Xenolpium, Antiolpium, Indolpium* et *Euryolpium* (Arachnides, Pseudoscorpions). *Rev. Suisse Zool.* **87**: 143–154 (distribution).

Xenolpium pacificum (With, 1907)

Olpium pacificum With, C.J. (1907). On some new species of the Cheliferidae, Hans., and Garypidae, Hans., in the British Museum. *J. Linn. Soc. Zool.* **30**: 49–85 [75]. Type data: holotype, BMNH *M., from Stewart Is., New Zealand.

Xenolpium pacificum norfolkense Beier, 1976

Xenolpium pacificum norfolkense Beier, M. (1976). The pseudoscorpions of New Zealand, Norfolk and Lord Howe. *N.Z. J. Zool.* **3**: 199–246 [211]. Type data: holotype, ANIC 9999 M., from Bumbora Reserve, Norfolk Is.

Distribution: Norfolk Is.; known only from type locality. Ecology: terrestrial, (predator); in cow dung.

Xenolpium squalidum Beier, 1966

Xenolpium squalidum Beier, M. (1966). On the Pseudoscorpionidea of Australia. *Aust. J. Zool.* **14**: 275–303 [291]. Type data: holotype, AM KS5861 M., paratype AM KS5861 F., from Kimberley Research Station, W.A.

Distribution: N coastal, W.A.; known only from type locality. Ecology: terrestrial, (predator), litter.

PSEUDOCHIRIDIIDAE

INTRODUCTION

Pseudochiridiids are small pseudoscorpions found in leaf litter or under the bark of trees. The Pseudochiridiidae has usually been considered a subfamily of the Cheiridiidae, but Muchmore (1982) reinstated it to full familial status. Two genera and 12 species currently are recognized. With the exception of four species, all are found in Africa and Madagascar.

Characterized by: coxae of leg IV expanded posteriorly, especially in females; monotarsate; heterofemorate; venom apparatus present in both chelal fingers; abdominal tergites and sternites chevron shaped.

References

Muchmore, W.B. (1982). Pseudoscorpionida. pp. 96–102 *in* Parker, S.P. (ed.) *Synopsis and Classification of Living Organisms.* New York : McGraw-Hill Vol. 2

Vitali-di Castri, V. (1969). Pseudochiridiinae (Pseudoscorpionida) du Muséum National d'Histoire Naturelle. Remarques sur la sous-famille et description de deux nouvelles espèces de Madagascar et d'Angola. *Bull. Mus. Natl. Hist. Nat. Paris (2)* **41**: 1175–1199

Pseudochiridium With, 1906

Pseudochiridium With, C.J. (1906). The Danish expedition to Siam 1899–1900. III. Chelonethi. An account of the Indian false-scorpions together with studies on the anatomy and classification of the order. *K. Danske Vidensk. Selsk. Skr. (7)* **3**: 1–214 [199]. Type species *Pseudochiridium thorelli* With, 1906 by original designation.

Afrocheiridium Beier, M. (1932). Pseudoscorpionidea II. Suborder Cheliferinea. *Das Tierreich* **58**: i–xxi 1–294 [5]. Type species *Pseudochiridium tragardhi* Tullgren, 1907 by original designation.

Taxonomic decision of Beier, M. (1955). Pseudoscorpionidea. pp. 263–328 *in* Hanstrom, B., Brinck, P & Rudebeck, G. (eds.) *South African Animal Life.* Results of the Lund University Expedition in 1950–1951. Stockholm : Almquist & Wiksells.

This group is also found in southeast Asia, Africa and U.S.A., see Vitali-di Castri, V. (1969). Pseudochiridiinae (Pseudoscorpionida) du Muséum National d'Histoire Naturelle remarques sur la sous-famille et description de deux nouvelles espèces de Madagascar et d'Angola. *Bull. Mus. Natl. Hist. Nat. Paris (2)* **41**: 1175–1199.

Pseudochiridium clavigerum (Thorell, 1889)

Chelifer claviger Thorell, T. (1889). Viaggio di Leonardo Fea in Birmania e regioni vicine. XXI. Aracnidi artrogastri Birmani raccolti da L. Fea nel 1885–1887. *Ann. Mus. Civ. Stor. Nat. Genova (2)* **7**: 521–729 pl V [591]. Type data: syntypes (probable), MCG *2 specimens, from Bhámo, Burma.

Pseudochiridium sundaicum Beier, M. (1953). Pseudoscorpionidea von Sumba und Flores. *Verh. Naturf. Ges. Basel* **64**: 81–88 [83]. Type data: syntypes, NHMB *2F., from Langgai, Sumba.

Taxonomic decision of Beier, M. (1965). Die Pseudoscorpioniden Neu-Guineas und der benachbarten Inseln. *Pac. Insects* **7**: 749–796 [765].

Distribution: Christmas Is.; also in Burma, India, Irian Jaya, Mayasia, Philippines and Sumba. Ecology: terrestrial, (predator), under bark, caves.

PSEUDOGARYPIDAE

INTRODUCTION

This is a family of medium sized pseudoscorpions found in habitats such as under bark of trees, under rocks, in leaf litter and in caves. Currently two genera and seven Recent species are recognized; three species are known from Baltic amber (Schawaller, 1978). *Neopseudogarypus scutellatus* Morris is Australia's only representative and is confined to Tasmania.

Characterized by: monotarsate; four corneate eyes on ocular tubercles, situated away from the anterior margin of the carapace; carapace sub-triangular, with postero-lateral alae and antero-lateral horns; spines present on coxa I.

References

Benedict, E.M. & Malcolm, D.R. (1978). The family Pseudogarypidae (Pseudoscorpionida) in North America with comments on the genus *Neopseudogarypus* Morris from Tasmania. *J. Arachnol.* **6**: 81–104

Schawaller, W. (1978). Neue Pseudoskorpione aus dem Baltischen Bernstein der Stuttgarter Bernsteinsammlung (Arachnida: Pseudoscorpionidea). *Stuttgarter Beitr. Naturk.* Series B No. 42: 1–22

Neopseudogarypus Morris, 1948

Neopseudogarypus Morris, J.C.H. (1948). A new genus of pseudogarypin pseudoscorpions possessing pleural plates. *Pap. Proc. R. Soc. Tasm.* **1947**: 43–47 [43]. Type species *Neopseudogarypus scutellatus* Morris, 1948 by original designation.

Neopseudogarypus scutellatus Morris, 1948

Neopseudogarypus scutellatus Morris, J.C.H. (1948). A new genus of pseudogarypin pseudoscorpions possessing pleural plates. *Pap. Proc. R. Soc. Tasm.* **1947**: 43–47 [44].

Type data: holotype, BMNH 1948.3.24.4 *F., paratypes AM KS5319 1M., BMNH 1948.3.24.2–3 *1F.,1M., AMNH *1F., MNHP*, from Launceston, Tas.

Distribution: Tas. Ecology: terrestrial, (predator), under stones.

STERNOPHORIDAE

INTRODUCTION

Sternophorids are small to medium sized pseudoscorpions that, with the exception of a single cavernicolous species, appear to be strictly corticolous. They are very pale in colouration and are flattened dorso-ventrally. Five genera and 21 species currently are recognized from most tropical and subtropical regions of the world, but several species are found in the temperate zones.

Characterized by: pseudosternum present between coxae of legs I to III; posteriorly angulate carapace; eyes absent; monotarsate; homofemorate; venom apparatus present in both chelal fingers.

References

Beier, M. (1932). Pseudoscorpionidea II. Suborder Cheliferinea. *Das Tierreich* **58**: i–xxi 1–294

Sternophorus Chamberlin, 1923

Sternophorus (Sternophorus) Chamberlin, 1923

Sternophorus Chamberlin, J.C. (1923). New and little known pseudoscorpions from the islands and adjacent shores of the Gulf of California. *Proc. Calif. Acad. Sci. (4)* **12**: 353–387 [371]. Type species *Sternophorus sini* Chamberlin, 1923 by original designation.

This group is also found in southeast Asia and North America; the other subgenus is Ethiopian, see Beier, M. (1951). Die Pseudoscorpione Indochinas. *Mém. Mus. Natl. Hist. Nat. Paris (ns)* **1**: 47–123; Chamberlin, J.C. (1932). On some false scorpions of the super-family Cheiridioidea. *Pan-Pac. Entomol.* **8**: 137–144; Murthy, V.A. & Ananthakrishnan, T.N. (1977). Indian Chelonethi. *Orient. Insect Monogr.* **4**: 1–210.

Sternophorus (Sternophorus) hirsti Chamberlin, 1932

Sternophorus (Sternophorus) hirsti hirsti Chamberlin, 1932

Sternophorus hirsti Chamberlin, J.C. (1932). On some false scorpions of the super-family Cheiridioidea. *Pan-Pac. Entomol.* **8**: 137–144 [143]. Type data: holotype, PUO JC–480.01001 M., from Barringun, N.S.W.

Distribution: Murray-Darling basin, N.S.W.; other subspecies found in Papua New Guinea. Ecology: terrestrial, (predator), under bark, woodland, low woodland. Biological references: Harvey, M.S. (1982). A parasitic nematode (Mermithidae) from the pseudoscorpion "*Sternophorus*" *hirsti* Chamberlin (Sternophoridae). *J. Arachnol.* **10**: 192.

TRIDENCHTHONIIDAE

INTRODUCTION

Members of this family are small to medium sized pseudoscorpions usually confined to leaf litter, especially in rainforests, but a few species have been collected from under bark of trees and under rocks. They are often referred to as the Dithidae, a synonym of Tridenchthoniidae.

Fifteen genera are currently recognized which include approximately 70 species. Of these species, most are confined to southeast Asia and Africa, but two species have been described from Australia. Two subfamilies have been erected, Tridenchthoniinae and Cecodithinae, but the status of the latter is doubtful because the only included genus and species, the Argentinian *Cecoditha parva* Mello-Leitao, was poorly described.

Characterized by: heterotarsate; spiracles obliquely orientated, and with sclerotic plates; genital area of female sclerotic; pedipalpal tibia pyriform; carapace usually parallel sided and relatively hirsute, possessing 50 or more setae.

References

Chamberlin, J.C. & Chamberlin, R.V. (1945). The genera and species of the Tridenchthoniidae (Dithidae): a family of the arachnid order Chelonethida. *Bull. Univ. Utah Biol. Ser.* **9**: 1–67

Anaulacodithella Beier, 1944

Anaulacodithella Beier, M. (1944). Über Pseudoscorpioniden aus Östafrika. *Eos, Madrid* **20**: 173–212 [175] [originally described as a subgenus of *Verrucadithella* Beier, 1931]. Type species *Chthonius mordax* Tullgren, 1907 by original designation.

Xenoditha Chamberlin, J.C. & Chamberlin, R.V. (1945). The genera and species of the Tridenchthoniidae (Dithidae) — A family of the Arachnid order Chelonethida. *Bull. Univ. Utah Biol. Ser.* **9**: 1–67 [17]. Type species *Chthonius mordax* Tullgren, 1907 by original designation.

Taxonomic decision of Beier, M. (1947). Zur Kenntnis der Pseudoscorpionidenfauna des südlichen Afrika, insbesondere der südwest- und südafrikanischen trockengebiete. *Eos, Madrid* **23**: 285–339 [287].

This group is also found in Africa and Pacific, see Beier, M. (1944). Über Pseudoscorpioniden aus Östafrika. *Eos, Madrid* **20**: 173–212; Beier, M. (1955). Pseudoscorpionidea. pp. 263–328 *in* Hanstrom, B., Brinck, P. & Rudebeck, G. (eds.) *South African Animal Life.* Results of the Lund University Expedition in 1950–1951. Stockholm : Almquist & Wiksells; Beier, M. (1966). Ergebnisse der österreichischen Neukaledonien-Expedition 1965. Pseudoscorpionidea. *Ann. Naturh. Mus. Wien* **69**: 363–371.

Anaulacodithella australica Beier, 1969

Anaulacodithella australica Beier, M. (1969). Neue Pseudoskorpione aus Australien. *Ann. Naturh. Mus. Wien* **73**: 171–187 [172]. Type data: holotype, M whereabouts unknown, paratypes 11M.,11F., 2 nymphs whereabouts unknown, NHMW *8M.,6F., from Lamington Natl. Park, Qld.

Distribution: NE coastal, SE coastal, Qld., N.S.W. Ecology: terrestrial, (predator), litter, closed forest. Biological references: Beier, M. (1974). Neue Pseudoskorpione aus Australien und Neu-Guinea. *Ann. Naturh. Mus. Wien* **78**: 203–213 (distribution).

Anaulacodithella plurisetosa Beier, 1976

Anaulacodithella plurisetosa Beier, M. (1976). The pseudoscorpions of New Zealand, Norfolk and Lord Howe. *N.Z. J. Zool.* **3**: 199–246 [200]. Type data: holotype, AM KS17 F., paratypes AM KS18–19 2F., NHMW *1F., from Boat Harbour, Lord Howe Is. (holotype and one paratype only).

Distribution: Lord Howe Is. Ecology: terrestrial, (predator), litter.

Heterolophus Tömosvary, 1884

Heterolophus Tömosvary, E. (1884). Data ad cognitionem Pseudoscorpionum. *Természetr. Füz.* **8**: 16–27 [24]. Type species *Heterolophus guttiger* Tömosvary, 1884 by subsequent designation, see Chamberlin, J.C. (1925). Notes on the status of genera in the chelonethid family Chthoniidae together with a description of a new genus and species from New Zealand. *Vidensk. Meddr. Dansk Naturh. Foren.* **81**: 333–338.

This group is circumtropical, see Chamberlin, J.C. & Chamberlin, R.V. (1945). The genera and species of the Tridenchthoniidae (Dithidae): a family of the arachnid order Chelonethida. *Bull. Univ. Utah Biol. Ser.* **9**: 1–67.

Heterolophus australicus Beier, 1969

Heterolophus australicus Beier, M. (1969). Neue Pseudoskorpione aus Australien. *Ann. Naturh. Mus. Wien* **73**: 171–187 [171]. Type data: holotype, F. whereabouts unknown, paratype 1 tritonymph whereabouts unknown, from Mt. Tamborine, Qld.

Distribution: NE coastal, Qld.; known only from type locality. Ecology: terrestrial, (predator), litter, closed forest.

WITHIIDAE

INTRODUCTION

The Withiidae contains medium sized pseudoscorpions distributed in most parts of the world. Approximately 30 genera and 150 species currently are recognized. Two subfamilies have been recognized, Withiinae and Paragoniochernetinae; only the Withiinae are recorded from Australia.

Characterized by: monotarsate; heterofemorate; venom apparatus present in both chelal fingers; sternites VI to IX usually with a discrete patch of specialized sensory(?) setae; junction between femur I and femur II of the first pair of legs usually vertical.

References

Beier, M. (1932). Pseudoscorpionidea II. Suborder Cheliferinea. *Das Tierreich* **58**: i–xxi 1–294

Beier, M. (1944). Über Pseudoscorpioniden aus Östafrika. *Eos, Madrid* **20**: 173–212

Metawithius Chamberlin, 1931

Metawithius Chamberlin, J.C. (1931). A synoptic revision of the generic classification of the chelonethid family Cheliferidae Simon. (Arachnida). *Can. Entomol.* **63**: 289–294 [293]. Type species *Chelifer murrayi* Pocock, 1900 by original designation.

Microwithius Redikorzev, V. (1938). Les pseudoscorpions des l'Indochine française receullis par M.C. Dawydoff. *Mém. Mus. Natl. Hist. Nat. Paris (ns)* **10**: 69–116 [106]. Type species *Microwithius yurii* Redikorzev, 1938 by monotypy.

Taxonomic decision of Beier, M. (1954). A second collection of Pseudoscorpionidea from Malaya. *Bull. Raffles Mus.* **25**: 38–46 [45].

This group is also found in southeast Asia, see Beier, M. (1951). Die Pseudoscorpione Indochinas. *Mém. Mus. Natl. Hist. Nat. Paris (ns)* **1**: 47–123; Murthy, V.A. & Ananthakrishnan, T.N. (1977). Indian Chelonethi. *Orient. Insect Monogr.* **4**: 1–210.

Metawithius murrayi (Pocock, 1900)

Chelifer murrayi Pocock, R.I. (1900). Chilopoda, Diplopoda, and Arachnida. pp. 153–162 pl XVI *in* Andrews, C.W. (ed.) *A Monograph of Christmas Island (Indian Ocean).* London : British Museum [156]. Type data: syntypes, BMNH 1898.10.14.7–9 *1M.,1F., from Christmas Is. Compiled from secondary source: With, C.J. (1906). The Danish expedition to Siam 1899–1900. III. Chelonethi. An account of the Indian false-scorpions together with studies on the anatomy and classification of the order. *K. Danske Vidensk. Selsk. Skr. (7)* **3**: 1–214.

Distribution: Christmas Is.; also Timor, Sumatra and the Nicobars. Ecology: terrestrial, (predator). Biological references: With, C.J. (1906). The Danish expedition to Siam 1899–1900. III. Chelonethi. An account of the Indian false-scorpions together with studies on the anatomy and classification of the order. *K. Danske Vidensk. Selsk. Skr. (7)* **3**: 1–214 (redescription, distribution).

Philomaoria Chamberlin, 1931

Philomaoria Chamberlin, J.C. (1931). A synoptic revision of the generic classification of the chelonethid family Cheliferidae Simon. (Arachnida). *Can. Entomol.* **63**: 289–294 [291]. Type species *Philomaoria novazealandica* Chamberlin, 1931 (= *Chelifer pallipes* White, 1849) by original designation.

This group is also found in New Zealand, see Beier, M. (1976). The pseudoscorpions of New Zealand, Norfolk and Lord Howe. *N.Z. J. Zool.* **3**: 199–246.

Philomaoria pallipes (White, 1849)

Chelifer pallipes White, A. (1849). Descriptions of apparently new species of Aptera from New Zealand. *Proc. Zool. Soc. Lond.* **17**: 3–6 [6]. Type data: holotype, BMNH*, from New Zealand.

Philomaoria novazealandica Chamberlin, J.C. (1931). A synoptic revision of the generic classification of the chelonethid family Cheliferidae Simon. (Arachnida). *Can. Entomol.* **63**: 289–294 [291] [originally described as "nova-zealandica"]. Type data: holotype, BMNH

JC-518.01001 *M., paratypes BMNH JC-518.01002 *F., BMNH JC-518.01003–26 *(24), from New Brighton, New Zealand.

Taxonomic decision of Beier, M. (1976). The pseudoscorpions of New Zealand, Norfolk and Lord Howe. *N.Z. J. Zool.* **3**: 199–246 [241].

Distribution: Lord Howe Is.; also in New Zealand. Ecology: terrestrial, (predator), litter.

Withius Kew, 1911

Withius Kew, H.W. (1911). A synopsis of the false scorpions of Britain and Ireland. *Proc. R. Irish Acad. (B)* **29**: 38–64 [49] [originally described as a subgenus of *Chelifer* Geoffroy, 1762]. Type species *Chelifer subruber* Simon, 1879 (= *Chelifer piger* Simon, 1878) by original designation.

This group is found world-wide, see Weygoldt, P. (1969). *The Biology of Pseudoscorpions.* Cambridge : Harvard Univ. Press xiv 145 pp.

Withius piger (Simon, 1878)

Chelifer piger Simon, E. (1878). Études arachnologiques. XI. Liste des espèces de la famille des Cheliferidae qui habitent l'Algérie et le Maroc. *Ann. Soc. Entomol. Fr. (5)* **8**: 144–153 [148]. Type data: holotype, MNHP 2379 *F., from Bou-Saada, Algeria.

Chelifer subruber Simon, E. (1879). *Les Arachnides de France, Chernetes.* Paris : Roret Vol. 7 vii 78 pp. [30]. Type data: syntypes, MNHP *M.,F., from Hyères, France.

Taxonomic decision of Heurtault, J. (1971). Chambre génitale, armature génitale et caractères sexuals secondaires chez quelques espèces de pseudoscorpions (Arachnida) du genre *Withius. Bull. Mus. Natl. Hist. Nat. Paris (2)* **42**: 1037–1053 [1050].

Distribution: SE coastal; also cosmopolitan. Ecology: terrestrial, predator; synanthropic. Biological references: Weygoldt, P. (1969). *The Biology of Pseudoscorpions.* Cambridge : Harvard Univ. Press xiv 145 pp. (biology).

AMBLYPYGI

M.S. Harvey

INTRODUCTION

The Amblypygi, or whip-spiders, are spectacular arachnids that are easily recognized by their sensorial first pair of legs, raptorial pedipalps and lack of a "tail". Nevertheless, for many years they were treated as a subfamily of the order "Pedipalpi", and it is only recently that they have been afforded full ordinal status, along with the Schizomida and Thelyphonida (or Uropygi). Over 80 species have been described and about half have been placed in the family Charontidae, the only family to have been recorded from Australia.

Amblypygids are found under rocks, logs, and loose sheets of bark, and when disturbed, run with extreme agility and speed. They are found in most tropical and subtropical regions of the world, but somewhat surprisingly, only one species is currently known from Australia.

The first description of an Australian amblypygid was by Lauterer (1895) who described *Charon annulipes* from Brisbane. This name is regarded as a *nomen dubium* because the holotype cannot be located, the description is inaccurate and seemingly faulty, and no amblypygid has since been recorded from southeast Queensland.

CLASSIFICATION

Four families are currently recognized: Charontidae, Phrynidae, Phrynichidae and Damonidae. The latter was only recently elevated to family level by Quintero (1976).

ACKNOWLEDGEMENTS

I wish to thank Mr K. Walker (NMV) for access to the holotype of *Charinus pescotti* and Dr R.G. Ordish (NMNZ) for information regarding a paratype of the same species. This work was undertaken whilst I was the recipient of an Australian Biological Resources Study grant. CSIRO, Division of Entomology, Canberra, kindly provided research facilities.

M.S.H.

References

Lauterer, J. (1895). An undescribed *Charon*, with notes on the metamorphis of the first pair of ambulatory legs into a physiological pair of feelers. *Rep. Australas. Ass. Adv. Sci.* **6**: 413–414

Quintero, D., Jr. (1976). *Trichodamon* Mello-Leitao and the Damonidae, new family status (Amblypygi : Arachnida). *Bull. Br. Arachnol. Soc.* **3**: 222–227

CHARONTIDAE

INTRODUCTION

Currently this family is recognized as including nine genera and approximately 40 species, and is characterized by possessing pulvilli on each of the tarsi of legs II, III and IV.

Charinus Simon, 1892

Charinus Simon, E. (1892). Arachnides des îles Philippines. *Ann. Soc. Entomol. Fr.* **61**: 35–52 [48]. Type species *Phrynus australianus* L. Koch, 1867 by original designation.

Lindosiella Kritscher, E. (1959). Ergebnisse der von Dr O. Paget und Dr E. Kritscher auf Rhodos durchgeführten zoologischen Exkursionen. II. Pedipalpi (Amblypygi). *Ann. Naturh. Mus. Wien* **63**: 453–457 [454]. Type species *Lindosiella ioannitica* Kritscher, 1959 by monotypy.

Taxonomic decision of Weygoldt, P. (1972). Charontidae (Amblypygi) aus Brasilien. *Zool. Jb. Abt. Syst.* **99**: 107–132 [123].

This group is also found in Africa, South America and islands of the southwest Pacific, see Weygoldt, P. (1972). Charontidae (Amblypygi) aus Brasilien. *Zool. Jb. Abt. Syst.* **99**: 107–132

Charinus pescotti Dunn, 1949

Charinus pescotti Dunn, R.A. (1949). New Pedipalpi from Australia and the Solomon Islands. *Mem. Natl. Mus. Vict.* **16**: 7–15 [8]. Type data: holotype, NMV F. K–153, paratypes NMNZ *1F., 3F. whereabouts unknown, from Barron Falls, Qld. (holotype only).

Distribution: NE coastal, Qld.; Solomon Ils. Ecology: terrestrial, (predator), under bark, closed forest; under stones. Biological references: Baehr, M. (1974). Wiederfund der einzigen bekannten australischen Geisselspinne *Charinus pescotti* Dunn 1949 (Chelicerata, Arachnida, Amblypygi). *Entomol. Zeit.* **84**: 101–105.

Incertae sedis

Charon annulipes Lauterer, J. (1895). An undescribed *Charon*, with notes on the metamorphosis of the first pair of ambulatory legs into a physiological pair of feelers. *Rep. Australas. Ass. Adv. Sci.* **6**: 413–414 [414] [the identity of this species is unknown]. Type data: holotype, whereabouts unknown, from Brisbane, Qld.

PALPIGRADI

M.S. Harvey

INTRODUCTION

The Palpigradi is one of the smallest arachnid orders, and only 50 or so species have been described in two families, Eukoeneniidae and Sternarthronidae. The latter is known only from a single, extinct species from the Jurassic of Germany.

Palpigrades are very small, pale, fragile animals confined to soil and to soil-rock interfaces. They have been described from most continents, but the majority are known from Europe and Africa. The only species known from Australia is the introduced *Eukoenenia mirabilis* (Grassi and Calandruccio).

ACKNOWLEDGEMENTS

I am grateful to the Australian Biological Resources Study for funds and to CSIRO, Division of Entomology, for providing research facilities.

M.S.H.

References

Beier, M. (1932). Pseudoscorpionidea II. Suborder Cheliferinea. *Das Tierreich* **58**: i–xxi 1–294

EUKOENENIIDAE

INTRODUCTION

The Eukoeneniidae is the only family of Recent Palpigradi, and currently contains five Recent genera and approximately 50 species. A sixth genus is known from the Pliocene(?) of Arizona, U.S.A. The family is characterized by a prosoma with five sternites.

Eukoenenia Börner, 1901

Koenenia Grassi, B. & Calandruccio, S. (1885). Intorno ad un nuovo Aracnide Artrogastro (*Koenenia mirabilis*) che crediamo rappresentante d'un nuovo ordine (Microteliphonida). *Naturalista Siciliano* **4**: 127–133, 162–168 [127] [*non Koenenia* Beushausen, 1884]. Type species *Koenenia mirabilis* Grassi and Calandruccio, 1885 by monotypy.

Eukoenenia Börner, C. (1901). Zur äusseren Morphologie von *Koenenia mirabilis* Grassi. *Zool. Anz.* **24**: 537–556 [551] [originally described as a subgenus of *Koenenia* Grassi and Calandruccio, 1885]. Type species *Koenenia mirabilis* Grassi and Calandruccio, 1885 by monotypy.

This group is also found in Europe, Asia, Africa and North America, see Rowland, J.M. & Sissom, W.D. (1980). Report on a fossil palpigrade from the Tertiary of Arizona, and a review of the morphology and systematics of the order (Arachnida, Palpigradida). *J. Arachnol.* **8**: 69–86.

Eukoenenia mirabilis (Grassi and Calandruccio, 1885)

Koenenia mirabilis Grassi, B. & Calandruccio, S. (1885). Intorno ad un nuovo Aracnide Artrogastro (*Koenenia mirabilis*) che crediamo rappresentante d'un nuovo ordine (Microteliphonida). *Naturalista Siciliano* **4**: 127–133, 162–168 [165]. Type data: syntypes, whereabouts unknown, from Catania, Sicily.

Distribution: S Gulfs, S.A.; southern Europe and northern Africa. Ecology: terrestrial, (predator), soil. Biological references: Remy, P.A. (1951). Nouvelles stations du Palpigrade *Koenenia mirabilis* Grassi. *Bull. Soc. Entomol. Fr.* **56**: 106–108 (distribution); Lee, D.C. & Southcott, R.V. (1979). Spiders and other Arachnids of South Australia. pp. 29–43 *in* Leonard, B.E. (ed.) *South Australian Year Book No. 14.* Adelaide : Australian Bureau of Statistics (distribution).

APPENDIX I

ABBREVIATIONS AND SYMBOLS

For definitions of nomenclatural terms which appear throughout the *Catalogue*, the reader should refer to the text and glossary of the 1961 edn. of the International Code of Zoological Nomenclature.

A.C.T	Australian Capital Territory
alt.	altitude
Art.	Article
Bd.	Band
E	east, eastern
ed./eds.	editor/editors
edn.	edition
emend.	emendation
F.	female
fasc.	fascicule
fig/figs	figure/figures
ft	feet
ICZN	International Code of Zoological Nomenclature
Is./Ils.	Island/Islands
km	kilometre
lf.	lief
lfg.	lieferung
livr.	livraison
m	metre
M.	male
mi	mile
ms	manuscript
Mt./Mts.	Mount, Mountain/Mountains
N	north, northern
Nat.	Natural
Natl.	National
no.	number
nom. nov.	*nomen novum*
nom. nud.	*nomen nudum*
ns	new series
N.S.W.	New South Wales
N.T.	Northern Territory
nov. comb.	new combination
pl/pls	plate/plates
pt/pts	part/parts
Qd./Qld.	Queensland
S	south, southern
S.A.	South Australia
ser.	series
sp.	species
Tas./Tasm.	Tasmania
Tom.	Tome/Tomus
var./Var.	variety
Verl.	Verlag
Vic./Vict.	Victoria
vol./Vol.	volume
W	west, western
W.A.	Western Australia
[*name*]	square brackets enclosing a valid or available name indicate a qualification of the use of that name in the context in which it appears.
*	appears only with reference to type specimen information and indicates that the author has not seen the specimen(s).

APPENDIX II

MUSEUM ACRONYMS

AIM	Auckland Institute and Museum, Auckland, N.Z.
AM	Australian Museum, Sydney, N.S.W., Australia
AMNH	American Museum of Natural History, New York, N.Y., U.S.A.
ANIC	Australian National Insect Collection, CSIRO Div. of Entomology, Canberra, A.C.T., Australia
BMNH	British Museum (Natural History), London, U.K.
BPBM	Bernice P. Bishop Museum, Honolulu, Hawaii, U.S.A.
BYM	B.Y. Main Collection, Dept. of Zoology, University of Western Australia, Nedlands, W.A., Australia
CMNZ	Canterbury Museum, Christchurch, N.Z.
CU	Cornell University, Ithaca, N.Y., U.S.A.
DSIR	Department of Scientific and Industrial Research, Auckland, N.Z.
EUQ	Queensland University, Dept. of Entomology, St. Lucia, Qld., Australia
FSM	Florida State Museum, Gainesville, Florida, U.S.A.
IZW	Institut Zoologii, Warsaw, Poland
MCG	Museo Civico di Storia Natural "Giacomo Doria", Genoa, Italy
MCZ	Museum of Comparative Zoology, Harvard University, Cambridge, Mass., U.S.A.
MGH	Museum Godeffroy, Hamburg, Federal Republic of Germany
MHNB	Museum d'Histoire Naturelle, Basel, Switzerland
MMUS	MacCleay Museum, University of Sydney, N.S.W., Australia
MNHP	Museum National d'Histoire Naturelle, Paris, France
NHMW	Naturhistorisches Museum, Wien, Austria
NHRM	Naturhistoriske Riksmuseum, Stockholm, Sweden
NM	Museum Wiesbaden, Wiesbaden, Federal Republic of Germany
NMV	Museum of Victoria, Melbourne, Vic., Australia
NMNZ	National Museum of New Zealand, Wellington, N.Z.
OMNZ	Otago Museum, Dunedin, N.Z.
OUM	Oxford University Museum, Oxford, U.K.
PUO	Portland University, Portland, Oregon, U.S.A.
QM	Queensland Museum, Fortitude Valley, Qld., Australia
QVM	Queen Victoria Museum, Launceston, Tas., Australia
RMNH	Rijksmuseum van Natuurlijke Histoire, Leiden, Netherlands
SAMA	South Australian Museum, Adelaide, S.A., Australia
SMF	Natur-Museum Senckenberg, Frankfurt-am-Main, Federal Republic of Germany
SMNS	Stadtliches Museum für Naturkunde, Stuttgart, Federal Republic Germany
TMH	Tasmanian Museum and Art Gallery, Hobart, Tas., Australia
WAM	Western Australia Museum, Perth, W.A., Australia
ZFMK	Zoologisches Forschungsinstitut und Museum "Alexander Koenig" und Universitaet Göttingen, Göttingen and Bonn, Federal Republic of Germany
ZMB	Museum für Naturkunde an der Universitaet Humbolt zu Berlin, German Democratic Republic
ZMH	Zoologisches Museum für Hamburg, Hamburg, Federal Republic of Germany
ZMLU	Museum, Universitetets Lund, Sweden

TAXONOMIC INDEX

MYGALOMORPHAE

TAXONOMIC INDEX

ARANEOMORPHAE

TAXONOMIC INDEX

PSEUDOSCORPIONIDA, AMBLYPYGI, PALPIGRADI

Publication Dates of Previous Volumes

Volume 1	29 November 1983
Volume 2	In Press

R83/758(2) Cat. No. 85 0402 1